THE
INTERNATIONAL SERIES
OF
MONOGRAPHS ON PHYSICS

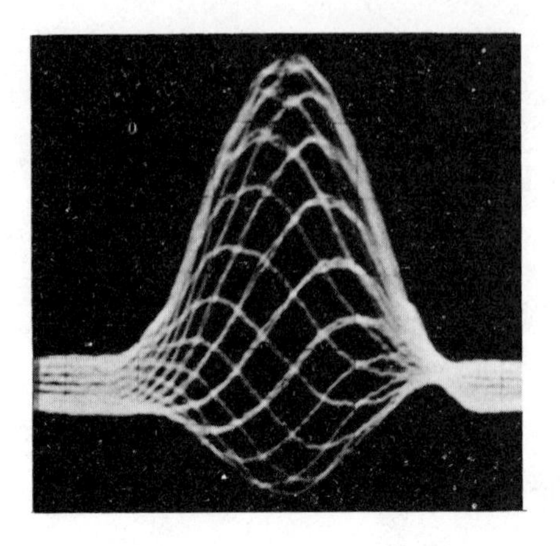

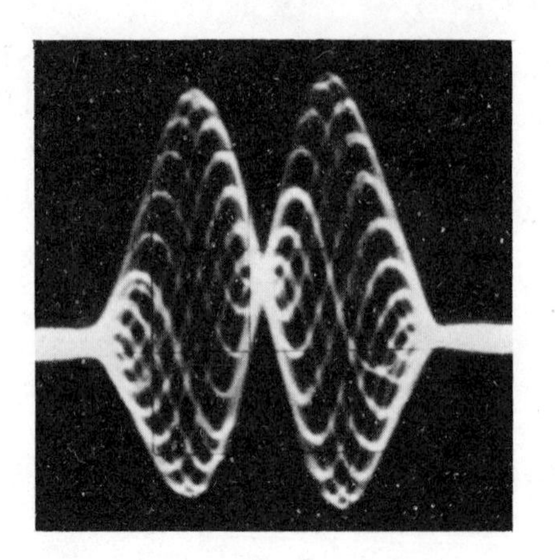

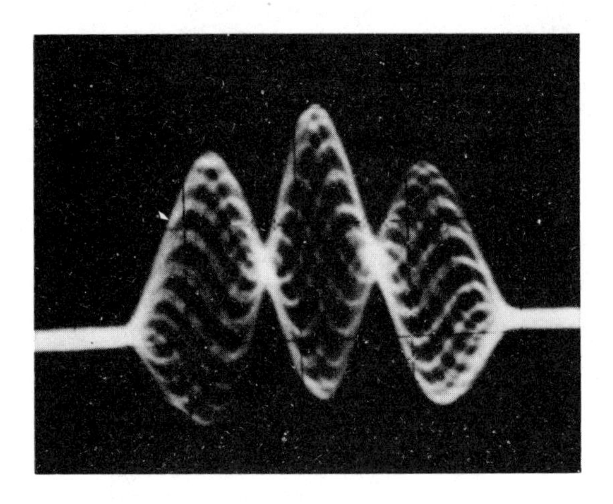

The figure illustrates three of the many possible modes of instability which can occur in cyclic particle accelerators. The circulating proton beam in the CERN 'booster' synchrotron, which is in the form of a bunch 15 m long and a few millimetres in diameter here develops a sinuous structure which grows in a few milliseconds until the particles strike the vacuum-chamber wall. Each trace records the appearance on one passage by an observation station. About 20 consecutive passages are superimposed to show how the different mode patterns evolve. This type of behaviour, known as the head-tail instability, is described in Chapter 6.

THE PHYSICS OF CHARGED-PARTICLE BEAMS

BY

J. D. LAWSON

CLARENDON PRESS · OXFORD

1977

Oxford University Press, Walton Street, Oxford OX2 6DP

OXFORD LONDON GLASGOW NEW YORK
TORONTO MELBOURNE WELLINGTON CAPE TOWN
IBADAN NAIROBI DAR ES SALAAM LUSAKA ADDIS ABABA
KUALA LUMPUR SINGAPORE JAKARTA HONG KONG TOKYO
DELHI BOMBAY CALCUTTA MADRAS KARACHI

ISBN 0 19 851278 3

Printed in Great Britain
by Thomson Litho Ltd.,
East Kilbride

PREFACE

The aim of this book is to present a broad synoptic view of the many varieties of charged-particle beams which have been studied in connection with a wide range of practical devices. Classification is in terms of essential physical features, rather than applications, and is related to underlying principles of optics, classical and statistical mechanics, and plasma physics. In this way diverse viewpoints are brought together, and secondary concepts which have emerged in connection with different applications are compared.

Formulae and curves of direct use to the designer of apparatus are not included, nor is there any discussion of computational methods. It is intended rather that the book should form a background text presenting the fundamental ideas, and showing how problems in a particular field of application relate to those in other fields. Information on more detailed and practical matters may be found by consulting the references, discussed in special sections of the text. These are denoted by the letter R, thus §4.R.1-3 denotes a section in which references relevant to Chapter 4, sections 1-3, are discussed.

Although there exist many books which separately cover such subjects as electron optics, microwave tubes, particle accelerators, and plasma physics, the present work is perhaps unique in bringing together material from all these fields. For this reason I have had to assimilate and re-present much with which I was unfamiliar; and the book owes much to the generous help given to me in this task by many colleagues who have given their time either for discussions, or for reviewing the manuscript at various stages.

The problem of proper acknowledgement in a field so wide as the one covered in this book is always difficult. A paper may be quoted where a better one, somehow missed by the author, remains unmentioned. I am particularly conscious that readers in the Soviet Union may feel that their very considerable contributions in this field are less well covered than those of Western Europe and the United States. This again is a question of choosing the familiar rather than perhaps the most appropriate. (At this point it is perhaps worth recording that my own enthusiasm for beams was largely engendered by the remarkable paper by Budker (1956) which contained the seeds of so much which was to follow.)

There is no name index in the book, and it must be emphasized that no attempt has been made to acknowledge by name the originator of every significant calculation. It is often more helpful to refer to later reviews or texts, in which the early contributions are discussed. In many cases pieces of work are cited as examples, this does not imply the absence of earlier, or comparable later work. Only references cited in the text are included in the list.

The choice of material naturally reflects my own judgement of what is interesting and important. I should welcome readers' views on any notable omissions or errors. It is too much to hope that some errors, both factual and typographical, will not remain.

The book has evolved over several years, during which the problem of completing it appeared at times to be insuperable. I am particularly indebted to Sir Denys Wilkinson, at whose suggestion the writing of the book was undertaken, and to Dr. G. H. Stafford, Director of the Rutherford Laboratory for their unfailing encouragement throughout the task.

On all matters pertaining to particle accelerators, I have received considerable help from Dr. H. G. Hereward. In a number of discussions, particularly emphasizing the physical basis of the subject, he has given me access to much informal written material, not acknowledged in detail. This has been followed up by a critical reading of the relevant parts of the manuscript. Dr. Martin Reiser provided many detailed comments at various stages, particularly on Chapter 2, and on a number of topics in later chapters, including especially the properties of relativistic ring beams. Dr. W. H. M. Clark helped with the clarification of a number of fundamental problems relevant to Chapter 6, and made valuable suggestions regarding presentation. Dr. A. Chao read an early draft of Chapters 2-4, and made many helpful suggestions. Dr. Sacherer helped with the section on instabilities in bunched beams, and contributed most of § 6.5.4. Many of the figures in Chapter 6 are from graphical print-outs of dispersion relations; others are sketches based on such print-outs. The programming and organization of this work was carried out by Mr. M. Newman. In addition I have received special help from the following, either in the form of discussions, direct help with particular problems, or critical reading of part of the manuscript: Dr. J. Connor, Dr. J. P. Davey, Dr. C. Lashmore-Davies, Mr. M. Donald, Dr. P. Hawkes, Dr. P. M. Lapostolle, Dr. L. Jackson Laslett, Mr. A. Reddish, Dr. R. Taylor, and Dr.A.J. Theiss. Many others have contributed on detailed points. On bibliographical matters I received much sympathetic help from Mrs. E. Marsh and her staff at the Rutherford Laboratory Library.

Formal acknowledgement is due to North-Holland Publishing Company for permission to re-use Figs. 4.6 and 4.9, which appeared in *Nuclear Instruments and Methods*, 139, 17, (1976), and to Dr. F. Sacherer for the photographs reproduced in the frontispiece, which appeared in the reference Sacherer (1974). No other figure has been reproduced directly, and where use has been made of published figures in preparing those used here acknowledgement is made locally.

Rutherford Laboratory, Chilton.

J. D. Lawson
April, 1977

CONTENTS

LIST OF SYMBOLS

A brief note on notation is given in section 1.4. The list below contains symbols in frequent use. Those which are only used for a few pages are defined locally and in general not listed. Where appropriate, the section in which the symbol is discussed, or equation where it is defined, is given.

Subscripts tend to have different meanings in different locations and the definition should be sought locally. One exception is the use of L, which (except in Ω_L) denotes that the quantity is measured in the Larmor frame. Note particularly the use of r discussed in section 1.4. The context should prevent confusion with the radial variable, sometimes used with the subscript a, b or 0, (as in section 3.4.1).

A	Vector potential
A	Amplitude, in phase-amplitude variable notation (§4.3.2)
A	Reduced beam radius, $a(\beta\gamma)^{\frac{1}{2}}$ (§4.3.7)
A	Atomic weight (Chapter 5)
a	Radius of cylindrical beam
a	Major radius of elliptical beam or radius along y-axis
a	Normal-mode amplitude (with various subscripts) (§6.3.7)
B	Magnetic induction (often referred to loosely as 'field')
B	Brightness
b	Radius of tube surrounding beam
b	Minor radius of elliptical beam or radius along y-axis
C	Circumference of accelerator orbit (§2.8.3)
C	Space-charge parameter in travelling-wave tube theory (§6.3.9)
C_c	Coefficient of chromatic aberration (§2.2.9)
C_s	Coefficient of spherical aberration (§2.2.8)
c	Velocity of light
c	Coupling constant in coupled-mode theory

D Collisionsless skin depth

E Electric field

F Force
$F(\ )$ Distribution function; argument and normalization defined locally
f Focal length of lens
f Neutralization fraction of partly neutralized beam
$f(\ )$ Distribution function (alternative to $F(\ )$)

H Hamiltonian function
h Harmonic number (§2.8)
h Planck's constant

I Current
I_A Alfvén current (§3.1)
I_B Beam current (§6.3.9)
I_C Circuit current (§6.3.9)
i Current density
i_c Current density at cathode

K Generalized perveance (equation (3.30))
K Coupling constant (§6.3.7)
k Wave number
k Perveance (§3.2.2.)
k Boltzmann's constant

L Lagrangian
L_0 One radiation length in metres (equation (5.10))
L In $(\theta_{max}/\theta_{min})$ for beam particle (§5.10)
L_0 Characteristic length associated with beam in a tube with conducting walls (equation (6.90))
L_1 As L_0 above, but with walls of finite impedance (equation (6.93))

M Transfer matrix (with appropriate subscripts, see note after equation (2.31))

m Matrix element (with two suffices)
m_0 Rest mass of particle
m^* Effective mass of particle (§2.7.4)

N Number of particles per unit length of beam
N_A Avogadro's number
N_a Number of particles per unit area of sheet beam
n Betatron field index (§2.5.3) (in §4.8.4 indices associated with magnetic and electric fields are distinguished as n_M and n_E).
n Particle density
n Azimuthal mode number (§6.3.5)
n_0 Unperturbed particle density

P Canonical momentum
p Momentum
P_θ, p_θ Components of angular canonical and mechanical momentum about axis
p Pressure, or components of pressure tensor

Q Normalized betatron oscillation frequency (§2.5.3)
q Charge of particle

R Reduced variable for radial co-ordinate, $(\beta\gamma)^{\frac{1}{2}}r$ (§2.2.2)
R Characteristic radius (defined locally)
R_0 Radius of equilibrium orbit (§2.5.3)
r Radial variable
r_0 Classical radius of particle (§1.4)

S Entropy
S Distance measured in radiation lengths (Chapter 5)
s Co-ordinate along beam, for beam with curved axis

T Temperature, always in the combination kT where k is Boltzmann's constant
T Value of k_r which satisfies the determinantal equation (§6.3.3)
t Time (as a variable)

U Energy density associated with wave

U Accelerator parameter for longitudinal stability analysis (equation (6.96))

U Heaviside step function

$U_\perp$ As above but relevant to transverse stability (equation (6.164))

u_0 A principal solution of equation (2.30)

V Volts

V Velocity of particle (§5.5)

$V, V_\perp$ Accelerator parameters (same references as for $U, U_\perp$)

$\tilde{V}_k, \tilde{V}_B$ Kinetic voltage (equation (6.126)

V_B Relativistically corrected beam voltage (§6.3.9)

v Velocity of particle (used rather than βc in non-relativistic calculations)

v_0 A principal solution of equation (2.30)

W Wronskian determinant (§2.2.2)

W Power

W, W_1 Variable related to orbital canonical momentum in accelerator (§2.8.2 and §6.3.5)

W Action function (§3.6)

w Variable related to phase-amplitude variables (§4.3.2)

w Scattering parameter (equation (5.14))

$\underline{\mathrm{w}}$ Vorticity (§3.6)

X Reduced variable, $(\beta\gamma)^{\frac{1}{2}}x$

X Transverse distance measured in radiation lengths (Chapter 5)

Z Atomic number

Z Complex function (§6.3.6)

Z_a, Z_b Impedances beam edge and wall (§6.3.4)

Z_B Beam impedance (equation (6.139))

Z_C Circuit impedance

Z_l Characteristic impedance of radial transmission line (equation (6.70))

Z_0 Impedance of free space = $(\mu_0/\varepsilon_0)^{1/2}$

$Z_{\parallel}$ Parallel impedance (equation (6.110))
$Z_{\perp}$ Transverse impedance (equation (6.171))

α Angle of arrival of ray on axis (§2.2.8)
α Momentum compaction constant (§2.7.4)
α_f Fine-structure constant
$\alpha_p(s)$ Momentum compaction function (§2.7.4)
$\alpha_s, \alpha_x, \alpha_y$ Damping coefficients in electron synchrotron

β Velocity of particle/c
β_a Value of β at edge of beam, $r=a$
β_w Phase velocity of wave/c
β_0 Velocity/c of steady component of beam (§6.2.5)
β_0 Lattice amplitude function (§4.3.4)

Γ See Table 2.1
γ Total energy of particle/m_0c^2, $(1-\beta^2)^{-\frac{1}{2}}$
γ_a Value of γ at edge of beam, $r=a$
γ_w Value of γ appropriate to particle moving with phase velocity of wave (equation (6.2))

δ Dirac δ-function

ε Emittance (§4.3.1)
ε_0 Permittivity of free space, $10^{-9}/36\pi$
ε_n Normalized emittance (§4.3.2)
$\bar{\varepsilon}$ Root-mean-square emittance (§4.3.8)
$\varepsilon_1, \varepsilon_2$ Image coefficients (§4.5.3)

η Accelerator lattice parameter (equation (2.145))

θ Angular co-ordinate
θ Angle of scattering (Chapter 5)

κ Constant specifying focusing strength
κ Dielectric constant
$\underline{\underline{\kappa}}$ Dielectric tensor

Λ	See Table 2.1
Λ	$\ln \Lambda$ = Coulomb logarithm for charged gas (§5.6)
λ	Wavelength
λ_D	Debye length (equation (4.88))
λ	$\lambda/2\pi$
μ	$2\cos\mu$ = trace of transfer matrix (equation (2.133))
μ	Optical magnification (§2.2.3)
μ_0	Permittivity of free space, $4\pi \times 10^{-7}$
ν	Budker's parameter (§3.1)
ν_y	Budker's parameter for cylindrical current sheet (§4.5.4)
ξ	$x + jy$
ξ	Chromaticity (equation (2.139))
ρ	Density in kg/m^{-3} (Chapter 5)
ρ	Dimensionless radial variable (§3.4.2)
ρ	Perturbation of beam radius (§3.3.1 and 6.4.7)
σ	Scattering cross-section
σ	Root-mean-square value of gaussian distribution
σ	Conductivity
τ	Relaxation time (§5.6)
Υ	See Table 2.1
Φ_0	Phase, in phase-amplitude variables (§4.3.2)
ϕ	Potential (§2.2.2)
ϕ_0	Potential corresponding to rest mass of particle, $-m_0c^2/q$, (§2.2.2)
ϕ	Synchrotron oscillation phase (§2.8.2)
χ	Phase parameter in head-tail effect (§6.5.4)
Ψ	Magnetic flux (§2.2.1)
Ψ_0	Flux linking circle through orbit when $\dot{\theta}=0$ (§2.2.1)

ψ	Variable related to phase-amplitude variables (§4.3.2)
Ω	Solid angle (Chapter 5)
Ω_L	Larmor frequency (§2.2.1)
Ω_s	Synchrotron frequency
ω	Angular frequency (often referred to simply as 'frequency')
ω_0	Orbital revolution frequency (§2.5.3)
ω_c	Cyclotron frequency (§2.2.1)
ω_f	Frequency of rotating frame of reference (§2.2.2)
ω_g	'Gap' frequency in accelerator (§2.8.2)
ω_p	Plasma frequency (§6.2.2, but see also equation (6.42))
ω_q	Reduced plasma frequency (§6.3.3)
ω_s	Synchronous frequency in accelerator (§2.8.2)

1
INTRODUCTION

1.1. Introduction

Charged-particle beams are widely used in scientific, industrial, and even domestic apparatus. The physical properties of such beams have been studied by workers in many different fields, often at different times and in rather different contexts. Thus, the essential requirements of a focused electron-microscope beam are different from those of a proton beam in a cyclotron or an electron beam in a low-noise travelling-wave tube. Nevertheless, despite differences in function and superficial appearance, beams used in a wide variety of applications may be shown to have many features in common. This is sometimes obscured by differences in approach or notation.

In this monograph the unity of beams encountered in different contexts is emphasized, and classification is in terms of physical features rather than application. A wide range of topics is surveyed, and for this reason the detail in many of them is restricted. Carefully selected references provide a starting point for more detailed study.

The approach is synthetic; simple examples are considered first, and the complications introduced gradually. Elementary methods are used wherever appropriate, and more sophisticated approaches introduced only as they are needed. This may not be as convenient for the problem solver as an initial general formulation, from which a number of examples follow as special cases. Nevertheless, it presents in a more vivid way the essential physical ideas, which form a set of conceptual building blocks which are useful when trying to think about new situations and to understand their inherent limitations.

1.2. The concept of a beam

It is not profitable to attempt a rigorous definition

of a beam; there is a reasonable consensus of what the term implies, though, as we shall see later, there is no abrupt transition between what can be thought of as a high-intensity beam and a collisionless plasma with a rather special velocity distribution function for the particles.

A beam is often cylindrical in shape (perhaps with cross-section varying along its length) and consists of charged particles moving in a direction approximately parallel to the axis of the cylinder. This axis may be curved if the beam passes through a transverse electric or magnetic field. Sometimes there is a background of 'neutralizing' particles of opposite sign to those with directed motion, which has little or no mean drift velocity. An alternative form is the sheet or strip beam, with transverse dimensions very large in one direction and relatively small in the other. Such beams can often be considered independent of one coordinate, with consequent simplification in their description. Yet again, the beam can form a hollow cylinder. If the thickness is small compared with the radius, then such a beam can sometimes be considered as a 'rolled-up' sheet beam.

In the definition it was stated that the particle motion is 'approximately parallel' to the axis. There is scope for argument here as to whether this is too restrictive. Clearly, to say that the charges of one sign have a net drift velocity down the cylinder is not restrictive enough; the current in a glow discharge is not a beam. To specify that no charge at any time moves backwards might be a better, though still arbitrary, criterion. In most practical beams the particle motion makes a small angle with the axis, and furthermore the spread of energy of the particles is small; sometimes it is a few per cent, and often considerably less.

The motion of the particles in a beam depends both on external fields and on the fields arising from other particles in the beam. Interaction with other particles in the beam is of two kinds. The first kind, typified by the 'space-charge' force in which the fields from a large number of particles combine to form a smooth electric field which only varies appreciably in a distance large compared with the interparticle distance, is independent of the particulate structure of the beam. The second type of force is essentially short range, and describes collisions between individual beam particles, or between such particles and stationary ions or atoms. The techniques for describing these two effects are somewhat different; we shall mostly be concerned with the first type.

In very dilute beams both these types of interaction are virtually absent and collective behaviour cannot occur. The properties of such beams can be found by summing the motion of individual particles in the external field. This is the province of charged-particle optics, considered in Chapter 2.

It is generally found that the kinetic energy of the beam particles exceeds the stored energy of any collective fields which they produce. Since these energies are proportional to the number of particles and the square of the number of particles respectively, it might be expected that the ratio of electric and magnetic stored energy to kinetic energy could be increased indefinitely by increasing the number of particles in the beam. It will be shown, however, that when these two energies become comparable the fields in general become sufficient to produce large transverse velocities so that a beam as discussed above no longer exists, and the system is perhaps best described as a plasma.

The term 'plasma' is used differently by different authors, as may be seen by looking up the definition in various texts. One view is that any system of charged particles in which the collective mutual interaction is not small compared with the effect of external fields constitutes a plasma, so that most of the beams considered in this book are plasmas, albeit rather dilute ones. A view representing the opposite extreme is that a plasma denotes a gas containing charged particles, but exhibiting overall charge neutrality, with characteristic dimensions large compared with the Debye shielding distance. There are good grounds for both views, which depend rather on one's field of interest. The connection between beam behaviour and that of plasmas as conventionally presented is particularly evident in Chapter 6.

1.3. Some applications of charged-particle beams

In later chapters the theory of beams is developed in a general way; the properties and limitations of various typical configurations are explored over a wide range of parameters. A number of examples relevant to particular applications are given, but no attempt is made to present an overall view of the operation of, for example, cyclotrons, electron microscopes, or travelling-wave tubes. Such treatment will be found in the more specialized books and reviews, which are referred to at appropriate points in the text.

Nevertheless, in order to give some indication of the wide variety of ways in which beams are now used in scientific research, industrial processes, and as ingredients of instruments and other devices, the present section is devoted to a brief synoptic survey. This is illustrative rather than systematic. Additional more specialized applications will be encountered in later sections. Uses connected with the more basic sciences will be described first, followed by a consideration of more routine and straightforward applications.

Beams in particle accelerators and storage rings have contributed greatly to our understanding of the physical world, ever higher energies being necessary to explore more deeply the structure of matter. At less extreme energies knowledge and understanding of nuclear structure have been obtained, and at lower energies still, detailed characteristics of atomic processes can be studied. Besides specifically nuclear and atomic studies, chemical and structural changes in materials can also be investigated.

The accelerated particles are not always used directly; instead they may create other particles in a target. Neutrons are produced in this way, typically from beams of electrons or deuterons. Electrons produce photons (X-rays), and mesons can be produced from sufficiently energetic protons; indeed some machines are designated as 'meson factories'. A recent development is the use of the ultraviolet synchrotron radiation generated by highly relativistic electron beams deflected by a strong magnetic field.

A substantial contribution to our understanding of the structure of materials and surfaces, and of the detailed structure of biological specimens, has been made with the electron microscope. The development of electron optics was stimulated by the requirements of this instrument, and a basic understanding of its performance and limitations was obtained in the 1930's. A great deal of technical development has occurred since then, however, and both transmission and scanning versions of considerable complexity and versatility are now commercially available. Collective beam effects are small, and limitations are now imposed by such factors as manufacturing precision and the provision of reliable high-brightness guns. The field continues to be a very active one.

A more recent development, employing finely focused electron beams for the study of surface materials, is electron microprobe analysis. A fine beam of electrons in the 10-30 keV energy range penetrates the sample to a depth of order 1 μm, and the characteristic X-rays of the materials on or near the surface are analysed to provide identification.

Surface properties can also be investigated by the scattering of low-energy beams of ions. In backscattering analysis a proton or helium beam is focused on to the material of interest; the atomic constitution of the material is then determined by analysing the distribution in energy of the backscattered particles. The bombarding energy is of the order of 100-1000 keV, and the method is well suited to the analysis of thin samples and the study of surfaces.

A somewhat different application of beams to the determination of the composition of small quantities of material is exemplified by mass spectrometry, a technique now in widespread use. In the mass spectrometer the material of interest is ionized and converted into a beam, which then passes through deflecting fields arranged so that particles with different charge-to-mass ratios are transported to different physical locations. Specialized designs have evolved for such instruments, now widely available commercially, though the basic principles of operation are straightforward.

Closely related to the mass spectrometer is the isotope separator, an instrument on a larger scale designed to separate macroscopic amounts of isotopes. A classic example of its use was in the separation of uranium isotopes for producing the first atomic weapon.

The technique of activation analysis, widely used with neutrons, is also sometimes useful with charged-particle beams in the range of some tens of MeV.

So far, various direct instrumental uses of beams have been described. Some of these have routine industrial applications, for example searching for vacuum leaks with mass spectrometers tuned to detect leaking helium gas. Other routine applications include the use of electron beams to produce X-rays for radiography, for cancer treatment, and for sterilization of materials by killing micro-organisms. More directly, electron beams are now widely used for welding and vacuum melting of metals in clean conditions.

Radioactive isotopes may be produced with beams of protons, deuterons, or α-particles with energies in the range of a few tens of MeV. The species which can be produced in this way complement those obtainable by neutron irradiation in reactors.

The most widespread use of electron beams is undoubtedly in the domestic television receiver. Here the beam is used as a generator of light, through the intermediary of phosphors of various kinds. The design of the electron-optical focusing and deflecting systems for television tubes is now a highly specialized field; ease of manufacture is an important feature, in addition to precision and reliability. Closely related to such tubes are those used in cathode-ray oscillographs, developed before television tubes and in widespread use both for routine instrumentation and for research purposes. The oscillograph has played an important part in the development of modern electronics technology. The phenomenon of gas focusing encountered in early 'soft' tubes is perhaps mainly of historical interest; the physics of the process is interesting, however, and will be treated briefly.

In some of the applications described so far the beam has been ultilized directly, as for example in a scattering experiment with an accelerator; in others its function is to produce some other physical manifestation such as X-radiation, heat, or light. An important class of device of this second type is the microwave tube, designed to generate or amplify power at microwave frequencies. The interaction with electromagnetic fields occurs along the length of the beam, and is not localized as in a target. A great deal of active development occurred during the period 1940-1960; this has died down now that the field is well understood and a few important types of device are established. (Others have been superseded by solid-state devices).

Continuous interaction between beams and a high-frequency structure is also a characteristic of many particle accelerators, and is especially evident in a linear accelerator. For such an accelerator powered by a microwave tube the circuit elements in the tube and accelerator can be considered as a sort of 'transformer', extracting energy from a high-current, low-voltage beam in the tube and transferring it to produce a high-voltage, low-current beam in the accelerator.

During recent years very intense particle beams have been developed in connection with ideas for collective particle accelerators and research into thermonuclear fusion. Since neither useful collective accelerators nor fusion reactors yet exist, this is clearly a field where continuing development is to be expected. It is in beams used in these fields that collective effects play a particularly important part. Intense electron ring beams have been studied in connection with the electron ring accelerator. Acceleration has been demonstrated, but as with other collective accelerator concepts the potential for development to a useful device has yet to be demonstrated.

In the field of fusion research cylindrical layers of relativistic electron beams have been studied in connection with the (now discontinued) 'Astron' project and similar devices. Proton and deuteron beams of very high power, tens of amperes with energy of tens of keV, are being developed for injection into magnetic confinement systems to heat plasmas and produce fusion reactions. (This requires neutralization by charge exchange after formation of the beam and subsequent stripping after it has penetrated into the magnetic field, otherwise confinement cannot be achieved.) A different approach requires the use of electron or ion beams of enormous instantaneous power (many terawatts) to heat and thereby compress pellets containing deuterium and tritium in the hope that 'inertial confinement' will be adequate to produce a net power yield from fusion reactions. Other suggestions for applications in this field will be encountered later.

More generally, in the field of plasma physics beam-plasma and beam-beam interactions have been studied under a wide range of conditions in an attempt to understand and elucidate more clearly the properties of waves in various beam and plasma configurations. In this widely explored area of study the subject of electron beams merges into what is conventionally known as plasma physics, since both a beam and plasma are involved. Furthermore, the techniques for dealing with waves and instabilities in either are very similar. Obviously this type of theory has some relevance to astrophysical phenomena such as solar flares, though excessive complication makes detailed comparison difficult.

Moving now away from Earth, artificial satellites have been propelled by ion motors, which eject a jet of heavy ions and also an equal number of electrons to preserve charge neutrality. The solar wind, though not a beam as defined in section 1.2, has by virtue of its streaming motion at least some of the characteristics of a beam. The van Allen belts deserve mention, though they are not beams but rather a special form of collisionless plasma. Although streaming cosmic rays can hardly be said to form a beam, it is interesting to note that some of the basic properties of high-current neutralized relativistic beams were first obtained in a paper by Alfvén (1939) relating to cosmic-ray streams. He identified an important characteristic current, which now bears his name.

1.4. Notation and style

Since our aim is to provide a unified discussion of a wide range of phenomena, we use a notation which is as general as possible, making use of characteristic dimensionless parameters wherever appropriate. The treatment is relativistic except where the contrary is stated. The letters N.R. and E.R. denote non-relativistic and extreme relativistic limits. Theoretical formulations are kept as direct and simple as possible, and outlines of the relevant background mechanical and electromagnetic theory are given when it is felt that this may be helpful.

For the analysis of cylindrical beams both cylindrical polar and cartesian co-ordinate systems are used. When the axis is curved, the z co-ordinate is replaced by s. For a system with axial symmetry, such as a betatron, where the beam axis is not the symmetry axis, confusion may arise. The simplest co-ordinate system is cylindrical polar, with z along the symmetry axis. To compare with systems which do not have axial symmetry, however, where the properties of the orbit vary periodically in θ, it is sometimes convenient to use (x,y,s) co-ordinates. For both these systems to be right-handed it is necessary to choose the signs such that on the beam axis

$$\theta = -s/R. \tag{1.1}$$

This is of opposite sign to θ measured in a polar system. These co-ordinate systems are illustrated in Fig.1.1. The confusion generated by this ambiguity is less than that which could arise from a fully consistent system; such a system would necessarily be opposite to convention in some applications.

Total energy, kinetic energy, momentum, and velocity will be expressed in terms of m_0, the rest mass, and c, the velocity of light, as $\gamma m_0 c^2$, $(\gamma-1)m_0 c^2$, $\beta\gamma m_0 c$, and βc res-

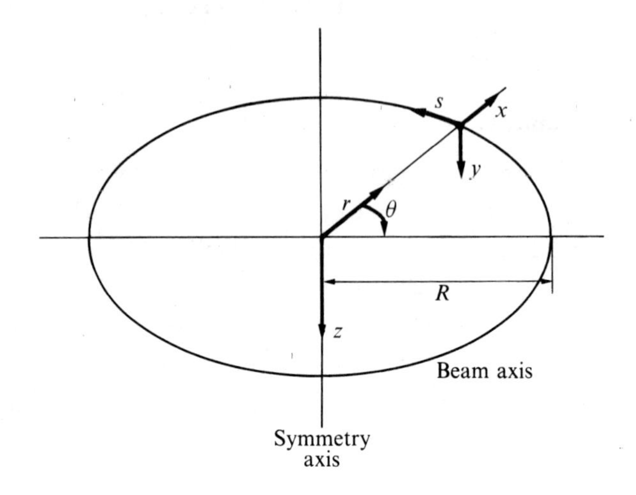

Fig.1.1. Alternative co-ordinate systems for curved orbits with axial symmetry.

pectively, where $\gamma^2 - 1 = \beta^2\gamma^2$. It is useful to note that $d\gamma/d\beta = \beta\gamma^3$ and $d(\beta\gamma)/d\beta = \gamma^3$. Where there is danger of ambiguity, the subscript on m will be made more specific; thus for electrons $m_0 = m_e$.

MKS units are used throughout. Electric charge is denoted by q, a positive number for positive charge. (In some books the electron charge e is positive, so that $e = -q$.) Other electrical symbols are conventional, and are shown in the list of symbols.

A parameter of frequent occurrence is the classical particle radius

$$r_0 = q^2/4\pi\varepsilon_0 m_0 c^2. \tag{1.2}$$

Where it is necessary to be specific to avoid ambiguity r_0 carries the same subscript as m; thus for protons

$$r_p = q^2/4\pi\varepsilon_0 m_p c^2. \tag{1.3}$$

Subscripts are frequently used; sometimes these denote direction, so that $\beta_z c$ denotes velocity in the z direction, and at other times they denote position, so that $\beta_{\theta a} c$ denotes the θ velocity at radius a in the beam. The symbols $\perp$ and $\parallel$ denote perpendicular and parallel to the beam axis.

The dot and prime respectively denote differentiation with respect to time and distance along the beam.

The quantity $\sqrt{(-1)}$ is denoted by j rather than i; this implies the definite convention $\exp \mathrm{j}(\omega t - kx)$ for harmonic quantities, whereas i is used by some authors as equivalent to j, and by others as -j. In the study of waves on beams all these usages are found; furthermore both MKS and gaussian units are both in common use. The exponential factor is often omitted; where this may cause confusion its presence is denoted by a tilde, thus $\tilde{E}_0$ denotes $E_0 \exp \mathrm{j}(\omega t - kx)$.

1.5. Note on background physics

The study of beams requires a knowledge of several branches of classical physics. In some instances adequate knowledge is assumed; elsewhere relevant concepts are summarized and references to more complete treatments are given. A brief discussion of the various disciplines involved is given in this section.

Much of the work requires a relatively straightforward application of relativistic particle dynamics combined with classical electromagnetic theory in the form of Maxwell's equations. Knowledge of Maxwell's equations and of simple relativistic kinematics is assumed. On the other hand, in the field of classical mechanics, an outline of Hamiltonian dynamics is given at the beginning of Chapter 4, prior to the introduction of Liouville's theorem and the concept of phase space. Later in the same chapter some of the more straightforward parts of the kinetic theory of gases are presented.

Well-known optical concepts are used in Chapter 2. Some results from particle scattering theory, essentially an application of classical dynamics, are required in Chapter 5. Some simple formulae from quantum scattering theory and the theory of radiation by charged particles are also quoted.

The earlier sections in Chapter 6 consist of material

treated in detail in many texts on plasma physics; for example, the discussion of wave motion in plasma and Landau damping. This is re-presented, sometimes in outline, in a form suitable for use later in the chapter.

2
BEAM OPTICS AND FOCUSING SYSTEMS

2.1. Introduction

The physics of beams without collective forces is in principle very simple. The trajectory of a particle with given charge-to-mass ratio is determined entirely by the external focusing fields and the three components of momentum at a specified point. Once these are known the orbits can be calculated from the equations of motion in terms of the known force on a moving charge. This is of course very laborious in general, and subsidiary concepts have been introduced, many from geometrical optics, to enable practical design to be carried out in a systematic manner.

In the design of many instruments it is necessary both to produce electric and magnetic fields of specified form and to calculate the motion of charged particles in such fields. Here we concentrate on the second of these problems, and assume that the form of the fields is given.

In the absence of interaction between the particles, a beam may be thought of as an ensemble of moving charges whose trajectories constitute a 'bundle', as illustrated in Fig.2.1. The diameter of this bundle is small compared with its length, and the trajectories through any cross-section generally make a fairly small angle with the 'axis', defined as a specially chosen trajectory within the bundle. Charged-particle optics deals with the determination of such trajectories and the collective properties of the bundles, hereafter referred to as beams. Trajectories will also be referred to as 'orbits' and 'rays'. Beams in which the particle velocity is a single-valued function of position are known as 'laminar'.

There are essentially two approaches to problems in charged-particle optics. In the first and most direct of these the equation of motion

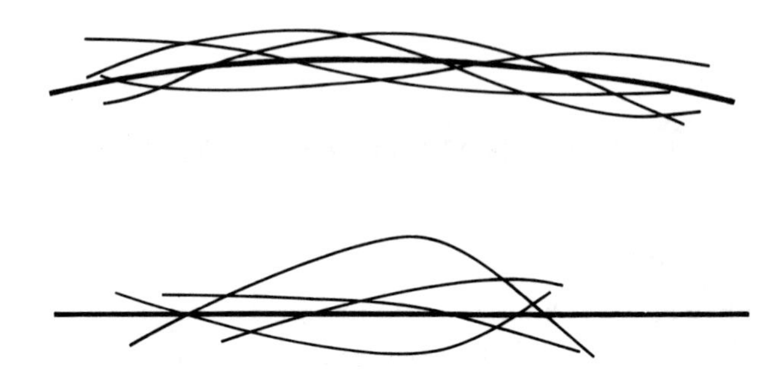

Fig.2.1. Bundle of trajectories in a charged-particle beam for two typical situations. The thick line represents the axis.

$$\frac{d\underline{p}}{dt} = q(\underline{E} + \underline{v}\times\underline{B}) \qquad (2.1)$$

is integrated in the given fields. In the second use is made of the variational principles of classical mechanics. For many purposes (for example, the study of lens aberrations) this approach is formally the more elegant; nevertheless for the purposes of the present book, the more direct method will be used.

The basic first-order optical equation which describes the motion of particles in a beam is known as the paraxial ray equation. A convenient axis for the beam is defined, and forces on the particles towards or away from the axis are expressed in terms of the fields on the axis, their derivatives, and the curvature of the axis. The fields actually experienced by the particles are expressed as first-order expansions of the field on the axis, and the angle between the particle trajectories and the axis is assumed small. This represents a good approximation in many situations. The axis is usually chosen to be a possible particle orbit, though this is not necessary, nor indeed always convenient. In cyclic accelerators for example it is sometimes convenient to choose an axis which is circular in form, even though the circle does not represent an orbit. This arises for example in systems with imperfections. Only in rectilinear systems can an orbit be found which is independent of the momentum of the particle; in all other systems the orbit is a function of the momentum, a property quantified in such concepts as 'dispersion' and 'momentum compaction'.

A paraxial equation appropriate to the most general configuration of fields which can occur would be exceedingly complicated; in this chapter we consider a number of restricted configurations which are of interest in practical devices. We examine them in such a way as to bring out both their common features and their essential differences. The effect of departing from paraxial conditions, characterized by the introduction of aberrations, is discussed in an illustrative rather than an exhaustive manner. The essential characteristics of different kinds of focusing will be indicated, with a brief discussion of some special applications such as focusing in fields which vary with time. We start by deriving the paraxial equation for systems with axial (or rotational) symmetry.

2.2. Systems with axial symmetry

2.2.1. Introduction; Busch's Theorem Many practical systems possess rotational symmetry about a rectilinear axis. It is evident that with such symmetry any electric and magnetic fields on the axis must be oriented along it, and the axis represents a possible particle trajectory. Away from the axis, radial components of field can exist; azimuthal components of electric field are not possible, nor of magnetic field unless there is a current flowing nearer the axis than the point in question.

Motion in such fields is described by the paraxial ray equation. This can be derived in several different ways, and presented in a number of different forms. In the present section a straightforward elementary approach is followed, and various forms of the equation are derived. Although the physical content of these is frequently the same, different formulations are appropriate to different problems; an attempt has been made here to bring together and compare all the commonly used versions.

As a first step in deriving the paraxial ray equation an important preliminary result known as Busch's theorem (Busch 1926) will be proved. This relates the angular velocity of a charged particle in an axially symmetrical magnetic field to the flux enclosed in a circle centred on the axis and passing through the particle. First, a straightforward physical derivation of the theory will be given.

We consider a charge q moving in an axially symmetrical magnetic field $\underline{B} = (B_r, 0, B_z)$ as shown in Fig.2.2. Then the

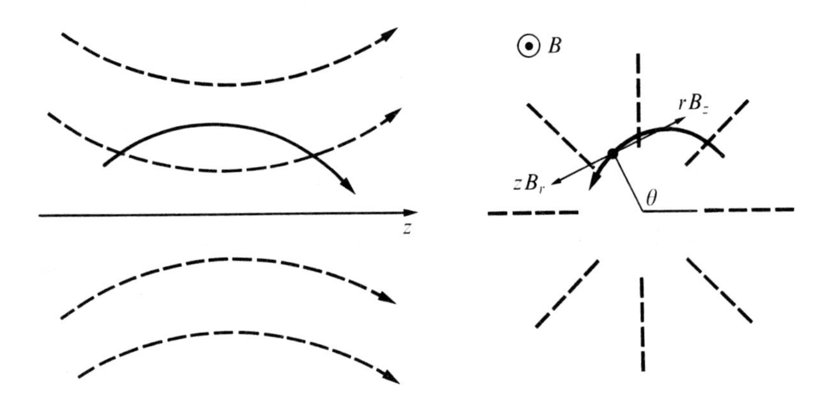

Fig.2.2. Forces on a particle moving in an axially symmetrical magnetic field. The broken lines represent the magnetic field, and the solid line the particle trajectory.

θ component of the Lorentz force can be equated to $1/r$ times the rate of change of angular momentum:

$$F_\theta = -q(\dot{r}B_z - \dot{z}B_r) = \frac{1}{r}\frac{\mathrm{d}}{\mathrm{d}t}(\gamma m_0 r^2 \dot{\theta}) \qquad (2.2)$$

Now the flux linking a circle C centred on the axis and passing through the charge is given by

$$\Psi = \int_0^r 2\pi r B_z \mathrm{d}r. \qquad (2.3)$$

As the particle moves from (r,z) to $(r + \mathrm{d}r,\ z + \mathrm{d}z)$ the rate of change of flux linking the circle can be found from the relation div $\underline{B} = 0$ to be

$$\frac{d\Psi}{dt} = 2\pi r(-B_r\dot{z}+B_z\dot{r}) . \tag{2.4}$$

Combining this with equation (2.2) and integrating with respect to time, we obtain

$$\dot{\theta} = \frac{-q}{2\pi\gamma m_0 r^2}(\Psi-\Psi_0) \tag{2.5}$$

where Ψ_0, the constant of integration, represents the flux linking the circle C when $\dot{\theta} = 0$.

Although this derivation illustrates in a direct way the physical basis of the theorem, more formally it merely expresses the fact that the canonical angular momentum P_θ about the axis is conserved:

$$P_\theta = qrA_\theta + \gamma m_0 r^2\dot{\theta} \tag{2.6}$$

where A_θ is the azimuthal component of the vector potential. Setting $\Psi = 2\pi rA_\theta$ and $\Psi_0 = 2\pi P_\theta/q$ gives equation (2.5).

As an example we may consider an electron launched with zero angular velocity from a cathode with axial symmetry; then Ψ_0 represents the flux threading the cathode at the radius at which the particle is emitted. Such cathodes are known as 'immersed'. If, on the other hand, the cathode is 'shielded', for example by placing the gun behind an iron plate with a small hole for the beam to pass through, then in a uniform field $(0, 0, B_z)$, $\Psi_0 = 0$ and $\Psi = \pi r^2 B_z$, whence

$$\dot{\theta} = -\frac{qB_z}{2\gamma m_0} = \tfrac{1}{2}\omega_c = \Omega_L \tag{2.7}$$

where ω_c and Ω_L are known respectively as the 'cyclotron' and 'Larmor' frequencies. At first sight it might be expected that $\dot{\theta}$ should equal ω_c; the paradox disappears, however, when it is realized that the particle orbit passes through the axis of the system. It may be seen from Fig.2.3 that

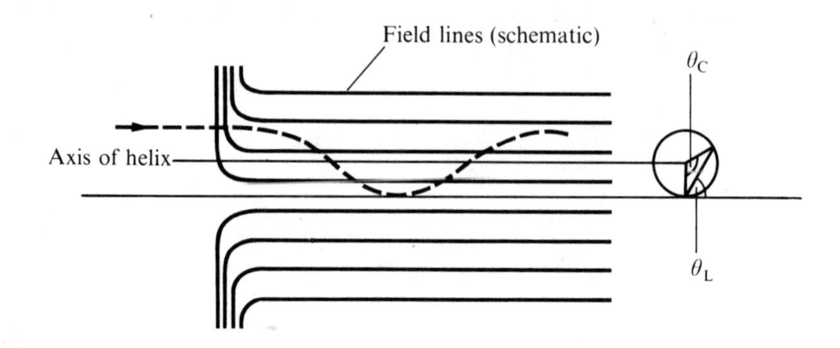

Fig.2.3. A particle moving in a circle which passes through O has angular velocity $\dot{\theta}_c$ measured about the centre of the circle, but $\dot{\theta}_L = \frac{1}{2}\dot{\theta}_c$ measured about O. Thus, a particle launched into a uniform magnetic field with zero canonical angular momentum has a frequency of ω_c about the axis of the helical orbit, but $\Omega_L = \frac{1}{2}\omega_c$ about the axis of the system.

the angular velocity about the axis is just half that about the centre of the circle formed by projecting the orbit on the (r,θ) plane.

2.2.2. The paraxial ray equation We now introduce the paraxial ray equation of form appropriate to a system with axial symmetry. This is a second-order differential equation, or pair of equations, which specifies the trajectory in the neighbourhood of the axis. The restriction to trajectories near the axis means that the angle which they make with it is small and also that only the first terms in the expansion of the fields are included. This implies that axial field components are essentially independent of r and radial components are proportional to r, where r is the radial distance from the z-axis. It is assumed also that the potential difference between the axis and the position of a particle is small compared with the kinetic energy of the particle. (This is almost, but not quite, implied in the previous assumptions.) The paraxial approximation, as will be seen, gives rise to gaussian optics. It does not embrace geometrical aberrations; these arise in the next approximation, when the above restrictions are not satisfied.

To derive the paraxial equation we start by equating the radial acceleration to the forces arising from the electric and magnetic fields; it is necessary to take into account the fact that in the presence of an axial electric field γ is a function of time:

$$\frac{d}{dt}(\gamma m_0 \dot{r}) - \gamma m_0 r \dot{\theta}^2 = q(E_r + r\dot{\theta}B_z). \quad (2.8)$$

From Busch's theorem, (equation (2.5)) and the condition that B_z is independent of r,

$$-\dot{\theta} = \frac{q}{2\gamma m_0}\left(B_z - \frac{\Psi_0}{\pi r^2}\right). \quad (2.9)$$

Eliminating $\dot{\theta}$, and writing $\dot{\gamma} \approx \beta q E_z / m_0 c$

$$\ddot{r} + \frac{\beta q E_z}{\gamma m_0 c}\dot{r} + \frac{q^2 B_z^2}{4\gamma^2 m_0^2} r - \frac{q^2 \Psi_0^2}{4\pi^2 \gamma^2 m_0^2}\frac{1}{r^3} - \frac{qE_r}{\gamma m_0} = 0 \quad (2.10)$$

This is the paraxial equation with time as independent variable. Although sometimes useful in this form, it is often more convenient to change the independent variable to z and to express the axial electric field in terms of $d\gamma/dz$; to do this the following equations are used, where primes denote derivatives with respect to z:

$$\begin{aligned} \dot{r} &= \beta c r' \\ \ddot{r} &= r''\dot{z}^2 + r'\ddot{z} \approx r''\beta^2 c^2 + r'\beta'\beta c^2 \end{aligned} \quad (2.11)$$

where βc has been written for $\dot{z}$; this is justified since $\dot{z} \gg \dot{r}$ in this approximation. Further, if there is no charge in the region occupied by the beam, we find, by taking the first term in the Taylor expansion of the field on the axis and using the condition that div $\underline{E} = 0$, that

$$E_r = -\tfrac{1}{2} r E_z' = -\tfrac{1}{2} r \gamma'' m_0 c^2 / q \quad (2.12)$$

Substituting equations (2.11) and (2.12) in equation (2.10) yields the required result:

$$r'' + \frac{\gamma' r'}{\beta^2\gamma} + \left\{\frac{\gamma''}{2\beta^2\gamma} + \left(\frac{qB_z}{2\beta\gamma m_0 c}\right)^2\right\} r - \left(\frac{q\Psi_0}{2\pi\beta\gamma m_0 c}\right)^2 \frac{1}{r^3} = 0 \quad (2.13)$$

This can be expressed more elegantly by introducing the Larmor frequency Ω_L and expressing Ψ_0 in terms of the canonical angular momentum P_θ:

$$r'' + \frac{\gamma' r'}{\beta^2\gamma} + \left(\frac{\gamma''}{2\beta^2\gamma} + \frac{\Omega_L^2}{\beta^2 c^2}\right) r - \left(\frac{P_\theta}{\beta\gamma m_0 c}\right)^2 \frac{1}{r^3} = 0. \quad (2.14)$$

The parameters β and γ are functions of the potential ϕ, and the equation is often expressed in terms of ϕ and $\phi_0 = -m_0c^2/q$. By making the substitutions

$$\gamma = 1 + \phi/\phi_0$$
$$\beta^2 = \frac{\phi(2\phi_0+\phi)}{(\phi+\phi_0)^2} \quad (2.15)$$

equation (2.13) may be written in the alternative form

$$\frac{\phi(2\phi_0+\phi)}{\phi_0+\phi} r'' + \phi' r' + \left\{\frac{\phi''}{2} - \left(\frac{qB_z}{2m_0c}\right)^2 \left(\frac{\phi_0^2}{\phi+\phi_0}\right)\right\} r + \left(\frac{q\Psi_0}{2\pi m_0 c}\right)^2 \left(\frac{\phi_0}{\phi+\phi_0}\right)\frac{1}{r^3} = 0. \quad (2.16)$$

At non-relativistic energies $\gamma \approx 1$ and $\beta^2 \approx 2\phi/\phi_0 \ll 1$. The quantity $\frac{1}{2}\beta^2\gamma^2\phi_0$, which is equal to $\phi(\phi_0+\frac{1}{2}\phi)/\phi_0$, is often called the 'relativistically corrected potential'.

Some properties of the orbits may be seen by looking at the form of the equation. The flux Ψ_0 may be set equal to $\pi B_0 r_1^2$, where r_1 represents the radius in a field B_0 at which the particle has zero angular velocity (equation 2.9). Substituting for Ψ_0 and dividing through by r_1 yields an equation in the variable (r/r_1). This implies that all trajectories for which Ψ_0 is proportional to r^2, for example those emerging normally from a plane cathode, are geometrically similar. Such trajectories cannot cross unless they pass through the axis, which can only occur when $\Psi_0 = 0$.

The paraxial equation gives the radial position of the

particle; for a given particle this depends on three initial conditions, corresponding essentially to the values of r, r', and (through Ψ or P_θ) θ'. A determination of the angular position requires a fourth condition, the initial angle θ_0. When this is known $\theta(z)$ can be found by integration of equation (2.9):

$$\theta - \theta_0 = -\int_0^z \frac{q}{2\beta\gamma m_0 c}\left(B_z - \frac{\Psi_0}{\pi r^2}\right) dz. \tag{2.17}$$

Setting $\Psi_0 = 2\pi P_\theta/q$ and expressing B_z in terms of Ω_L, this may be expressed in the alternative form

$$\theta - \theta_0 = \int_0^z \frac{1}{\beta c}\left(\Omega_L + \frac{P_\theta}{\gamma m_0 r^2}\right) dz. \tag{2.18}$$

It is often convenient, as will be seen below, to work in a frame rotating at the Larmor frequency, known as the 'Larmor frame'. This makes an angle

$$\theta_L = \int_0^z \Omega_L \, dz/\beta c \tag{2.19}$$

with the stationary frame. In the Larmor frame the angle of a particle is given by

$$\theta - \theta_L = \int_0^z \frac{P_\theta dz}{\beta\gamma c m_0 r^2}. \tag{2.20}$$

When P_θ (or Ψ_0) is zero, motion in this frame is in a plane through the axis (known as the 'meridional plane') and may be found from equations (2.13) or (2.16) alone. For non-meridional motion equations (2.13) or (2.16) must be solved simultaneously with equation (2.20). In the Larmor frame the particle has *mechanical* angular momentum P_θ.

Transformation to the Larmor frame is often useful and will frequently be made in later chapters. It represents a particular case of a more general transformation to rotating co-ordinates discussed in Appendix 1. When a transformation to a rotating frame is made, the apparent axial magnetic field is modified and an additional radial electric field appears. Denoting by subscripts 1 and 2 the stationary and rotating frames and by ω_f the rotation frequency, the values of B_z and E_r, the additional electric field, seen in the rotating frame by a particle with Larmor frequency Ω_L are

$$\begin{aligned} B_{z2} &= (1-\omega_f/\Omega_L)B_{z1} \\ E_{r2} &= r\omega_f(1-\omega_f/2\Omega_L)B_{z1} \end{aligned} \qquad (2.21)$$

The transformation is only valid for systems in which transverse velocities are non-relativistic. This is always so in paraxial systems, even though the particles are moving with relativistic velocities in the z direction. When $\omega_f = \Omega_L$, the transformation becomes simply

$$B_{z2} = 0, \qquad E_{r2} = \tfrac{1}{2}\, r\,\Omega_L B_1. \qquad (2.22)$$

The electric field represents an inward force.

The paraxial equation, essentially in the form represented by equation (2.16), was developed in the early days of electron microscopy. It has been derived here in an elementary and straightforward manner, introducing the approximations *ab initio*. We could equally well have used the Lagrangian as a starting point to find formally the equations of motion in the three directions; elimination of time between the three equations, making suitable approximations, gives the trajectory equation (2.13). For systems with axial symmetry the angle θ is a cyclic co-ordinate; the θ equation of motion yields Busch's theorem from which equation (2.17) is obtained.

Even more generally, one can arrive directly at the trajectory equations by making use of the principle of least action. This principle provides an elegant and powerful starting point for a general study of the science of electron optics. It was first applied to the subject by Glaser (1952) and has been developed in the book by Sturrock (1955). A non-relativistic derivation of the paraxial ray equation is given by Zworykin *et al.* (1945); an outline of the derivation for relativistic particles is presented in Appendix 2.

It is often convenient to work in rectangular co-ordinates and to obtain equations in terms of x and y, describing the projected orbits on two perpendicular planes. The equations in this form are linear, but coupled. We start with the equations of motion in cartesian coordinates:

$$\begin{aligned} \frac{\mathrm{d}}{\mathrm{d}t}(\gamma m_0 \dot{x}) &= q(E_x + B_z \dot{y} - B_y \dot{z}) \\ \frac{\mathrm{d}}{\mathrm{d}t}(\gamma m_0 \dot{y}) &= q(E_y - B_z \dot{x} + B_x \dot{z}). \end{aligned} \tag{2.23}$$

Substituting, as before (c.f. equations (2.11) and (2.12)), the relations

$$\ddot{x} = x''\beta^2 c^2 + x'\beta'\beta c^2 \tag{2.24}$$

$$\begin{aligned} E_x &= -\tfrac{1}{2}\,\gamma'' x\, m_0 c^2/q \\ B_x &= -\tfrac{1}{2}\left(\frac{\partial B_z}{\partial z}\right) x \end{aligned} \tag{2.25}$$

and expressing B_z in terms of the Larmor frequency

$$\Omega_{\mathrm{L}} = -\frac{qB_z}{2\gamma m_0}, \qquad \Omega_{\mathrm{L}}' = -\frac{q}{2\gamma m_0}\frac{\partial B_z}{\partial z} + \frac{\gamma'}{\gamma}\Omega_{\mathrm{L}} \tag{2.26}$$

gives rise to the pair of equations

$$\begin{aligned} x'' + \frac{\gamma' x'}{\beta^2\gamma} + \frac{\gamma'' x}{2\beta^2\gamma} + \frac{2\Omega_{\mathrm{L}} y'}{\beta c} + \frac{\Omega_{\mathrm{L}}' y}{\beta c} + \frac{\gamma'\Omega_{\mathrm{L}} y}{\beta\gamma c} &= 0 \\ y'' + \frac{\gamma' y'}{\beta^2\gamma} + \frac{\gamma'' y}{2\beta^2\gamma} - \frac{2\Omega_{\mathrm{L}} x'}{\beta c} - \frac{\Omega_{\mathrm{L}}' x}{\beta c} - \frac{\gamma'\Omega_{\mathrm{L}} x}{\beta\gamma c} &= 0. \end{aligned} \tag{2.27}$$

These can be combined to form a fourth-order linear equation in x or y, with solution depending on the initial values of x and its first three derivatives. (An instructive example is to consider a particle in a uniform magnetic field; the equation in this case reduces to third order, with independent solutions

$$
\begin{aligned}
x &= x_1 \sin(2\Omega_L z/\beta c)\\
x &= x_2 \cos(2\Omega_L z/\beta c)\\
x &= x_3.
\end{aligned}
\tag{2.28}
$$

The first two solutions represent projections of the spiralling motion and the third expresses the fact that the position of the axis is arbitrary.)

Equations (2.27) may be more compactly written by introducing the complex variable $\xi = x+jy = re^{j\theta}$. This enables them to be combined as

$$
\xi'' + \left(\frac{\gamma'}{\beta^2\gamma} - \frac{2j\Omega_L}{\beta c}\right)\xi' + \left\{\frac{\gamma''}{2\beta^2\gamma} - \frac{j}{\beta c}\left(\Omega_L' + \frac{\gamma'}{\gamma}\Omega_L\right)\right\}\xi = 0 \tag{2.29}
$$

with a corresponding equation for ξ^*. Real and imaginary coefficients represent forces parallel and perpendicular respectively to the corresponding velocities. This equation has of course the same physical content as (2.13) and (2.16). It is linear, however, and Busch's theorem was not used in the derivation, so that Ψ_0 (or P_θ) does not appear explicitly. Both x and y can be zero, though not of course simultaneously, unless $\Psi_0 = 0$. In this form of the equation the physical information contained in a knowledge of Ψ_0 must be used in specifying the initial conditions; x, y, x', y' need to be specified in place of r, r', θ, Ψ_0. In the absence of a magnetic field B_z there is no coupling between the motion in the x and y directions, and equation (2.27) reduces to identical equations in x and y.

By making a transformation to the rotating Larmor frame we have seen that the paraxial ray equation can be expressed as two identical equations in x and y of the form

$$
x'' + g_1(z)x' + g_2(z)x = 0. \tag{2.30}
$$

Alternatively, if we wish to use the laboratory frame the equation has the same form but with complex coefficients and complex variable $\xi = x + jy$, where x and y now refer to stationary axes.

Equation (2.29) is linear, and, as we see later, other forms of the paraxial equation can be put in similar form. It is convenient at this point to summarize a number of mathematical properties of equations of this type. The variable x may be considered either as real or complex in what follows. Since equation (2.30) is linear and of second order it therefore has two independent solutions $u(z)$ and $v(z)$, the general solution being of the form $Au(z) + Bv(z)$. It is sometimes convenient to take u and v as those solutions with initial conditions (1,0) and (0,1) at some value of z; these are known as the principal solutions and will be denoted by u_0 and v_0. (In some treatments, using time as independent variable, these solutions are written $M(t)$ and $S(t)$, corresponding to 'magnification' and 'stiffness'.) Alternatively, in the study of lenses, the solutions with values (1,0) at $-\infty$ and $+\infty$ respectively are often useful.

The values of x and x' at a point z_1 may be related to those at z_2 by means of a transfer matrix

$$\begin{pmatrix} x_2 \\ x_2' \end{pmatrix} = \begin{pmatrix} m_{11} & m_{12} \\ m_{21} & m_{22} \end{pmatrix} \begin{pmatrix} x_1 \\ x_1' \end{pmatrix} . \tag{2.31}$$

If M_{21} is the matrix relating (x,x') at z_2 to its value at z_1 then $M_{31} = M_{32}M_{21}$. The matrix M_{21} is sometimes written $M(z_2|z_1)$. In terms of the principal solutions $u_0(z)$ and $v_0(z)$ the matrix evidently has the form

$$M_{21} = \begin{pmatrix} u_0(z) & v_0(z) \\ u_0'(z) & v_0'(z) \end{pmatrix} . \tag{2.32}$$

An important quantity associated with equation (2.30) is the Wronskian determinant, constructed from a pair of independent solutions

$$W = uv' - vu'. \tag{2.33}$$

Differentiating and substituting in equation (2.30) gives

$W' = -g_1 W$, whence by integration

$$W(z) = W_1 \exp\left(-\int_{z_1}^{z} g_1(z)\,dz\right). \tag{2.34}$$

It follows that for a system in which $g_1 = 0$ the Wronskian is a constant. By taking the determinant of both sides of the relation

$$\begin{pmatrix} u_2 & v_2 \\ u_2' & v_2' \end{pmatrix} = \begin{pmatrix} m_{11} & m_{12} \\ m_{21} & m_{22} \end{pmatrix} \begin{pmatrix} u_1 & v_1 \\ u_1' & v_1' \end{pmatrix} \tag{2.35}$$

and using the fact that the determinant of the product matrix is equal to the product of the two component determinants, we find that

$$W_2 = |M_{21}|W_1 \tag{2.36}$$

from which the important result follows that, if $g_1 = 0$ then $|M| = 1$.

The coefficient g_1 of z' in equation (2.30) can be made zero by introducing the 'reduced' variable

$$X = \left(\frac{W_1}{W}\right)^{\frac{1}{2}} x = x \exp\left(\tfrac{1}{2}\int_{z_1}^{z} g_1(z)\,dz\right) \tag{2.37}$$

to give

$$X'' + (g_2 - \tfrac{1}{4}g_1^2 - \tfrac{1}{2}g_1')X = 0 \tag{2.38}$$

When working with reduced quantities, the determinant of the transfer matrix M_{21} is always unity.

Equation (2.37) may be used to put equation (2.13), with $P_\theta = 0$, into reduced form. Substitution in equation (2.33) gives

$$W = W_1 \exp\left(-\int \frac{\gamma'}{\beta^2\gamma}\,dz\right) = \frac{W_1}{\beta\gamma} \tag{2.39}$$

so that $R = r(\beta\gamma)^{\frac{1}{2}}$. It will be seen that W is inversely proportional to the momentum of the particle; the significance of this result will become evident later. In terms of the potential ϕ

$$\begin{aligned} R/r &= \{(\phi/\phi_0)^2 + 2\phi/\phi_0\}^{1/4} \\ &\approx (2\phi/\phi_0)^{1/4} \qquad \text{(N.R.)} \end{aligned} \tag{2.40}$$

In terms of these quantities the polar form of the paraxial equation may be written

$$R'' + \left\{\frac{\gamma'^2(\gamma^2+2)}{4\beta^4\gamma^4} + \frac{\Omega_L^2}{\beta^2c^2}\right\} R - \left(\frac{P_\theta}{m_0c}\right)^2 \frac{1}{R^3} = 0 \tag{2.41}$$

or

$$R'' + \frac{1}{4}\left\{\frac{\phi'^2(\phi^2+2\phi\phi_0+3\phi_0^2)}{(\phi^2+2\phi\phi_0)^2} + \frac{\phi_0^2}{\phi(2\phi_0+\phi)}\frac{q^2B^2}{m_0^2c^2}\right\} R - \left(\frac{q\Psi_0}{2\pi m_0c}\right)^2 \frac{1}{R^3} = 0 \tag{2.42}$$

which, in the non-relativistic limit, becomes

$$R'' + \left\{\frac{3}{16}\left(\frac{\phi'}{\phi}\right)^2 + \frac{\phi_0q^2B^2}{8\phi m_0^2c^2}\right\} R - \left(\frac{q\Psi_0}{2\pi m_0c}\right)^2\frac{1}{R^3} = 0 \quad \text{(N.R.)}. \tag{2.43}$$

The reduced variables for the complex form of the paraxial equation (2.30) may be found in the same way:

$$\begin{aligned} \Xi &= \xi \exp\left(-j\int_0^z \frac{\Omega_L}{\beta c} + \frac{\gamma'}{\beta^2\gamma}\right)dz \\ &= (\beta\gamma)^{1/2}\ \xi \exp(-j\theta_L) \end{aligned} \tag{2.44}$$

where θ_L is the angle through which the Larmor frame has rotated. Thus the change from ξ to Ξ represents not only a change of amplitude which is a function of $\beta\gamma$, but also a rotation such that the co-ordinates of the particle are specified in the Larmor frame. Equation (2.29) in reduced form becomes

$$\Xi'' + \left(\frac{\gamma'^2(\gamma^2+2)}{4\beta^4\gamma^4} + \frac{\Omega_L^2}{\beta^2c^2}\right)\Xi = 0 \tag{2.45}$$

which is identical in form to equation (2.41) with $P_\theta = 0$, except that Ξ now represents a complex variable, and the equation is valid when $P_\theta \neq 0$. The variable Ξ can be written as $X_L + jY_L$, where X_L and Y_L are projections on cartesian co-ordinates orthogonal to the z axis, the x axis being oriented at the Larmor angle $\theta_L(z)$.

It is often convenient to work in the laboratory frame. This may be done by defining a further variable

$$\Lambda = X + jY = (\beta\gamma)^{1/2}\xi \tag{2.46}$$

so that, from equation (2.44),

$$\begin{aligned}\Lambda &= \Xi \exp(j\theta_L) \\ &= (X_L + jY_L)\exp(j\theta_L).\end{aligned} \tag{2.47}$$

From equation (2.47)

$$\Lambda' = (\Xi' + j\Omega_L \Xi/\beta c)\exp(j\theta_L) \tag{2.48}$$

which, together with equation (2.47) may be written in matrix form

$$\begin{aligned}\begin{pmatrix}\Lambda \\ \Lambda'\end{pmatrix} &= \exp(j\theta_L)\begin{pmatrix}1 & 0 \\ j\Omega_L/\beta c & 1\end{pmatrix}\begin{pmatrix}\Xi \\ \Xi'\end{pmatrix} \\ \begin{pmatrix}\Xi \\ \Xi'\end{pmatrix} &= \exp(-j\theta_L)\begin{pmatrix}1 & 0 \\ -j\Omega_L/\beta c & 1\end{pmatrix}\begin{pmatrix}\Lambda \\ \Lambda'\end{pmatrix}\end{aligned} \tag{2.49}$$

We now introduce the two principal solutions of equation (2.45). Since the coefficients in this equation are real, these are real functions for which

$$\begin{aligned}\Gamma(0) &= 1, \quad \mathrm{T}(0) = 0 \\ \Gamma'(0) &= 0, \quad \mathrm{T}'(0) = 1\end{aligned} \quad . \tag{2.50}$$

Now making use of equation (2.32) (reading Γ,T for u_0,v_0),

$$\begin{pmatrix} \Lambda \\ \Lambda' \end{pmatrix} = \exp(j\theta_L) \begin{pmatrix} 1 & 0 \\ j\Omega_L/\beta c & 1 \end{pmatrix} \begin{pmatrix} \Gamma & \mathrm{T} \\ \Gamma' & \mathrm{T}' \end{pmatrix} \begin{pmatrix} 1 & 0 \\ -j\Omega_{L0}/\beta c & 1 \end{pmatrix} \begin{pmatrix} \Lambda_0 \\ \Lambda'_0 \end{pmatrix} \tag{2.51}$$

where subscripts zero denote conditions at $z = 0$. Multiplying the matrices yields

$$\begin{pmatrix} \Lambda \\ \Lambda' \end{pmatrix} = \begin{pmatrix} M & N \\ M' & N' \end{pmatrix} \begin{pmatrix} \Lambda_0 \\ \Lambda'_0 \end{pmatrix} \tag{2.52}$$

where

$$\begin{aligned} M &= (\Gamma - j\Omega_{L0}\mathrm{T}/\beta c)\exp(j\theta_L) \\ N &= \mathrm{T}\exp(j\theta_L). \end{aligned} \tag{2.53}$$

It may readily be verified that the determinant of this matrix is $\exp(2j\theta)$.

A simple example is provided by a uniform magnetic field of length $L = \beta c\phi/\Omega_L$. For such a field

$$\begin{pmatrix} \Gamma & \mathrm{T} \\ \Gamma' & \mathrm{T}' \end{pmatrix} = \begin{pmatrix} \cos\phi & (\beta c/\Omega_L)\sin\phi \\ -(\Omega_L/\beta c)\sin\phi & \cos\phi \end{pmatrix}. \tag{2.54}$$

Using the transformation to find the matrix in equation (2.52),

$$\begin{pmatrix} \Lambda \\ \Lambda' \end{pmatrix} = \begin{pmatrix} 1 & j\beta c\{\exp(-2j\phi)/\Omega_L^2\} \\ 0 & \exp(-2j\phi) \end{pmatrix} \begin{pmatrix} \Lambda(0) \\ \Lambda'(0) \end{pmatrix}. \tag{2.55}$$

This may also be expressed as a real 4×4 matrix (see section 2.7.6):

$$\begin{pmatrix} X \\ X' \\ Y \\ Y' \end{pmatrix} = \begin{pmatrix} 1 & (L\sin 2\phi)/2\phi & 0 & L(1-\cos 2\phi)/2\phi \\ 0 & \cos 2\phi & 0 & \sin 2\phi \\ 0 & -L(1-\cos 2\phi)/2\phi & 1 & (L\sin 2\phi)/2\phi \\ 0 & -\sin 2\phi & 0 & \cos 2\phi \end{pmatrix} \begin{pmatrix} X(0) \\ X'(0) \\ Y(0) \\ Y'(0) \end{pmatrix}. \tag{2.56}$$

When $\phi = \pi$ this becomes a unit matrix, which represents one complete rotation at the cyclotron frequency.

Many forms of the paraxial equation with axial symmetry have now been introduced; the notation for the variables is given in Table 2.1. The equation will be extended in later chapters to include self-fields of the beam. It will also be used as a basis for finding the envelope equation for a beam with crossing trajectories.

For systems with axial symmetry the use of complex notation avoids the need for 4 × 4 matrices. When axial magnetic fields and quadrupole fields are both present, however, a 4 × 4 matrix is essential to describe the motion. Such a situation is discussed in section 2.7.7.

2.2.3. The formation of images; magnification. Several forms of the paraxial ray equation have been presented; these describe the motion of a single particle in a rectilinear system in which the focusing is a function of z, the distance along the axis.

The optical concept of an image depends on the properties of a bundle of orbits. We consider an object, in the first instance lying in a plane perpendicular to the axis, which emits particles over a range of angles in such a way that the current emerging from a small element of area varies across the surface. At an image plane the particles originating at a point P_o on the object converge again to a point P_i, provided that only particles making a small angle with the axis are accepted so that the orbits are paraxial. This behaviour is illustrated in Fig.2.4. To prove that all rays emerging from a point converge on a point and to find the magnification of the system, we use the reduced complex variable Ξ. Let Ξ_0 be the co-ordinates of P_o; then the general solution of the paraxial equation may be written

$$\Xi = \alpha_1\Gamma(z) + \alpha_2\mathrm{T}(z) \tag{2.57}$$

where Γ and T are the principal solutions with initial conditions (1,0) and (0,1). The coefficients α_1, α_2 are complex, and the same equation relates complex conjugates of the

TABLE 2.1

Notation for various forms of the paraxial ray equation

Variables	Frame	Co-ordinate system	Type	Relations between variables	Principal solutions
x, y	Stationary	Cartesian, real	Natural	$x=r\cos\theta$, $y=r\sin\theta$	u_0, v_0
r, θ	Stationary	Polar, real	Natural		
R, θ	Stationary	Polar, real	Reduced	$R=(\beta\gamma)^{1/2}\, r$	
ξ, ξ^*	Stationary	Complex	Natural	$\xi=x+jy=r\exp(j\theta)$	
Ξ, Ξ^*	Larmor	Complex	Reduced	$\Xi=(\beta\gamma)^{1/2}\,\xi\,\exp(-j\theta_L)$	Γ, T
X_L, Y_L	Larmor	Cartesian, real	Reduced	$X_L=(\beta\gamma)^{1/2}\,(x\cos\theta_L+y\sin\theta_L)$ $Y_L=(\beta\gamma)^{1/2}\,(y\cos\theta_L-x\sin\theta_L)$	
Λ, Λ^*	Stationary	Complex	Reduced	$\Lambda=(\beta\gamma)^{1/2}\,\xi$	

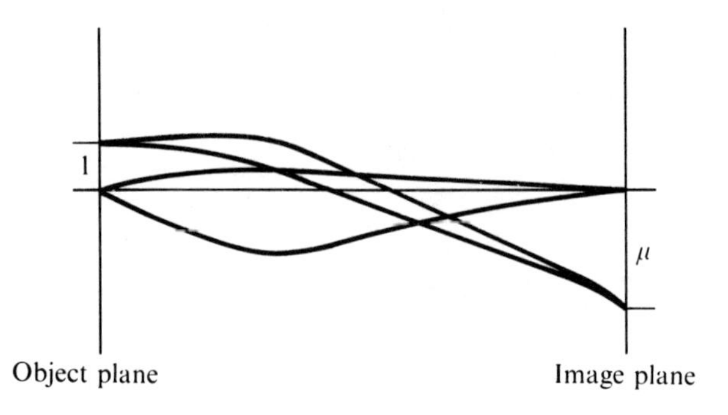

Fig.2.4. Formation of paraxial image with magnification μ (angles are exaggerated).

coefficients and variables. Whenever T is zero, Ξ depends on α_1 only. At such values of z, therefore, all particles emanating from a point on the object again pass through a point; this is the condition for an image.

The transfer matrix between object and image plane then has the form given by equation (2.31), with $m_{12} = 0$,

$$M_{oi} = \begin{pmatrix} \Gamma & 0 \\ \Gamma' & T' \end{pmatrix} \tag{2.58}$$

In particular, $\Xi_i = \Gamma\Xi_o$. The quantity Γ, which is a real number, may be termed the reduced magnification, and the actual magnification is given by

$$\mu = \frac{r_i}{r_o} = \Gamma\left(\frac{\beta_o\gamma_o}{\beta_i\gamma_i}\right) \tag{2.59}$$

where subscripts i and o , as elsewhere in this section, refer to image and object. For a trajectory starting on the axis Ξ_o and Ξ_i are zero; since $|M_i| = 1$ it follows from equation (2.58) that $\Gamma = (T')^{-1} = \Xi_o'/\Xi_i'$, whence, from equation (2.59)

$$\mu = \left|\left\{\frac{\Xi_o'(\beta_o\gamma_o)^{1/2}}{\Xi_i'(\beta_i\gamma_i)^{1/2}}\right\}\right| . \tag{2.60}$$

Furthermore, the image is rotated through an angle

$$\theta = \tan^{-1}(\mathrm{Im}\,\Xi_i/\mathrm{Re}\,\Xi_i) - \tan^{-1}(\mathrm{Im}\,\Xi_o/\mathrm{Re}\,\Xi_o). \quad (2.61)$$

If the object and image are in field-free regions equation (2.60) becomes

$$\mu = \frac{r_o'\beta_o\gamma_o}{r_i'\beta_i\gamma_i} \quad (2.62)$$

This will be seen later (section 5.3) as an example of a more general result. In dynamical terms it states that, as a particle moves from object to image, its transverse momentum changes in inverse proportion to μ. It is not difficult to show further that the longitudinal magnification associated with an object of finite extent in in the z-direction in paraxial approximation is equal to μ^2.

2.2.4. Some general properties of lenses. Lenses play an important part in the formation and control of beams; in the next few sections we describe briefly some of their essential properties.

We study first lenses having either an electrostatic or magnetic field configuration with axial symmetry, confining attention to the region near the axis where the behaviour is paraxial. The fields are produced by electrodes away from the axis, so that their axial extent is not short compared with the beam diameter. Frequently, however, the lenses can nevertheless be considered 'thin' in the sense that the change in r of a trajectory passing through the lens is small.

Five parameters are needed for the specification of the properties of a thick lens; these are the position of the two principal planes, the two focal lengths, and the Larmor rotation angle. If we work in the Larmor frame, this angle can be decoupled from the other parameters; we do not therefore consider it further. The other parameters we discuss in terms of two solutions $u_1(z)$ and $u_2(z)$ of the paraxial ray equation which have initial and final values of (r,r') respectively outside the field region given by $(1,0)$. In Fig.2.5 the two planes at z_a and z_b represent

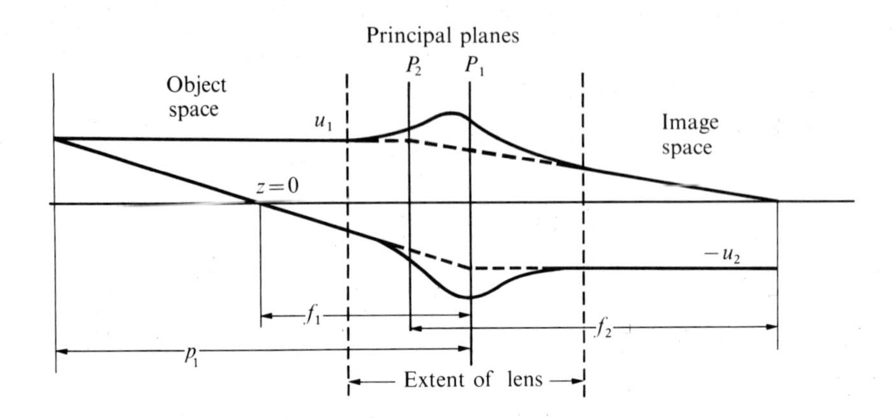

Fig.2.5. Parameters associated with a thick lens: P_1 and P_2 are the principal planes, f_1 and f_2 the focal lengths, u_1 and u_2 solutions of the paraxial equation with initial and final conditions (1,0), and the broken lines represent the extent of the lens.

the boundaries of the region in which the focusing field exists. The regions, $z < z_a$ and $z > z_b$ may be termed the 'object space' and 'image space'. Projections of the end sections of the two trajectories, which are straight lines, meet at points which determine the principal planes P_2 and P_1, as shown in the figure. Distances from these planes to the points where the rays cross the axes define the focal lengths f_1 and f_2. The position of the two principal planes and the values of f_1 and f_2 define completely the optical properties of the lens. The formulae relating object and image distance and magnification to the lens parameters can be found from simple geometrical considerations. Proofs are given in any standard text on optics; here the resulting formulae are quoted with the notation defined in Fig.2.5. The origins are taken to be at the principal planes, distances measured to the right being positive and to the left negative:

$$\frac{f_1}{p_1} + \frac{f_2}{p_2} = 1$$
$$\frac{h_1}{h_2} = -\frac{f_1}{f_2}\frac{p_2}{p_1} \qquad (2.63)$$

where h_1, h_2 are the heights of the object and image and p_1 and p_2 their respective distances from the corresponding principal plane. The 'power' of the lens is defined as $(f_1 f_2)^{-1/2}$. It is negative when f_1 and f_2 are negative. Fig.2.5 refers to converging lenses; we shall see later that all electrostatic and magnetic lenses with no charge or current in the beam space are of this type. The simplifications implied in equation (2.63) are not applicable to the location and characterization of images which are formed in the region where the fields do not vanish. Objects and images in this region are said to be 'immersed', and their relation to images and objects outside the region can only be found by ray tracing through the lens. The image illustrated in Fig.2.4 is of this type. Because of the possibility of such images the more general terms 'incident space' and 'emergent space' are to be preferred to 'object space' and 'image space' introduced above.

In general, the parameters f and p can conveniently be assigned to a lens only after trajectory calculations have been made; when two trajectories are known, relations between the input and output conditions yield enough information to find the four constants. Explicit evaluation in terms of integrals of fields along the axis is in general too complicated to be useful, though some approximate results may be obtained in this way.

It is often convenient to represent the lens system by an equivalent matrix. The form of the matrix depends of course on where the boundaries of the lens are considered to be. Particularly simple is the matrix for transit through a thin lens, for which the principal planes coincide. This is readily verified to be

$$M = \begin{pmatrix} 1 & 0 \\ -\dfrac{1}{f_2} & \dfrac{f_1}{f_2} \end{pmatrix} = \left(\frac{f_1}{f_2}\right)^{1/2} \begin{pmatrix} \left(\dfrac{f_2}{f_1}\right)^{1/2} & 0 \\ \dfrac{-1}{(f_1 f_2)^{1/2}} & \left(\dfrac{f_1}{f_2}\right)^{1/2} \end{pmatrix}. \quad (2.64)$$

The determinant of the first matrix can only be unity if $f_1 = f_2$. This implies that for a lens in which $f_1 \neq f_2$ the particle momentum is different on both sides of the lens. The second matrix has unit determinant, so that the ratio of natural to reduced variables changes by $(f_1/f_2)^{1/2}$ in passing through the lens. Since from equation (2.39) this ratio is $(\beta\gamma)^{-1/2}$, it follows that

$$f_1/f_2 = \beta_1\gamma_1/\beta_2\gamma_2. \quad (2.65)$$

This result, closely related to equation (2.62), corresponds to the Helmholtz-Lagrange relation in light optics relating the focal lengths to refractive indices on both sides of the lens. The angular divergence of the rays at the focus is inversely proportional to $\beta\gamma$, a result to be encountered in more general form in Chapter 4.

In the following sections some basic properties of the more common types of electrostatic and magnetic lenses are described.

2.2.5. Electrostatic lenses In Fig.2.6 a number of commonly used configurations for electrostatic lenses are sketched. Starting with cylinders, planes containing circular holes, and grids, a large number of combinations can be devised. These can be further increased by the use of conical, spherical, and other curved shapes. Most of these have been extensively studied, their advantages and disadvantages for different applications evaluated, and their properties exhibited in the form of suitable graphs and tables. This information is spread widely throughout the literature (for references see section 2.R.2). Here we deduce some of the simpler general properties of these lenses. To do this we

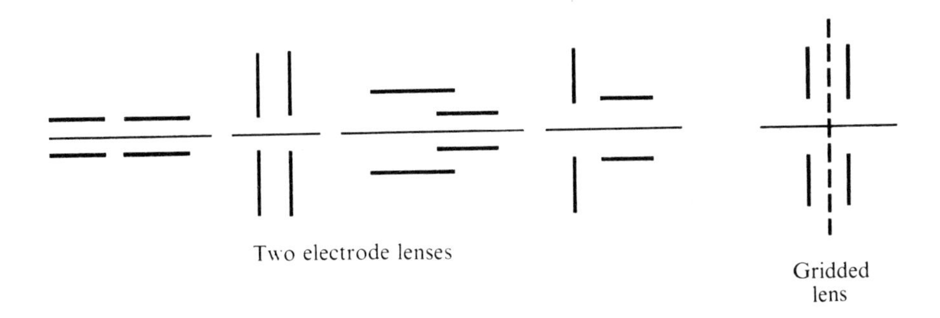

Fig.2.6. Basic configurations for two-electrode electrostatic lenses. In practice the variety of configurations is very large; apertures and cylinders have finite thickness and are rounded. Hemispherical and conical elements are also used. Three-electrode lenses, across which there is no net change of potential, can readily be formed by adding a further cylinder or plate. A gridded lens is also shown.

start with the paraxial equation, omitting the magnetic terms. To harmonize with the normal usage in electron optics we use the notation of equation (2.16) with $\underline{B} = 0$:

$$\frac{\phi(2\phi_0+\phi)}{\phi_0+\phi}\, r'' + \phi' r' + \frac{\phi''}{2}\, r = 0 \qquad (2.66)$$

where ϕ is a function of z determined by the lens geometry. The problem is to determine the lens parameters as a function of ϕ. Solution of equation (2.66) is facilitated by the use of reduced variables (equation (2.40))

$$R'' + T(z)R = 0 \qquad (2.67)$$

where $T(z)$ is defined in equation (2.42) with B and Ψ_0 set equal to zero. For a 'weak' lens, defined as a lens in which R remains sensibly constant along a trajectory originally parallel to the axis, the focal length f may be readily found. It follows that, if R changes little through the lens, then f must be much greater than thickness of the lens, which we may therefore regard as thin. If subscripts 1 and 2 refer to conditions before entering and after leaving the lens, then for a trajectory with $r_1' = 0$

$$\frac{1}{f_2} = -\frac{r_2'}{r_1} = -\left(\frac{\phi_1^2+2\phi_1\phi_0}{\phi_2^2+2\phi_2\phi_0}\right)^{1/4}\frac{R_2'}{R_1} \tag{2.68}$$

For a thin lens, with $|\phi_2-\phi_1| \ll \phi_2$,

$$\frac{R_2'}{R_1} \approx \int_{z_1}^{z_2}\frac{R''\mathrm{d}z}{R} = -\int T(z)\,\mathrm{d}z \tag{2.69}$$

leading to the final result

$$\frac{1}{f_2} = \left(\frac{\phi_1^2+2\phi_1\phi_0}{\phi_2^2+2\phi_2\phi_0}\right)^{1/4}\int_{-\infty}^{\infty}\frac{\phi'^2(\phi^2+2\phi\phi_0+3\phi_0^2)\,\mathrm{d}z}{4(\phi^2+2\phi\phi_0)^2} \tag{2.70}$$

which in the non-relativistic and extreme relativistic limits becomes

$$\frac{1}{f_2} = \frac{3}{16}\left(\frac{\phi_1}{\phi_2}\right)^{1/4}\int\left(\frac{\phi'}{\phi}\right)^2\mathrm{d}z \quad \text{(N.R.)} \tag{2.71}$$

$$\frac{1}{f_2} = \frac{1}{4}\left(\frac{\phi_1}{\phi_2}\right)^{1/2}\int\left(\frac{\phi'}{\phi}\right)^2\mathrm{d}z \quad \text{(E.R.)} \tag{2.72}$$

In both these limiting forms the integral is independent of ϕ. Evidently f_2 must be positive; it would appear therefore that all lenses are convergent. (This observation is not true for lenses containing grids or foils, for which Poisson's equation is not valid in the beam space, nor for incomplete lenses such as the aperture lens; both of these types are considered later.)

The validity of the thin-lens formula calculated using equations (2.67) and (2.68) can be estimated by considering the idealized uniform lens, where T is constant from $z = 0$ to $z = L$ and zero elsewhere and TL is constant. In this illustrative calculation the potential variation across the lens is taken as zero, so that $r/r' \approx R/R'$. The transfer matrix for such a lens is readily seen to be

$$M_{21} = \begin{pmatrix} \cos LT^{1/2} & T^{-1/2}\sin LT^{1/2} \\ -T^{1/2}\sin LT^{1/2} & \cos LT^{1/2} \end{pmatrix} \tag{2.73}$$

(see section 2.4.2 below). For the principal solution initially parallel to the axis

$$\begin{pmatrix} r_2 \\ r'_2 \end{pmatrix} = M_{21}\begin{pmatrix} 1 \\ 0 \end{pmatrix} = \begin{pmatrix} \cos LT^{1/2} \\ -T^{1/2}\sin LT^{1/2} \end{pmatrix} \tag{2.74}$$

where the subscripts 1 and 2 refer to conditions before and after the lens.

Defining f as the distance from the lens centre to the point where the initially parallel ray meets the axis,

$$f = \frac{L}{2} - \frac{r_2}{r'_2} = \frac{L}{2} + \frac{\cos LT^{1/2}}{T^{1/2}\sin LT^{1/2}} . \tag{2.75}$$

Writing $(LT)^{-1} = f_t$, the focal length of the thin lens where $L \to 0$, the second term can be expanded to give

$$f \approx f_t + \frac{L}{3} . \tag{2.76}$$

The focal length of the thick lens exceeds that of the thin lens by one-third of its length.

If there is a change of potential in crossing a lens, then from equation (2.41) $f_1 \neq f_2$. By symmetry

$$\frac{f_2}{f_1} = \left(\frac{\phi_2^2+2\phi_2\phi_0}{\phi_1^2+2\phi_1\phi_0}\right)^{1/2} = \frac{\beta_2\gamma_2}{\beta_1\gamma_1} , \tag{2.77}$$

in agreement with equation (2.65).

Equation (2.70) shows that, for a lens in which $\phi_1 \approx \phi_2$, the strength is proportional to the square of the electric field. This implies that focusing is a second-order effect, a fact which may be seen physically as follows. From the shape of the electric field, it is evident that the direction of the force on the charge changes sign from inward to outward (or vice versa) as the particle passes through the lens. If the particle direction and velocity remained unchanged, then the net inward impulse received by a particle moving at a distance r_1 from the axis would be

$$F = \frac{\int qE_r(r_1)\,\mathrm{d}z}{\beta c}\,. \tag{2.78}$$

Now

$$\frac{\partial E_z}{\partial z} = -\frac{1}{r}\frac{\partial}{\partial r}(rE_r) \tag{2.79}$$

so that to paraxial approximation, in which $\partial E_r/\partial r$ is constant,

$$\int E_r\mathrm{d}z = -\frac{r_1}{2}\int\frac{\partial E_z}{\partial z}\,\mathrm{d}z = 0. \tag{2.80}$$

An actual particle receives a net acceleration by virtue of the fact that both r and the particle velocity vary; this unbalances the two opposing forces, and the effects of the variation of both radius and velocity increase the inward force and decrease the outward force. The force is proportional both to the field and to the amount of unbalance, which is itself proportional to the field, resulting in an overall quadratic dependence which is always of such a sign as to produce focusing.

In the discussion so far it has been assumed that the rays are meridional. The properties of skew rays may be conveniently described as before in terms of the complex variables

$$\xi = x+\mathrm{j}y = r\exp(\mathrm{j}\theta)$$

and its complex conjugate.

We have so far considered a lens as being bounded by field-free regions. It is sometimes, however, convenient to consider the focusing properties between two regions, one of which is not field free; in electron guns for example the beam is emitted by a cathode into a region of non-zero field. For configurations of this sort, which represent only part of a lens as defined earlier, the action can be either converging or diverging. A particularly simple and important lens of this form is the aperture lens, consisting of a hole in a conducting sheet which separates two regions of uniform field. The behaviour of such a lens can be estimated quite easily if it is assumed that the region of non-uniform field is confined to a range of z comparable with the aperture radius and short compared with the focal length.

Let ϕ_1' and ϕ_2' be the potential gradients on the two sides of the aperture, and ϕ the potential corresponding to the energy of the particle; then, if the focal length is large compared with the aperture, the r' term in the paraxial ray equation (2.66) is small for a trajectory initially parallel to the axis. Under these conditions it may be integrated to give

$$-\frac{r}{r_2'} = f_2 = \frac{2\phi(2\phi_0+\phi)}{(\phi_0+\phi)(\phi_2'-\phi_1')} . \tag{2.81}$$

In the non-relativistic limit this takes the simple form

$$f_2 = \frac{4\phi}{\phi_2'-\phi_1'} . \tag{2.82}$$

The sign of f depends on whether the strength of the field increases or decreases across the aperture, so that, in contrast to a complete lens, it can be negative. Because the field is not zero in the region of the focus the trajectories are curved and an initially parallel beam will not pass through the focus defined by equation (2.81). As indicated earlier, the concepts outlined in Fig.2.5 are not applicable to this type of situation. The strength of the lens depends on the ratio of the fields, but not on the size of the aperture hole.

So far only lenses without charges or currents in the beam space have been considered. An example of a lens which does not have this restriction is the gridded lens. Such a lens consists essentially of two aperture lenses back to back separated by a plane at a different potential from the apertures, as shown in Fig.2.6. The plane is composed of a fine mesh or grid, so that it is effectively transparent to the charges. Although it has technical limitations, such as a limited capacity for handling intense beams, the focusing provided is much stronger than that in the corresponding three-aperture lens and spherical aberration is reduced. The discontinuous reversal of slope of the field lines at the grid, combined with the change of direction of field, provides a radial force which, in contrast to other lenses, is of the same sign throughout the lens. Depending on the potential of the grid, the lens can be either focusing or defocusing. The focal length depends on ϕ rather than on ϕ^2, the lens exhibits first order focusing in contrast to those discussed earlier where there are no charges in the beam space. The expression for the focal length is not a simple function of the geometry, and the reader is referred to the references for details.

This brief outline contains the essential physical ideas relating to the linear properties of electrostatic lenses. Aberrations are considered in section 2.2.7.

2.2.6. Magnetic lenses To determine the focal length of a magnetic lens we start with the paraxial equation for a system with magnetic fields only and consider a trajectory originally parallel to the axis, so that $P_\theta = 0$:

$$r'' = -\left(\frac{qB_z}{2\beta\gamma m_0 c}\right)^2 r = -\left(\frac{\Omega_L}{\beta c}\right)^2 r \tag{2.83}$$

with

$$\theta = \theta_0 + \int \left(\frac{\Omega_L}{\beta c}\right) dz. \tag{2.84}$$

This equation is relativistic, and the expression for r'' is of the same form as the reduced equation (2.67) used to evaluate the focal length of an electrostatic lens.

Using the same approximations as in the electrostatic calculation for the thin lens, it may readily be shown that

$$\frac{1}{f_1} = \frac{1}{f_2} = \int \left(\frac{\Omega_L}{\beta c}\right)^2 dz. \tag{2.85}$$

As with electrostatic lenses, f is always positive. The focusing is again a second-order effect, but for apparently different physical reasons. The force towards the axis is proportional to the product of the field and the circumferential velocity; the circumferential velocity which is produced is itself proportional to the field, giving a net effect proportional to the square of the field. Two essential differences between magnetic and electrostatic focusing are that the dependence on particle energy is very different, and also for magnetic focusing skew motion is produced in trajectories originally parallel to the axis. These differences disappear if the system is viewed in the Larmor frame. Motion of a typical trajectory through such a lens is shown in Fig.2.7.

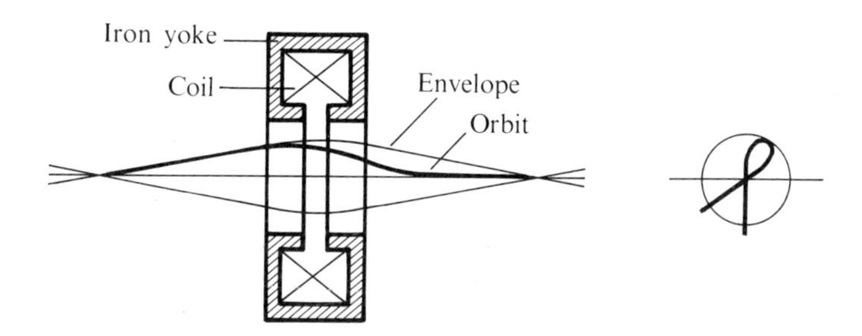

Fig.2.7. Basic elements of a magnetic lens showing a typical orbit. In practical designs the iron is often shaped in the neighbourhood of the gap.

In the previous section it was remarked that much stronger electrostatic lenses can be produced by using grids, which effectively introduce charge into the beam region. By introducing current into the beam region of magnetic lenses, an analogous enhancement of strength can be produced. This arises because such lenses exhibit first-order focusing, whereas axially symmetric lenses with a clear beam space only produce a second-order effect. Magnetic lenses with current in the beam region have been produced for special applications requiring the focusing of very high-energy particles, which are not stopped or appreciably scattered by the conductor which carries the current. The 'magnetic horn' and a 'plasma lens', both designed for focusing very high-energy π-mesons, which subsequently decay to μ-mesons and neutrinos, are illustrated in Fig.2.8.

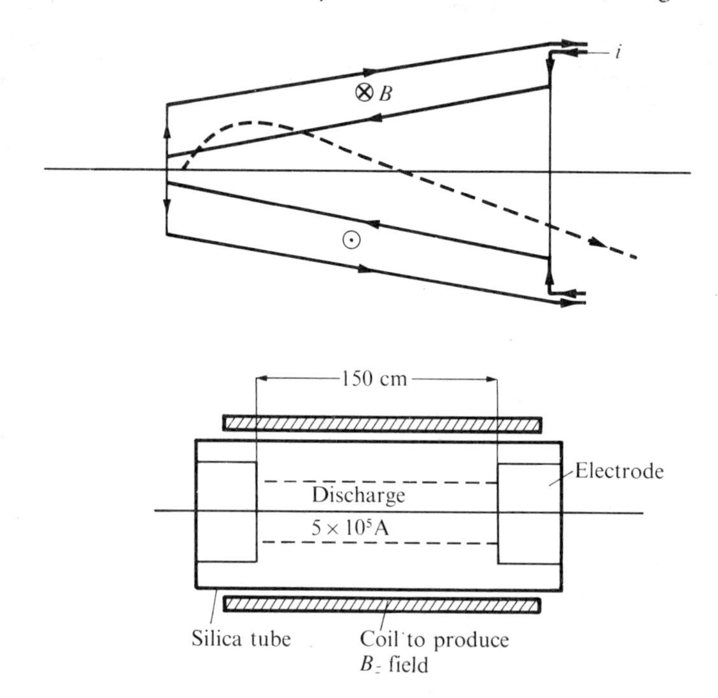

Fig.2.8. Schematic diagram of pulsed magnetic horn (van der Meer, 1961) and plasma lens (Forsyth, Lederman, and Sunderland 1965), both used to focus π-mesons in experiment to produce μ-neutrinos.

The first-order theory of thin electrostatic and magnetic lenses with axial symmetry is, as we have seen, relatively straightforward, though detailed calculations for particular lenses can be lengthy. More complicated is an analysis of the various types of aberration and distortion which arise from mechanical imperfections, essential non-linearities in the optics, energy spread in the beam, thermal velocities, particle diffraction, and space-charge effects. The topics of space charge and the effects of thermal-velocity distributions will be discussed later in a more general context; spherical aberration is analysed in the next section, and other forms of aberration are briefly described. The way in which aberrations limit the size of spot to which a beam can be focused is discussed in Chapter 4.

Lenses exhibiting axial symmetry find wide application in electron guns, cathode-ray tubes, electron microscopes, ion sources, and d.c. electron and ion accelerators, as well as other more specialized instruments. After a discussion of aberrations we treat systems with a different type of symmetry.

2.2.7. Aberrations in axially symmetrical lenses The lens properties studied so far have been derived from the paraxial ray equation; objects in a plane perpendicular to the beam axis can be perfectly imaged in a plane where points map to points and the image is geometrically similar to the object.

The paraxial equation is only an approximation, however, and higher-order terms cause the image to be imperfect in various ways. In general the image is distorted, it does not lie in a plane, and rays from a point on the object do not pass through a single point on the image. In a different category from these effects is chromatic aberration, which does not imply any non-linear terms in the ray equation; it can be thought of as producing a continuum of perfect images, each one corresponding to a particular particle energy present in the beam. Its characteristics may readily be evaluated in terms of the variation with energy of the coefficients of the paraxial ray equation. For the moment we confine attention to axially symmetrical lenses. The most important defect is the spherical aberration, and this is the only one which will be considered in detail.

Because of the fundamental constraints on the field shapes in the beam region imposed by the conditions curl $\underline{B} = 0$ and div $\underline{E} = 0$, the measures that can be taken to correct aberrations are limited; in particular the spherical aberration cannot be made zero. The situation is more difficult than in light optics, where there is a great deal of freedom in choosing lens shapes and materials. To achieve the corresponding degree of freedom in electron optics it would be necessary to place shaped charge or current distributions within the beam region. Indeed, spherical aberration in electrostatic lenses can in principle be reduced or removed entirely by the use of foils, grids, or charge clouds in the focusing region; unfortunately the scattering and interception introduced is unacceptable for most applications. A concise summary of some of these possibilities with references is given by Ash (1964); more details may be found in the review by Septier (1966).

We consider now aberrations which arise from including terms in r^3 in the paraxial ray equation (2.14) or (2.15). (Terms in r^2 are excluded by symmetry, since changing the sign of r must change the direction of the force.) Since the equation is no longer linear, the amount of distortion is affected not only by the coefficients of the third-order terms in the equation, but also by the allowed radial excursion of the orbits.

In order to characterize the aberrations, we consider a particle leaving a point (r_o, θ_o) on the object which arrives at a point $(r_i + \Delta r, \theta_i + \Delta\theta)$ on the image plane, where Δr and $\Delta\theta$ are zero in a perfect lens. Now Δr and $\Delta\theta$ depend on the path taken by the ray from (r_o, θ_o), this can be defined by specifying the values of r and θ in any arbitrarily chosen defining plane. (This is conveniently located at the point where the principal paraxial ray with zero initial slope, projected back from the image point, crosses the axis.) The co-ordinates in this plane will be denoted by r_d and θ_d, as shown in Fig.2.9.

It is convenient to work in complex variables $\xi = re^{j\theta}$. The effect of imperfections can then be expressed as the quantity

$$\Delta\xi_i = \xi_i - M\xi_o \tag{2.86}$$

where M is the (complex) magnification of the lens without imperfections. The complex form of M takes into account the rotation of the image as well as its size, (equation (2.60)

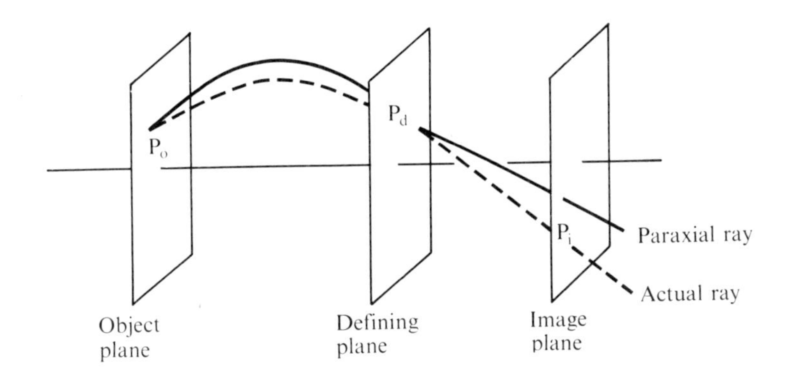

Fig.2.9. Diagram to illustrate aberration from the paraxial ray. The co-ordinates of the points P_o, P_d, and P_i where the actual ray crosses the three planes, are given by the real and imaginary parts of ξ_o, ξ_d, and $\xi_i + \Delta\xi_i$.

without the modulus sign). The quantity $\Delta\xi_i$ is a function of ξ_o, ξ_d and their complex conjugates. Constraints on the form of the function can be found by noting that, if ξ_i and ξ_o are both changed by a factor $e^{j\theta}$, then this represents a pure rotation of the co-ordinate system so that $\Delta\xi_i$ must change by θ also. This eliminates all even products in the expansion of $\Delta\xi_i$, such as $\xi_o\xi_d$, and some odd products, for example $\xi_o\xi_d^{*2}$. The lowest-order non-zero terms are third order, and those which satisfy the required constraints are retained in equation (2.86):

$$\Delta\xi_i = A\xi_d^*\xi_d^2 + B\xi_o^*\xi_d^2 + C\xi_o\xi_d\xi_d^* + D\xi_o^2\xi_d^* + E\xi_o\xi_o^*\xi_d + F\xi_o^2\xi_o^*. \qquad (2.87)$$

This expression contains 12 real coefficients, the two associated with A being $\frac{1}{2}(A+A^*)$ and $-\frac{1}{2}j(A-A^*)$. From purely geometrical considerations it is possible to show, using the theorem of Malus and Dupin, that some of these coefficients are related (Born and Wolf 1965). This theorem can be stated in the form that the imaginary part of $\partial\Delta\xi_i/\partial\xi_d$ is zero. Since ξ_o and ξ_d are independent, this condition may be applied separately to each term of equation (2.87). Applying it yields

$$\begin{aligned} \operatorname{Im} A &= \operatorname{Im} E = 0 \\ \operatorname{Re} B &= \tfrac{1}{2}\operatorname{Re} E \\ \operatorname{Im} B &= -\tfrac{1}{2}\operatorname{Im} C \,. \end{aligned} \tag{2.88}$$

There are therefore eight coefficients which characterize the distortion and aberration of the image. In purely electrostatic systems with no image rotation only the real coefficients remain. There are five of these, as in geometrical optics.

For details of the classification of the forms of aberration, the type of distortion they produce and the methods of calculation the reader is referred to texts on electron optics and the references quoted. A great deal of effort has been expended on the study of aberrations and methods of reducing them. This is a highly specialized subject; we confine attention to spherical and chromatic aberration to be discussed in the next two sections.

2.2.8. Spherical aberration Of the classes of aberration implied by the various coefficients in equation (2.87), only one applies to points on the lens axis, where $\xi_0 = 0$. This is the spherical aberration, with coefficient A, which from equation (2.88) is real. Rays leaving the object point at different angles to the axis cross again at different values of z. From equation (2.87) it is evident that rays strike the image plane for a perfect lens at radial distances proportional to r_d^3, which in turn is proportional to the cube of the angle to the axis at which particles leave the object point (Fig.2.9). The spherical aberration coefficient C_s is defined by the equation

$$\Delta r_i = C_s \alpha_i^3 \tag{2.89}$$

where α_i is the angle of arrival to the axis of a ray originating at the object point. It represents a property of the lens which does not involve the defining aperture. It is evident from equation (2.87), however, that, if the image point is in a field-free region and a defining aperture of diameter $r_d = |\xi_d|$ is placed at a distance z_o from the image point, then $C_s = Az_o^3$.

The structure of the beam near the cross-over is shown in Fig. 2.10. Although the circle of confusion on the image

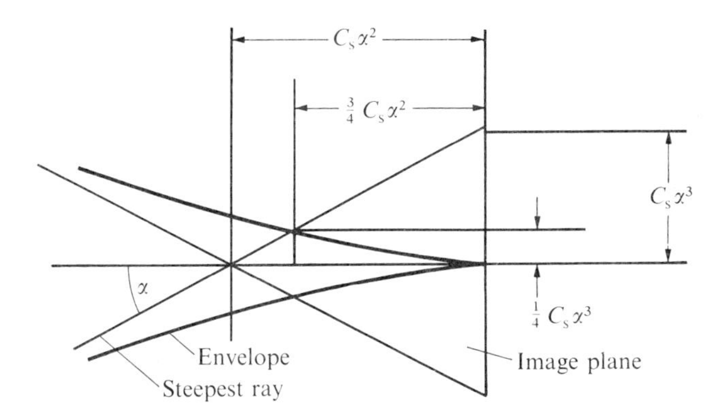

Fig.2.10. Structure near the focus of a lens with spherical aberration. The equation of the envelope of the rays is $r = 2(z^3/27C_s)^{1/2}$ where z is measured backwards from the image plane.

plane corresponding to a perfect lens has radius $C_s\alpha_i^3$, all rays pass through a circle of radius $\frac{1}{4}C_s\alpha_i^3$ at a distance $\frac{3}{4}C_s\alpha_i^2$ in front of the image plane. The coefficient C_s may be evaluated in various ways. A straightforward approach is to retain the cubic terms in the paraxial ray equation and evaluate the difference between the accurate and the paraxial trajectory originating at the object point. If the former is known, C_s can be expressed in terms of an integration involving a knowledge of the paraxial trajectory and the values of the magnetic field and electric potential on the axis. An alternative, and more elegant, approach is the variational method of characteristic functions introduced by Glaser (1952) and developed by Sturrock (1951, 1952). Both are somewhat lengthy, however, and they will not be described here. Details can be found in the references cited in section 2.R.2. For several common types of lens the value of the dimensionless parameter C_s/f has been evaluated and is available in the form of graphs and tables.

The sign of the spherical aberration is always such that rays remote from the axis are focused more strongly than rays close to it. Scherzer (1936) has shown that, in the absence of currents or charges in the region through which the trajectories pass, the formula for the aberration can be expressed as the sum of squared terms, so that the sign cannot be changed; in particular the aberration cannot be made zero. In addition to its obvious importance in electron microscopy, spherical aberration represents an important constraint in the design of high-quality electron guns and in the production of fine, bright electron spots for such purposes as electron machining (section 4.4.8). An interesting historical review of the subject is given in the article by Septier (1966). In addition to the discussion of the use of space charge referred to earlier, the possibilities are examined of using time-varying fields, and departures from axial symmetry to overcome the essential constraints found by Scherzer.

2.2.9. Chromatic aberration The phenomenon of chromatic aberration will now be introduced with the aid of a simple example, and a suitable coefficient defined.

If a parallel beam is incident on a lens and leaves it with particle momentum p, then for a perfect lens there will be a focal point on the axis at some point z_i. Particles with momentum $p + \Delta p$ on the other hand are focused at $z = z_i + (\partial f/\partial p)\Delta p$. If the angle of convergence of the rays is α, then the radius of least confusion is evidently given by

$$r_c = \alpha\left(\frac{\partial f}{\partial p}\right)\Delta p = \alpha f\left(\frac{p}{f}\frac{\partial f}{\partial p}\right)\frac{\Delta p}{p}. \tag{2.90}$$

It is convenient to define the coefficient of chromatic aberration C_c for a lens as

$$r_C = 2C_C \alpha \frac{\Delta p}{p} = 2C_C \alpha \frac{\Delta\gamma}{\beta^2\gamma} \qquad (2.91)$$

where $C_C/f = \frac{1}{2}(p/f)(\partial f/\partial p)$. In the non-relativistic limit this may be written

$$r_C = C_C \alpha \frac{\Delta\phi}{\phi} \qquad (2.92)$$

where ϕ is the potential through which the particles have been accelerated to reach the momentum p. It should be noted that $\Delta\phi$ refers to half the total energy spread in the beam.

For first- and second-order thin magnetic lenses $(p/f)(\partial f/\partial p)$ is equal respectively to -1 and -2; this follows from the fact that f is proportional to $p/\int B\mathrm{d}z$ and $p^2/\int B^2\mathrm{d}z$ for such lenses. The same is true for electrostatic lenses provided that the variation of potential energy across the lens is small compared with the particle energies. General expressions for thick lenses have been obtained in terms of the field or potential along the axis. These are complicated.

This very simplified discussion illustrates the physical nature of the effect, and provides a definition of C_C suitable for later use.

2.R.2. Notes and references Most of the material in these sections was developed in the 1930's in the context of electron microscopy. It is well covered in a number of texts, in which references to original papers and derivations may be found. Much of the treatment is non-relativistic, however, relativistic effects being sometimes incorporated as a 'correction' in which the relativistically corrected potential (section 2.2.2) is used.

Early classic texts include those of Zworykin *et al* (1945) and Glaser (1952). More formal is the theoretical text by Sturrock (1955) with emphasis on variational methods. Recent examples of straightforward and practically oriented texts are those of Klemperer and Barnett (1971), Grivet and Septier (1972), and Hawkes (1972). Sections on the paraxial ray equation and cylindrical lenses may also be found in the two-volume compilation *Focusing of charged particles* edited by Septier (1967a). The paraxial ray equation is also discussed at length by Pierce (1954a) and Kirstein, Kino, and Waters (1967); the emphasis in these texts is more on applications to microwave tubes than microscopy. Pierce's treatment emphasizes physical rather than mathematical features and forms a good introduction to applications in microwave tubes. The treatment of Kirstein *et al.* is more advanced; our discussion in terms of complex variables is based on theirs. Both these books consider only non-relativistic beams; both consider lenses, but only to first order since aberration theory is barely relevant to microwave tubes.

Although the books mentioned in the previous paragraph contain much information on lenses, a great deal more detailed information is scattered throughout the literature in journals and conference proceedings. A comprehensive account is that of Mulvey and Wallington (1973).

Recent texts with application to other fields such as microprobe analysis, electron-beam welding, and ion implantation also contain material on the topics under review. They derive fundamental ideas from earlier texts reviewed here.

2.3. Two-dimensional systems

2.3.1. Strip beams In this section the paraxial equation appropriate to strip beams is calculated and the properties of strip lenses evaluated. The methods follow closely those of section 2.2 with axes oriented such that z is the beam direction and all parameters are independent of y. The yz plane is referred to as the 'axis plane'. Although strip beams in practice always have 'ends' in the y direction, the results can be applied also to hollow cylindrical systems in which the beam thickness is small compared with the radius. For this situation $x \approx r-a$, $y \approx a\theta$, where $r = a$ represents the axis plane of the beam. The magnetic field is of such a form that $B_y = 0$, and $\partial B_z/\partial x = 0$.

Derivation of the paraxial equation is straightforward;

equivalent to Busch's theorem is the statement that $P_y = \gamma m_0 \dot{y} + qA_y$ is constant. In terms of B this is

$$\dot{y} = \frac{q}{\gamma m_0}(\Psi_0 - B_z x) \tag{2.93}$$

where Ψ_0 is the flux of B_z per unit length in the y direction, bounded on one side by the $x = 0$ plane and on the other by a parallel plane through the point at which $\dot{y} = 0$ for the particle in question. The paraxial equation is found by a method analogous to that used to deduce equation (2.14). The balance of forces in the x direction involves y, which is eliminated by use of equation (2.93) to give an equation in x, its time derivatives, and the fields. Instead of a term in $1/r^3$, finite Ψ_0 gives rise to a term independent of x. This implies motion which is not symmetrical about the symmetry plane; we therefore omit this term and confine attention to the case when $\Psi_0 = 0$. In notation analogous to that of equation (2.13) the paraxial equation for a strip beam with $\Psi_0 = 0$ is

$$x'' + \frac{\gamma' x'}{\beta^2 \gamma} + \left(\frac{\gamma''}{\beta^2 \gamma} + \frac{4\Omega_L^2}{\beta^2 c^2}\right) x = 0. \tag{2.94}$$

Expression in terms of ϕ, or with time as independent variable, may be obtained by comparison with equations (2.16) or (2.10).

The geometry for which equation (2.94) was calculated is of limited practical interest, at least for magnetic focusing. Of more interest is the situation with a uniform magnetic field in the y direction and components of both electric and magnetic field in the z and x directions. For such a system the axis plane becomes curved, and for a general analysis curvilinear coordinates must be used. For this reason we defer discussion of it until section 2.5, after the introduction of curved axes.

2.3.2. Strip lenses The focal length for an electric strip lens may be found in a manner analogous to that for an axially symmetrical lens. Writing equation (2.94) in terms of ϕ and ϕ_0 and setting $\Omega_L = 0$,

$$\frac{\phi(2\phi_0+\phi)}{\phi_0+\phi}\,x'' + \phi' x' + \phi'' x = 0. \qquad (2.95)$$

By methods analogous to those of section 2.2.5 we obtain

$$\frac{1}{f_2} = \left(\frac{\phi_1^2+2\phi_0\phi_1}{\phi_2^2+2\phi_0\phi_2}\right)^{1/4}\int\left\{\frac{\phi'^2(\phi^2+\phi\phi_0+3\phi_0^2)}{4(\phi^2+2\phi\phi_0)^2} + \frac{(\phi+\phi_0)\phi''}{2(\phi^2+2\phi\phi_0)}\right\}dz \qquad (2.96)$$

This is more complicated than for the cylindrical lens because of the appearance of the term ϕ'', which does not cancel in deriving the reduced equation. In the non-relativistic limit, however, the integral simplifies to

$$\int\left\{\frac{3}{16}\left(\frac{\phi'}{\phi}\right)^2 + \frac{\phi''}{4\phi}\right\}dz = \int\frac{7}{16}\left(\frac{\phi'}{\phi}\right)^2 dz \quad \text{(N.R.)} \qquad (2.97)$$

This identity may be proved by splitting the r.h.s. in the ratio of 3:4 and integrating the second term by parts. One part of the integral then equals the second term on the l.h.s. and the other vanishes at both limits. For the focal length we have therefore

$$\frac{1}{f_2} = \frac{7}{16}\left(\frac{\phi_1}{\phi_2}\right)^{1/4}\int\left(\frac{\phi'}{\phi}\right)^2 dz \quad \text{(N.R.)} \qquad (2.98)$$

For a slot lens the formula analogous to equation (2.81) is

$$f_2 = \frac{\phi(2\phi_0+\phi)}{(\phi_0+\phi)(\phi_2'-\phi_1')}\,. \qquad (2.99)$$

These formulae are of the same form as for the cylindrical lens, but the focusing is weaker.

Magnetic strip lenses are of limited use, since y components of velocity are produced, causing shear and spread in the y direction. In annular systems, however, this may not be a disadvantage. It is readily shown from equation (2.94) that the focal length of a strip lens is four times as great as that of a cylindrical lens with the same field on the axis.

2.R.3. Notes and references Strip beams are of little interest in electron microscopy, and the properties of strip lenses are not normally included in texts. Strip beams are discussed by Pierce (1954a) and Kirstein *et al* (1967). Strip lenses are analysed by Pierce in non-relativistic approximation.

2.4. Systems with two planes of symmetry

2.4.1. Introduction As explained in sections 2.2.5 and 2.2.6, both electrostatic and magnetic lenses in axially symmetrical systems produce second-order focusing. This observation is true only in the absence of charges and currents in the beam region; a uniform charge or current centred on the axis produces first-order focusing, as will be seen in Chapter 3. The plasma lens illustrated in Fig.2.8 falls into this class, though such lenses are of limited practical interest.

An alternative way of providing first-order focusing is to abandon circular symmetry. This can be done by curving the z axis, or alternatively by keeping the axis straight and introducing quadrupolar fields. For the moment we consider the second of these possibilities.

Cylindrical multipole fields in free space satisfy the conditions that div $\underline{E}$ and curl $\underline{B}$ = 0. It is readily verified that a '2n-pole' variation of the radial field proportional to $f(z)r^{n-1}\cos 2(n-1)\theta$ satisfies these equations. In particular, a pure electric quadrupole field has the form

$$E_r = \frac{E_0 r}{a}\cos 2\theta\,, \qquad E_\theta = \frac{-E_0 r}{a}\sin 2\theta$$

or (2.100)

$$E_x = E_0\, x/a\,, \qquad E_y = -E_0\, y/a$$

where a is the radius at which $E_r^2 + E_\theta^2 = E_0^2$. A quadrupolar magnetic field is described by

$$B_r = \frac{B_0 r}{a} \sin 2\theta\,, \qquad B_\theta = \frac{B_0 r}{a} \cos 2\theta$$

or (2.101)

$$B_y = B_0 x/a \quad , \quad B_x = B_0 y/a \,.$$

The axes have been chosen such that a particle moving in the z direction and lying in the zx or xy plane experiences a force towards or away from the axis. A sketch of the field, produced by electrodes or magnets, is shown in Fig.2.11. Such fields form the basis of quadrupole lenses, described in the next section.

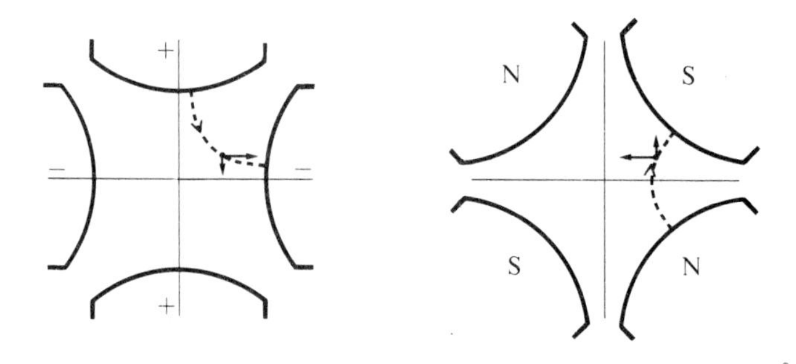

Fig.2.11. Electric and magnetic quadrupoles, showing a typical field line, and the direction of the forces on a particle coming out of the paper.

2.4.2. Quadrupole lenses For a magnetic quadrupole with the axes shown in Fig.2.11, a charge q moving in the z-direction experiences a force towards the yz plane proportional to x, and a force away from the xz plane proportional to y. Using again the paraxial assumption of small angle to the axis, the equation of motion of the charge may be written

$$\begin{aligned} \gamma m_0 \ddot{x} + q\beta c B_0 x/a &= 0 \\ \gamma m_0 \ddot{y} - q\beta c B_0 y/a &= 0 \end{aligned} \qquad (2.102)$$

where B_0/a is the radial field gradient. For a field uniform in the z-direction this represents sinusoidal (focusing) motion in the xz direction and exponential (defocusing) motion in the direction at right angles. Equation (2.102) may be written as trajectory equations

$$x'' + \kappa x = 0, \qquad y'' - \kappa y = 0 \tag{2.103}$$

where $\kappa = qB_0/\beta\gamma m_0 ca$. For an electrostatic lens the same equations are obeyed, except that the axes have to be taken where the circumferential rather than the radial field is zero, and κ is given by $qE_0/\gamma m_0 a\beta^2 c^2$ (equation 2.100).

A quadrupole field extending for a finite axial length L constitutes a quadrupole lens. In such lenses the length L is often greater than the radius but less than $\kappa^{-1/2}$. The quadrupole field does not end abruptly at the end of the lens; the 'end effect' depends on the details of the design and also the field distribution. For L much greater than the radius the effect is small and can partly be compensated by defining a suitable effective length. For many applications the aberrations introduced by additional multipole effects associated with the ends can be ignored. Focusing and defocusing lenses are illustrated in Fig.2.12 and expressions for the cardinal distances are shown in the table.

In contrast to axially symmetrical systems, where, in the absence of charges and currents in the beam space, the focusing is a second-order effect, quadrupoles produce first-order focusing in one plane and defocusing in the other. In axially symmetrical systems the power is proportional to E^2 or B^2 (equations (2.70) and (2.85)), whereas in a quadrupole the power is directly proportional to the field strength. It is readily shown that for electric and magnetic focusing respectively, for a thin lens,

$$\frac{1}{f_e} = \pm \frac{q}{\gamma m_0 \beta^2 c^2} \int \frac{E_0}{a}\,dz$$
$$\frac{1}{f_m} = \pm \frac{q}{\gamma m_0 \beta c} \int \frac{B_0}{a}\,dz \tag{2.104}$$

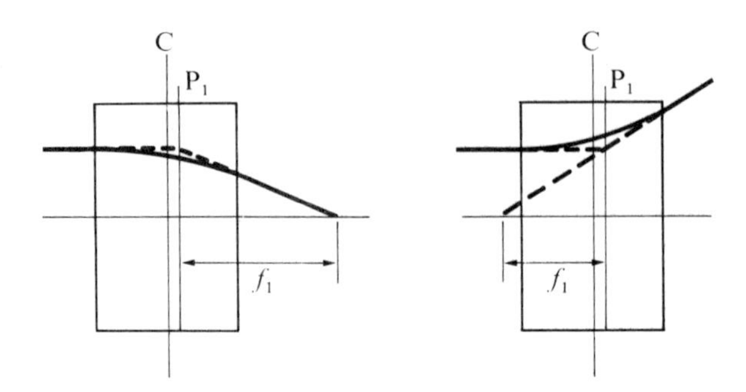

Fig.2.12. Cardinal points of focusing and defocusing planes of quadrupole lens. C denotes lens centre, P_1 and f_1 are the principal plane and focal length respectively

	Distance CP_1	Focal length f_1
Focusing plane	$\frac{L}{2} - \frac{(1-\cos\kappa^{1/2}L)}{\kappa^{1/2}\sin\kappa^{1/2}L}$	$\frac{1}{\kappa^{1/2}\sin\kappa^{1/2}L}$
Defocusing plane	$\frac{L}{2} - \frac{(\cosh\kappa^{1/2}L-1)}{\kappa^{1/2}\sinh\kappa^{1/2}L}$	$\frac{-1}{\kappa^{1/2}\sinh\kappa^{1/2}L}$

In axially symmetrical lenses the focusing action depends on both longitudinal and transverse fields; in the quadrupole, however, interaction with longitudinal field components, which necessarily occur at the ends of the quadrupole, produces only second order effects.

Although a single quadrupole focuses in one plane and defocuses in the other, two lenses arranged as a focusing-defocusing pair have a net focusing effect. Such a quadrupole pair is much stronger than an axially symmetrical lens of comparable size and field strength, but the focusing is again a second-order effect representing the difference between the focusing and defocusing actions of the two lenses;

that the former is stronger is seen to be plausible from the fact that the particle is on average further from the axis in the focusing region, so that the net force towards the axis is stronger. Detailed calculations of the focusing properties of quadrupole doublets (and more complicated arrays) can be found with the aid of the transfer matrices for focusing and defocusing lenses of length L:

$$M_f = \begin{pmatrix} \cos \kappa^{1/2} L & \kappa^{-1/2} \sin \kappa^{1/2} L \\ -\kappa^{1/2} \sin \kappa^{1/2} L & \cos \kappa^{1/2} L \end{pmatrix} \qquad (2.105a)$$

$$M_d = \begin{pmatrix} \cosh \kappa^{1/2} L & \kappa^{-1/2} \sinh \kappa^{1/2} L \\ \kappa^{1/2} \sinh \kappa^{1/2} L & \cosh \kappa^{1/2} L \end{pmatrix} \qquad (2.105b)$$

and for a drift space between lenses of length

$$M_o = \begin{pmatrix} 1 & D \\ 0 & 1 \end{pmatrix} . \qquad (2.106)$$

A different sign convention is often used. The two equations (2.103) are both written with positive sign, with $\kappa = \kappa_f$ in a focusing section and $\kappa = -\kappa_d$ in a defocusing section.

It is instructive to calculate the focal length of a pair of thin lenses separated by a distance D. From equations (2.105) and (2.106), using the small-angle approximation and neglecting L in comparison with D, the overall transfer matrix for such a system is

$$M = \begin{pmatrix} 1 & 0 \\ -\kappa L & 1 \end{pmatrix} \begin{pmatrix} 1 & D \\ 0 & 1 \end{pmatrix} \begin{pmatrix} 1 & 0 \\ \kappa L & 1 \end{pmatrix}$$

$$= \begin{pmatrix} 1+\kappa DL & D \\ -\kappa^2 DL^2 & 1 - \kappa DL \end{pmatrix} . \qquad (2.107)$$

The distance from the second lens at which the principal ray $(x, x') = (1, 0)$ crosses the axis is given by the ratio of the matrix elements $-m_{11}/m_{21}$ as

$$z_2 = \frac{1 + \kappa DL}{\kappa^2 DL^2} \quad . \tag{2.108}$$

The focal length of the lens combination, which is in general thick, is given by simple geometry (Fig.2.5) as $z_2/(1 + \kappa DL)$, which is equal to $1/\kappa^2 DL^2$. The lens power is thus proportional to the *square* of κ, and indicates second-order focusing. In the yz plane a similar argument holds, except that the focusing quadrupole comes first, so that κL has the opposite sign. This does not affect the focal length, but the principal plane is in a different place; the lens is therefore necessarily astigmatic.

It is readily verified that the ratio of focal length of the quadrupole pair just analysed to that of a solenoid of length $2L$, with a value of B equal to that at radius a in the quadrupole, is $a^2/2LD$. When the two lenses are adjacent this ratio is $a^2/2L^2$; this may be found by expanding the. functions in equation (2.105). Symmetrical triplet combinations can be designed which do focus from point to point. This property is restricted to a particle pair of points, however, and the combination is in general astigmatic.

The most important application of magnetic quadrupoles is in accelerator and beam-handling installations, where beams of very high-energy particles have to be precisely guided. Properties of various lens combinations, particularly doublets and triplets, have been fully investigated. An extensive review confined to linear systems has been given by Regenstreif (1967) and aberrations are treated by Hawkes (1966, 1972). Electrostatic quadrupoles are less used in focusing systems, though they have been used for partial correction of certain types of aberration in electron-optical systems for microscopes.

Higher-order multipoles, in particular sextupoles and octupoles, are used for introducing or controlling non-linear effects in large accelerator lattices; their use in this respect is discussed later (section 2.7.5). In systems where magnetic focusing interacts with the magnetic moment of a particle rather than its charge sextupoles play the same role as quadrupoles do in the focusing of charged beams. This property is made use of in the design of polarized ion sources and lenses for deflecting cold neutrons. In an analogous manner quadrupole fields act on such particles in the same way that dipole fields act on charged particles; they produce a deflection in a plane parallel to the magnetic moment.

2.5. Systems with one plane of symmetry and curved axes

2.5.1. Introduction The focusing systems considered so far have had rectilinear axes; when curved axes are introduced, new features appear. In the first place only one plane of symmetry exits; we find later that with such symmetry first-order focusing in both planes is possible. Secondly, dispersion occurs; particles with different energies cannot have a common axis. It is often convenient to specify an axis corresponding to a particular energy, and to express axes for neighbouring energies in terms of a linear expansion; an example of this is studied in section 2.5.4. A very important class of system with curved axes is represented by the closed orbits in particle accelerators and storage rings. The periodicity of the focusing structure and the very large number of revolutions emphasize features absent or unimportant in 'single-transit' devices.

When using curvilinear axes we shall denote the direction along the axes as s, the direction perpendicular to s but in the plane of the axis as x, and the direction perpendicular to both of these as y. The direction of x is measured away from the average centre of curvature. An alternative convention often used is to denote y by $-z$. The system here, however, prevents confusion with the convention for rectilinear axes used in earlier sections.

In the next sections we consider edge focusing, where an orbit enters a magnet at an angle, and betatron focusing, which arises from the combination of curvature and a gradient of magnetic or electric field.

2.5.2. Edge focusing When a beam passes into a region of uniform magnetic field perpendicular to the direction of motion, the particles are deflected. The beam axis becomes curved and its shape is determined by the momentum of the particles. In addition, particles of the same momentum off the axis experience a focusing (or defocusing) effect which may be associated with the transition region at the edge of the magnet. This is known as 'edge focusing', and will now be analysed; the 'momentum resolution' is considered later (sections 2.5.4 and 2.7.2).

The focusing in a plane perpendicular to the field B_y may be understood from Fig.2.13 where, for simplicity, we

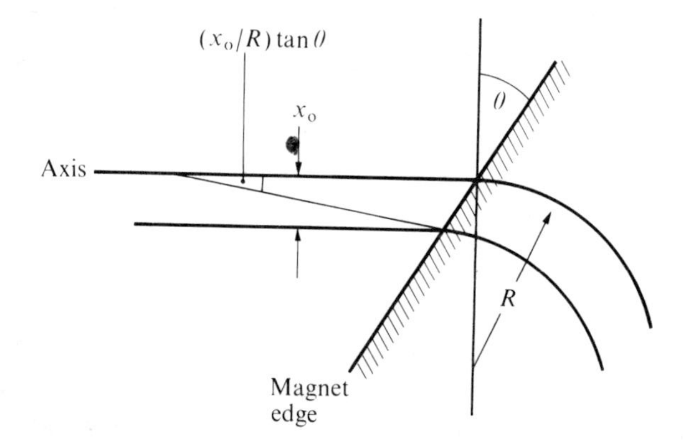

Fig.2.13. Focusing at a hard edge. Two rays initially parallel cross the magnet edge. Unless $\theta = 0$ one of these enters before the other, and is therefore deflected by a larger amount. From the geometry of the figure it is evident that the focal length is $R/\tan\theta$, where θ is taken as negative if the rays bend towards the normal at the interface, as illustrated above, and positive otherwise.

consider a 'step' field at which the radius of curvature of the particles jumps from infinity to R. This is of course an approximation, but it is good if the gap between magnet poles is small compared with the radius of curvature of the orbit in the magnet. A particle off the axis by an amount x_0 enters the field earlier or later than a particle on the axis. If θ is the angle which the axis makes with the normal to the magnet surface, then the off-axis particle suffers an additional deflection $(x_0/R)\tan\theta$, giving rise to a focal length

$$f = R/\tan\theta . \qquad (2.109)$$

Evidently the sign of $\tan\theta$ is such that if the rays bend towards the normal (as in the figure) it is negative. To find the vertical focusing, we note that a particle on a trajectory at distance y_0 from the symmetry plane receives an impulse $q\beta c \sin\theta \int B_\perp dt$, where $B_\perp$ is the fringe-field component perpendicular to the surface. This deflects it through an angle

$$y' = q\sin\theta \int B_\perp ds/\beta\gamma m_0 c . \qquad (2.110)$$

Now $\cos\theta \int B_\perp ds = B_y y_0$ and $R = \beta\gamma m_0 c/qB_y$, (equation (2.112)), so that the focal length $-y_0/y'$ is given by

$$f_y = -R/\tan\theta . \qquad (2.111)$$

This is equal numerically to the radial focusing but opposite in sign.

Besides being of use in analysing and beam-handling systems, this type of focusing is employed in the sector-focused cyclotron to overcome the vertical defocusing associated with the radially increasing field, as described later in section 2.7.6. With radial sectors the vertical focusing contributed by each edge is positive but small. With spiral sectors the force is larger but alternates in sign; the net effect, however, is greater. Orbits in such machines are normally established by computer, using a step-by-step integration through fields found by measurement.

2.5.3. Betatron-focusing In betatrons, classical cyclotrons, synchrocyclotrons, and some of the older and smaller synchrotrons, the particles move azimuthally in axially symmetrical fields. In larger and more modern machines, where use is made of the alternating-gradient and sector-focusing principles, the field varies azimuthally, with three-fold or higher symmetry, and the orbit shape departs from a circle. Oscillations in the radial and axial directions about the 'closed orbit', are known as 'betatron oscillations', and one of the important parameters in the design of an accelerator is the frequency Q of these oscillations normalized to the particle rotation frequency. A discussion of some general properties of the closed orbit follows in section 2.7.3. In systems with axial symmetry it is evidently a circle.

We start with a simple first-order analysis of betatron oscillations in a field with axial symmetry. First, we define the equilibrium orbit of a particle of momentum $\beta\gamma m_0 c$ as the radius R_0 at which it would move in a circular orbit centred on the axis. This orbit is in the 'median plane' where there is no radial component of B. The equilibrium radius R_0 is found by balancing the Lorentz force, equal to $q\beta_s cB_y$, and the centrifugal force, $\gamma m_0 \beta_s{}^2 c^2/R_0$. This yields

$$R_0 = \beta_s \gamma m_0 c / qB_{y0} \tag{2.112}$$

where B_{y0} is the field at radius R_0. (Note that R_0 is positive in the (x,y,s) co-ordinate system. In cylindrical polar co-ordinates it would have been negative, see section 1.4.)

Having found the equilibrium orbit, we now calculate the motion of a particle displaced slightly from it. A particle of the same momentum at radius R_0+x where $x \ll R_0$ experiences a net radial force

$$\gamma m_0 \ddot{x} = \frac{\gamma m_0 \beta^2 c^2}{R_0+x} - q\beta c\left(B_{y0} + x\frac{\partial B_y}{\partial r}\right). \tag{2.113}$$

Expanding to first order, making use of equation (2.112) and denoting $-(R_0/B_{y0})(\mathrm{d}B_y/\mathrm{d}r)$ by n, the 'field index', this equation becomes

$$\ddot{x} + (1-n)\omega_0^2 x = 0 \tag{2.114}$$

where ω_0 is the angular rotation frequency of the particle. It corresponds (numerically but not in sign) to the cyclotron frequency in the field B_{y0}. Focusing in the axial direction may be found by calculating the force towards the orbit which arises from the radial component of field B_x;

$$\gamma m_0 \ddot{y} = q\beta c B_x . \tag{2.115}$$

Since curl $\underline{B} = 0$, it follows that to first order

$$B_x = y \frac{\partial B_y}{\partial r} = - \frac{nB_0 y}{R_0}$$

whence

$$\ddot{y} + n\omega_0^2 y = 0. \tag{2.116}$$

Expressing these results in terms of derivatives measured along the equilibrium orbit,

$$x'' + \frac{(1-n)x}{R_0^2} = 0$$
$$y'' + \frac{ny}{R_0^2} = 0 . \tag{2.117}$$

For stability in radial and axial directions, therefore, B_y must decrease with radius, but not more rapidly than $1/r$. The field lines for such a distribution, and suitable magnet poles for producing them are sketched in Fig.2.14.

The Q values associated with the focusing, defined in the first paragraph of this section, are

$$Q_x = (1-n)^{1/2} , \qquad Q_y = n^{1/2} . \tag{2.118}$$

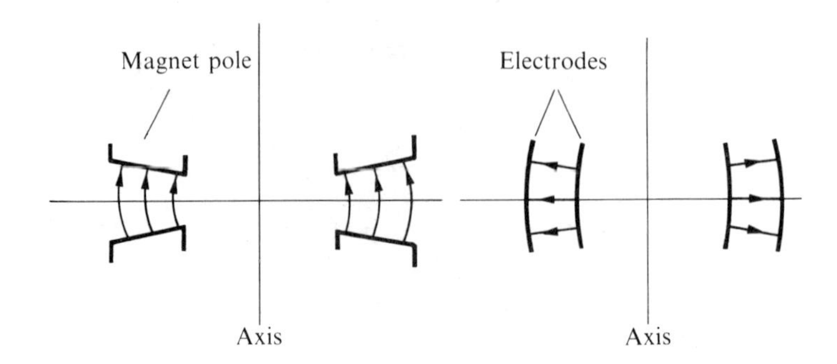

Fig.2.14. Pole and electrode configurations for producing stable circular orbits.

Subscripts x and y denote corresponding directions for the oscillations. Sometimes subscripts R and V, for 'radial' and 'vertical' are used. Commonly cylindrical polar co-ordinates are chosen, with corresponding subscripts r and z. In Europe the symbol Q is used, but in the United States this quantity is denoted by ν.

Stability in both planes requires that $0<n<1$, so that Q_x and Q_y are both less than unity; the particles make less than one oscillation per revolution in the machine. Such focusing is sometimes known as 'weak focusing'. When $n = \frac{1}{2}$, $Q_x = Q_y = 1/\sqrt{2}$. This is first-order focusing, made possible in both planes simultaneously by the curved equilibrium orbit. Setting $n = 0$ corresponds to a uniform field. Vertically there is neither focusing nor defocusing, and the particle moves with $y' =$ constant. In a horizontal plane all orbits are circles; if one such orbit is considered as the equilibrium orbit, then it is evident that a slightly displaced orbit has $Q = 1$.

Although electric fields with axial symmetry are not used in accelerators, they have been used in spectrometers. The Q values for such a configuration can be determined from a similar type of argument to that used for magnetic fields. The result is different, however; this arises essentially because in a magnetic field the energy of a particle is constant whereas in the electric field the energy is a function of radial position, and it is the angular momentum rather than energy which is conserved. In the vertical plane the condition that div $\underline{E}$ rather than curl $\underline{B}$ is zero is used; this again makes a difference. The results of such a calculation are

$$Q_x^2 = 3-\beta^2-n \qquad Q_y^2 = n-1. \tag{2.119}$$

Here the vertical focusing disappears when $n = 1$; this represents a field decreasing as $1/r$ radially and uniform in the axial direction. The radial Q-value depends on the particle energy, varying from $\sqrt{2}$ to 1 from the non-relativistic to extreme relativistic region.

These results apply also to a gravitational field; for an inverse-square law $n = 2$, so that $Q_x = (1-\beta^2)^{1/2} = \gamma^{-1}$ and $Q_y = 1$. For small β, $Q_x = 1$, and this implies (for a planet) an orbit that repeats from year to year with fixed perihelion; for $\gamma > 1$ a relativistic precession frequency of $(\gamma-1)\omega_0$ is obtained, which for small β is $\frac{1}{2}\beta^2\omega_0$. This differs from the correct value $3\beta^2\omega_0$ obtained from general relativity. When both electric and magnetic fields are present the calculation is slightly more tedious, but still straightforward. The basic equation expressing the radial balance of forces is

$$\frac{\mathrm{d}}{\mathrm{d}t}(\gamma m_0 \dot{x}) = \frac{\gamma m_0 \beta^2 c^2}{r} + qE_x - q\beta c B_y = 0 \tag{2.120}$$

where $r = R_0 + x$. This is expanded by writing

$$\gamma = \gamma_0 + \Delta\gamma,$$

$$\beta = \beta_0 + \Delta\beta,$$

$$E_x = E_{x0}\,(1 - n_{\mathrm{E}}x/R_0)$$

and

$$B_y = B_{y0}\,(1 - n_{\mathrm{M}}x/R_0).$$

Both β and γ can be expressed in terms of x, using the energy relation for small x,

$$qE_{x0}x = \Delta\gamma m_0 c^2 .$$

These substitutions yield a linear second-order equation in x, from which the oscillation frequency can be found. The corresponding frequency for the y-direction can be found by a slight extension of the methods used to obtain equation (2.116). Corresponding to equations (2.118) and (2.119) we have the more general relations

$$Q_x^2 = 1 + \frac{\beta_0(1-n_E) - \beta n_M}{\beta_0 + \beta} + \frac{\beta_0^2}{\gamma^2(\beta_0+\beta)^2}$$

$$Q_y^2 = \frac{n_M\beta + (n_E-1)\beta_0}{\beta_0 + \beta} \tag{2.121}$$

where $\beta_0 = - E_{x0}/B_{y0}c$. These formulae have been obtained by Reiser (1973), (with a different sign convention).

In the analysis presented here only first-order terms have been considered. A more complete treatment shows a very large number of second-order terms arising essentially from the curvature of the orbit and finite amplitude of the oscillations. Although most of these are unimportant, terms which couple the axial and vertical motion can act in a resonant manner in transferring energy between these two modes, when an integral relationship $N_1Q_x = N_2Q_y$ exists between the Q-values. We return to this topic in section 2.7.8.

Although in the discussion so far we have tacitly assumed systems with axial symmetry and a circular equilibrium orbit, it is evident that a sector magnet or electrostatic deflector system of finite azimuthal extent can be used as a thick lens. Thus, an '$n = \frac{1}{2}$' magnetic spectrometer has two foci separated by $\sqrt{2}\pi$. Likewise, an electrostatic system consisting of concentric hemispheres focuses (in the non-relativistic limit) between points at an angular separation of π. Focusing in one plane only can be achieved by a uniform magnetic field extending over an angle π, or an electric field between concentric cylinders proportional to r^{-1} and extending over an angle $\pi/(2-\beta^2)^{1/2}$.

If the source contains particles of varying energy or momentum, then the radial position of the focus shows a first-order variation. This dispersive property, essential to the operation of spectrometers, is discussed with illustrative examples in the next section.

2.5.4. Systems with dispersion The emphasis in this chapter has been so far on the focusing characteristics of the lenses and magnet systems which have been considered. These properties vary with particle energy. For some applications such variations are a disadvantage; in other systems, however, the dispersion is made use of in order to separate beams of mixed energy or (as in mass spectrometers) of mixed mass.

The design of spectrometers is a highly specialized art, though the basic principles are straightforward. Many types have evolved which are very different in detail depending on the particle energy, energy range, and resolution requirements. Most of them depend on the angular separation of orbits of particles moving in transverse electric or magnetic fields. At high energies, magnetic fields are normally used, whereas for electrons at non-relativistic energies electric fields are generally more convenient. Crossed fields can also be used to separate beams of mixed energy. The electric field exerts the same force on all particles, whereas the magnetic field acts more strongly on the faster ones; it is straightforward to arrange that the wanted particles move in a straight line whereas the unwanted ones are deflected. Such a device is known as a 'Wien filter'. The same principle has been used at high energies to separate mixed beams of high-energy particles which have already been analysed in momentum. Chromatic aberration of a lens is used in the classical beta-ray spectrometer to separate particles of different energy from a point source.

An important ingredient of many spectrometer assemblies is the magnetic prism. As an illustrative example the first-order properties of a simple prism will be calculated, though the general method can readily be extended to more complicated systems. For second-order effects (geometrical aberrations) the specialized literature should be consulted.

We first examine the general relation between the input and output rays, including the effect of a deviation $\Delta p/p$ in momentum from the central value $p = \beta\gamma m_0 c$. The magnet has a plane of symmetry $y=0$, and co-ordinates along and perpendicular to the central orbit are denoted by s and x. The central orbit, or beam axis, is taken as being in the symmetry plane. If entry and exit points are denoted by the subscripts s_1 and s_2, the quantities, x_2, x_2', y_2, y_2' are in general functions of x_1, x_1', y_1, y_1', and $\Delta p/p$. If we include only first-order terms (paraxial approximation), the symmetry about the orbit plane implies no coupling between x and y motion, so that x_2 and x_2' are functions of x_1, x_1', and $\Delta p/p$, and similarly for y_2 and y_2'. It is convenient to introduce a 3×3 matrix to describe the transformation; since $\Delta p/p$ is unchanged in the magnet, this is of the form

$$M = \begin{pmatrix} m_{11} & m_{12} & m_{13} \\ m_{21} & m_{22} & m_{23} \\ 0 & 0 & m_{33} \end{pmatrix}. \tag{2.122}$$

The various elements can be evaluated once the field configuration is known. The four elements in the top left corner correspond to those encountered already in section 3.2.2. The elements m_{13} and m_{23} correspond to the values of x_2 and x_2' appropriate to a particle starting on the central orbit, per unit fractional deviation of momentum $\Delta p/p$.

As an example, we consider a sector magnet with a betatron field, the central ray being taken as the equilibrium orbit of radius R_0 for the corresponding momentum. Elements m_{11}, m_{12}, m_{21}, m_{22} will be as in M_f of equation (2.105), with $\kappa = (1-n)/R_0^2$. A particle starting at $s = 0$ with momentum $\Delta p + p_0$ can be considered as at the minimum of a betatron oscillation of amplitude $\alpha R_0 \Delta p/p_0$ about an orbit of radius $R_0(1+\alpha\Delta p/p_0)$, where $\alpha = (\Delta R/R_0)/(\Delta p/p_0)$. (This is essentially the 'momentum compaction' defined and discussed later in section 2.7.4). For a betatron field $\alpha = (1-n)^{-1}$, as may be verified by setting $n = -(r/B)(dB/dr)$ and $\Delta p/p_0 = \Delta B/B_0$, with $p_0 \propto Br$. From the definition of α, $\Delta R = R_0\alpha(\Delta p/p_0) = R_0(\Delta p/p_0)/(1-n)$. The equation for the trajectory of the 'off-momentum' particle measured with respect to the equilibrium orbit R_0 of the particle with momentum p_0 is thus

$$x = \frac{\Delta p}{p}\frac{R_0}{1-n}\left[1-\cos\left\{\frac{(1-n)^{1/2}s}{R_0}\right\}\right]. \qquad (2.123)$$

From this x' can be found, giving for the matrix (2.123) corresponding to a sector magnet

$$M_s = \begin{pmatrix} \cos\kappa^{1/2}s & \kappa^{-1/2}\sin\kappa^{1/2}s & (\kappa R_0)^{-1}(1-\cos\kappa^{1/2}s) \\ -\kappa^{1/2}\sin\kappa^{1/2}s & \cos\kappa^{1/2}s & \kappa^{-1/2}R_0^{-1}\sin\kappa^{1/2}s \\ 0 & 0 & 1 \end{pmatrix}$$

(2.124)

where $\kappa = (1-n)/R_0^2$.

For elements with a rectilinear beam axis, or of no axial extent (such as a magnet edge or a thin lens), m_{13} and m_{23} are zero. The first-order properties of a sector magnet with hard edges such that the central axis enters and leaves at angles θ_1 and θ_2 can be represented by the matrix $M_{E2}M_sM_{E1}$, where M_s is given by equation (2.124) and M_{E1} by

$$M_{E1} = \begin{pmatrix} 1 & 0 & 0 \\ (\tan\theta_1)/R & 1 & 1 \\ 0 & 0 & 1 \end{pmatrix} \qquad (2.125)$$

where θ_1 is negative if the beam is bent towards the normal when entering the magnet. The appropriate matrix is evaluated in more detailed treatments, such as those of Enge (1967) or Livingood (1969).

An interesting property of a uniform-field sector-focusing magnet (n = 0) with normal incidence is that it focuses in such a way that the object, vertex of the magnet, and image are collinear. This property, sometimes known as 'Barber's rule', can be deduced by considering the matrix $M_b M_s M_a$, where M_a and M_b are the matrices corresponding to the drift distances a and b between object and image respectively and the magnet face. Setting $n = 0$ in M_s (equation 2.124), the condition for focusing is found by setting m_{12} in the matrix $M_b M_s M_a$ equal to zero (see section 2.2.3). This gives the condition

$$(a+b)\cos\left(\frac{s}{R_0}\right) + \left(R_0 - \frac{ab}{R_0}\right)\sin\left(\frac{s}{R_0}\right) = 0. \qquad (2.126)$$

This may be arranged as

$$\tan\left(\frac{s}{R_0}\right) = -\frac{(a+b)R_0}{R_0^2 - ab} \qquad (2.127)$$

from which $\phi_a + \phi_b + \phi_s = \pi$, where $\tan\phi_a = a/R_0$, and $\tan\phi_s = s/R_0$. This implies that the object, image, and vertex points of the magnet are collinear (Fig.2.15).

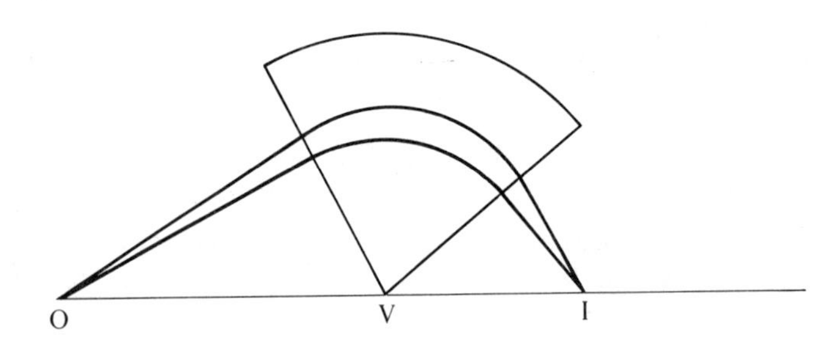

Fig.2.15. In a focusing system consisting of a sector magnet with normal entry and exit and uniform magnetic field the object, vertex, and image are collinear (equation (2.127)).

The transfer matrix for a system in which this condition is met, (with lengths normalized to R_0), is

$$M_b M_s M_a = \begin{pmatrix} \cos\phi_s - (b/R_0)\sin\phi_s & 0 & 1 - \cos\phi_s + (b/R_0)\sin\phi_s \\ -\sin\phi_s & \cos\phi_s - \sin\phi_s & \sin\phi_s \\ 0 & 0 & 1 \end{pmatrix} \tag{2.128}$$

Such a system has no vertical focusing and is of limited practical value as it stands. More practical (and more complicated) arrangements are described in the specialized literature discussed in section 2.R.4-5.

Many other devices in common use involve deflection of particles; examples are cathode-ray tubes and photomultiplier tubes. These are rather specialized and are not treated here; references are given in section 2.R.4-5.

2.R.4-5. Notes and references Quadrupole lenses did not come into use in focusing systems until after the application of the strong focusing principle in 1952 (section 2.7.2). Since that time they have been studied in considerable detail. A general treatment with a fair amount of detail may be found in the text of Grivet and Septier (1972). (Note that the material in earlier editions is much less.) An extended theoretical treatment, which includes aberrations, appears in the monograph of Hawkes (1966) and a later review (Hawkes 1970). The optics of combinations of quadrupoles is considered by Regenstreif (1967) and King (1964). Practical details (including the design of superconducting quadrupoles) may be found in the proceedings of accelerator and magnet conferences. These also contain papers on computer-aided design.

Although the mechanism of betatron focusing was already recognized in the 'thirties and the role of vertical focusing in classical cyclotrons was understood, the subject was clarified by the introduction of the field index n by Kerst and Serber (1941). The use of a similar index for the electrostatic field was a natural extension of this. The theory of betatron focusing is explained in all the texts on cyclic accelerators discussed in section 2.R.6-7.

The theory of edge and sector focusing applied to prisms and deflecting magnets is presented in contributions by Wollnik and Enge in the book edited by Septier (1967a). Dipole magnets are also studied in detail by Livingood (1969).

The properties and use of both dipole and quadrupole magnets in beam-transport systems are extensively discussed in the texts by Steffen (1965) and Banford (1966).

2.6. Curvilinear strip beams

In the last few sections systems with curved axes have been analysed. The properties of the axes have been rather special and simple. The general problem, including acceleration along the axis, is extremely complex; indeed, there is little point in attempting a general formulation since specific problems are normally best tackled individually by numerical methods.

The corresponding problem for a strip beam is just about tractable analytically, but leads in general to a lengthy and not very useful form of the paraxial equation. The non-relativistic equation was calculated using a variational method by Sturrock (1959) and by a direct method starting from the Lagrangian by Kirstein *et al*. (1967). In Appendix 3 the relativistic version is derived for a system in which the magnetic field has a single component in the direction in which all quantities are uniform (the y-direction). The electric field has components along the beam (the s-direction in curvilinear co-ordinates) and perpendicular to it (the x-direction). The co-ordinate system is shown in Fig.2.16; the term 'axis surface' denotes the curved surface which corresponds to the axis in a system with focusing in both planes, or (more closely) to the 'axis plane' defined in the planar strip beam in section 2.3.1. It should be noted that the beam studied there is not included in the present analysis, since it includes a component of magnetic field along the direction of motion. In curvi-

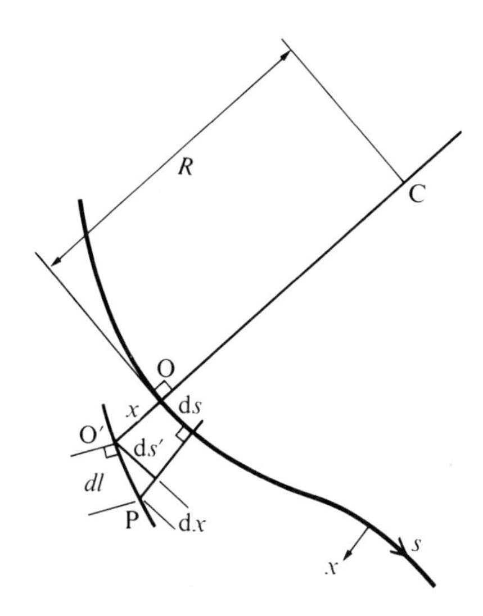

Fig.2.16. Co-ordinates for curvilinear axes. The curved line represents a section of the axis surface.

linear co-ordinates it is not realistic to include B_s without including B_x also; this is done by Kirstein *et al.*, but the final result contains terms explicitly involving the vector potential $\underline{A}$ which are difficult to express in terms of $\underline{B}$. The arbitrariness implied by the appearance of $\underline{A}$ can only be removed by specifying initial conditions for trajectories lying on the axis surface. Even in the absence of B_s and B_x the axis surface is not in general uniquely defined. In the trivial limiting case of only B_y present, representing motion in a uniform magnetic field, the location of the cylindrical axis surface is clearly arbitrary.

The paraxial equation is only useful when the equation defining the axis surface is known also. In Appendix 3 these two equations are shown to be

$$x'' + \frac{\gamma' x'}{\beta^2\gamma} + \left\{ \frac{\gamma''}{\beta^2\gamma} + \frac{2}{R^2} - \frac{2q}{\beta\gamma m_0 c^2 R}(cB_y - \beta E_x) + \frac{q^2}{\beta^2\gamma^2 m_0^2 c^4}(cB_y - \beta E_x)^2 \right\} x = 0 \tag{2.129}$$

and

$$\frac{\gamma m_0 \beta^2 c^2}{R} - qB_y \beta c + qE_x = 0. \quad (2.130)$$

In these equations E_x, R, and γ are in general functions of s, the distance along the orbit. If E_s is zero, however, they define a circular orbit, and substitution into equation (2.129) yields the relation for Q for motion in electric and magnetic fields with $n_{\mathrm{M}} = 0$ and $n_{\mathrm{E}} = 1$ (section 2.5.3).

Identification of specific terms with specific focusing effects is not immediately obvious; for example, the last three terms in the braces, together with equation (2.130), are needed to establish the simple case of focusing in a uniform magnetic field, for which the term in the braces reduces to $1/R^2$ implying that $Q = 1$. In the presence of an electric field only, with R constant, it may be verified that the equations yield $Q = (2 - \beta^2)^{1/2}$, as expected from equation (2.119) with $n = 1$. With a general non-circular curved axis this type of focusing is known as 'deflection focusing'. When R is an oscillating function of s, so that the axis surface is 'wavy', it has been termed 'slalom focusing'. This is discussed by Kirstein *et al.* (1967), who give references to the original papers.

2.7. Accelerator orbit theory

2.7.1. Introduction In this section an attempt is made to summarize the essential features of orbit theory in cyclic machines. The emphasis is on general concepts; some of these are closely related to ideas discussed already, whereas others, for example those associated with exact periodicity, introduce essentially new features. Only single-particle orbits are considered here; steady collective effects are analysed in Chapter 3, concepts associated with ensembles of orbits are developed in Chapter 4, and collisional and radiation effects are discussed in Chapter 5.

2.7.2. Alternating-gradient ('strong') focusing This important principle, used in all large accelerators, was not applied before its rediscovery by Courant, Livingston, and Snyder (1952). It had previously been expounded in a limited publication by Christofilos (1950), but the significance of his work was not generally realized until 1952. It has already been remarked that the combination of a pair of quadrupole lenses produces net focusing in both planes; this fact was not, however, generally appreciated until 1952.

The first application of the principle was to the design of synchrotrons, where it was shown that a magnet divided azimuthally into a number of equal segments, with n alternating between a large positive value and a large negative value, could produce strong focusing with $Q \gg 1$ in both vertical and horizontal planes. Such focusing not only reduces the oscillation amplitudes which arise from a given angular displacement of the particles from the equilibrium orbit, but also enables a larger energy spread to be contained in a given radial aperture width.

The characteristics of the orbits in periodic focusing lattices have been studied in great detail in connection with the design of large synchrotrons. For large values of Q the behaviour of the particles is a very sensitive function of the parameters, especially misalignments and errors which perturb even slightly the periodicity of the lattice. In the presence of localized errors in the gradient for example, Q becomes complex near integral and half-integral values, so that the amplitude of the motion (in linear approximation) increases indefinitely (see section 2.7.3).

As a specific example we demonstrate the focusing action of the very simplest form of 'alternating-gradient' machine, using results already obtained for betatron focusing. From equations (2.117) it is readily verified that the transfer matrix for a length $s = R_0\theta$ of the orbit is of the same form as equation (2.105a) for a quadrupole, with $\kappa = n/R_0^2$ for vertical motion and $\kappa = (1-n)/R_0^2$ for horizontal motion. If now n is made very large compared with unity, the matrix remains the same in respect of vertical motion but for horizontal motion n is replaced by $(1-n)^{1/2} \approx jn$, and the matrix then takes the form given by equation (2.105b).

If we consider now a segment in which n has a large negative value, so that the field increases with radius, the first matrix applies to radial motion and the second to vertical motion. The matrix for traversing a sector pair is therefore $M_f M_d$ in the radial direction and $M_d M_f$ in the vertical direction. Specifically

$$M_f M_d = \begin{pmatrix} CC_h + SS_h & \kappa^{-1/2}(CS_h + SC_h) \\ \kappa^{1/2}(CS_h - SC_h) & CC_h - SS_h \end{pmatrix} \quad (2.131)$$

where C and S refer to the cosine and sine of $n^{1/2}s/R_0$ and the subscript h denotes the corresponding hyperbolic function. This will be denoted by M_0, the matrix corresponding to a 'unit cell' of the machine; since the length of the cell is $2s$ there are altogether $\pi R_0/s = N_0$ cells.

In order to demonstrate the focusing action we write the matrix as

$$M_0 = \cos\mu \begin{pmatrix} 1 & 0 \\ 0 & 1 \end{pmatrix} + \sin\mu \begin{pmatrix} m_{11} & m_{12} \\ m_{21} & -m_{11} \end{pmatrix} \quad (2.132)$$

where

$$\cos\mu = \cos(n^{1/2}s/R_0)\cosh(n^{1/2}s/R_0) = \tfrac{1}{2}\,\mathrm{Tr}\,M_0 \quad (2.133)$$

and Tr denotes the trace of the matrix. By repeated multiplication of equation (2.132) and use of the fact that the determinant of M_0 is unity (see section 2.2.3) it is found that

$$M_0^N = \cos N\mu \begin{pmatrix} 1 & 0 \\ 0 & 1 \end{pmatrix} + \sin N\mu \begin{pmatrix} m_{11} & m_{12} \\ m_{21} & -m_{11} \end{pmatrix} \quad (2.134)$$

The motion remains bounded therefore if $\cos\mu$ is real; the condition for this is that the right-hand side of equation (2.133) should be less than unity, which it is if $n^{1/2}s/R_0$ is less than about $0{\cdot}6\pi$. If $n^{1/2}s/R_0$ slightly exceeds this value the motion consists of an exponentially growing oscillation; the beam is 'overfocused'. For even larger values, focusing is again possible; as $n^{1/2}s/R_0$ increases further, unstable and progressively smaller stable regions alternate.

The quantity μ may be regarded as the phase shift of the betatron oscillation across one cell. The Q for the complete machine is therefore given by

$$Q = \frac{N_0\mu}{2\pi} = \frac{N_0}{2\pi}\cos^{-1}\left\{\cos\left(\frac{n^{1/2}\pi}{N_0}\right)\cosh\left(\frac{n^{1/2}\pi}{N_0}\right)\right\}. \quad (2.135)$$

By making n and N_0 sufficiently large, Q can in principle be increased without limit. Values used in large accelerators lie typically in the range 6 for a 25 Gev proton accelerator to 20 for machines of energy of a few hundred GeV. Practical limitations to the attainable Q values are imposed by tolerances.

Large synchrotrons consist of a sequence of magnets arranged in a ring separated by 'straight sections', which may not all be of the same length. Between the magnets is placed such equipment as accelerating cavities, injection electrodes, and diagnostic devices. The design of magnet lattices for such machines has become a highly specialized art. Sophisticated computer programmes have been constructed to obtain and optimize orbit characteristics in terms of the lattice parameters. A variant on the type of focusing element just described is provided by a 'separated-function' lattice in which both uniform-field bending magnets and quadrupoles are used. (The transfer matrix for a bending magnet in which the beam enters normally is given by the matrix (2.105*a*) with $n = 1$, and hence $\kappa = 1/R_0^2$. Compared with the quadrupole focusing this effect is very weak.) For focusing and transporting extracted beams arrays of quadrupoles and bending magnets are used; here again detailed design of such beams is normally accomplished with the aid of computer programmes.

Periodic focusing systems with axial symmetry are used in microwave tubes; since they are generally employed in situations where space charge is important, treatment of them is deferred until the next chapter (section 3.5).

2.7.3. The existence of closed orbits; resonances In magnetic field systems with axial symmetry, as considered for example in section 2.5.3, it is obvious that circular closed orbits exist. Motion about such orbits need not, however, be stable, as has already been seen. It can be shown quite generally that, in a magnetic field with a plane of symmetry to which the field is everywhere normal, at least one closed orbit can always be found. A proof of this fact has been given by Courant and Snyder (1958). For electric fields, and magnetic fields without a plane of symmetry, such an orbit does not necessarily exist.

In almost all cyclic particle accelerators the closed orbit is designed to be planar. If there is a spread of energy in the beam, particles with different energies have different closed orbits; the term 'equilibrium' orbit is used to denote the closed orbit of a particle at some specified energy. In betatrons and synchrotrons the position of the equilibrium orbit remains fixed, but both field and energy increase with time; in synchrocyclotrons, on the other hand, the field is fixed and the radius of the equilibrium orbit increases with time. In classical cyclotrons, with fixed field and orbital frequency, there are particles at all radii simultaneously, so that there is no unique equilibrium orbit radius.

In the types of field studied so far the closed orbit is a circle, or, for a machine with straight sections, may be constructed from circles and straight lines. In sector-focused cyclotrons and fixed-field alternating-gradient (FFAG) machines, on the other hand, the field varies continuously along the equilibrium orbit. If the field configuration is specified analytically, for example in terms of the product of azimuthal harmonics and polynomials in the radial direction, the equilibrium orbit can be found by expanding about an arbitrary closed curve (such as a circle) and regarding the deviations of the field from those appropriate to the chosen orbit as forcing terms in the equation of motion. How this may be done, with application to a class of fields of interest, is explained for example by

Symon *et al.* (1956) and in the review article by Laslett (1967). The calculations are somewhat laborious, and in the practical design of sector-focused cyclotrons, which use such fields, closed orbits are generally found by trial and error by means of computer calculations rather than analytically.

Betatron frequencies about the closed orbit have already been calculated for machines with a circular orbit (section 2.5.3); Symon *et al* show that for small oscillations equation (2.117) may still be used, when n and R_0 are functions of s, though the oscillations are no longer sinusoidal. For small oscillations the equation is necessarily linear, and a transfer matrix M_{c} of the form of equation (2.132) may be specified for one transit of the closed orbit,

$$\begin{pmatrix} x(C) \\ x'(C) \end{pmatrix} = M_{\text{c}} \begin{pmatrix} x(0) \\ x'(0) \end{pmatrix} \tag{2.136}$$

where C is the circumference of the orbit. The value of Q is given by $\cos 2\pi Q = \cos\mu$; this expression does not define Q uniquely, but the integral part can be found by inspection. For Q between N and $N+1$, where N is an integer, the oscillating orbit crosses the closed orbit either $2N$, $2N+1$ or $2N+2$ times per revolution.

The properties of the closed orbit of a particle with energy slightly different from that of a particle on the equilibrium orbit is characterized by the 'momentum compaction' or 'off-energy' function $\alpha_{\text{p}}(s)$. If x is the distance of a particle with momentum $p_0 + \Delta p$ from the equilibrium orbit, then

$$\alpha_{\text{p}}(s) = \frac{\Delta x}{\Delta p/p} \; . \tag{2.137}$$

This function is sometimes denoted by η or X_{p}. For a machine with axial symmetry Δx is independent of s, and $2\pi\Delta x = \Delta C$, the additional distance travelled by the off-momentum particle. In a non-circular machine the average value of α_{p} round the orbit, defined as

$$\alpha = \frac{p}{C}\frac{\Delta C}{\Delta p} \tag{2.138}$$

is often used. (Note, however, that α is dimensionless, whereas α_p is a length.) This factor is further discussed in the next section.

Not only does the size and shape of the equilibrium orbit vary with particle momentum, but the Q value varies also. This is expressed by the 'chromaticity' of the lattice:

$$\xi = \frac{p}{Q}\frac{\Delta Q}{\Delta p} . \tag{2.139}$$

For a weak focusing synchrotron the field index $n = -(r/B)(\mathrm{d}B/\mathrm{d}r)$ is a function of r if $\mathrm{d}B/\mathrm{d}r$ is constant, and it is readily verified (making use of equation 2.139) that

$$\xi_x = -\frac{n(1+n)}{2(1-n)^2}, \qquad \xi_y = \frac{1+n}{2(1-n)} . \tag{2.140}$$

In complicated lattices where Q is large and has to be specified within narrow limits to avoid the many possible resonances (see for example section 2.7.5), a large value of chromaticity is undesirable; on the other hand, at high currents chromaticity can be helpful in suppressing transverse instability (sections 6.4.4 and 6.5.3). A general expression in terms of the lattice parameters α_0 and β_0, defined later in section 4.3.4, of a lattice composed of linear focusing elements has been calculated by Beck *et al.* (1967). For strong-focusing lattices the numerical value of ξ_x and ξ_y are found to be of order -1 and +1. The chromaticity is also affected by introducing a non-linear field gradient in the focusing field, so that $\mathrm{d}B/\mathrm{d}r$ is a function of r. A special example of this is the 'scaling field' $B/B_0 = (r/R_0)^{-n}$. Orbits with different momentum, and hence different values of R_0 are geometrically similar, so that $\xi = 0$. To first order the field at R_0 is the same as for the betatron field defined in section 2.5.3, and in this sense the chromaticity is a second-order effect. Non-linear effects are further considered in section 2.7.5.

In practical machines many problems are caused by imperfections of various kinds. For example, if Q is near an integral value, the closed orbit is very sensitive to slight errors in field which arise from magnet imperfections and misalignments. As a very simple illustrative example we consider a machine with $2\pi Q = \mu$, $0<Q<1$. Imagine a slight field imperfection at $\theta = 0$ such that the field is slightly below its design value over a very small range of azimuth. Such an error causes a radial 'kick' to be experienced by the particle, which is deflected through an angle $\Delta\theta$. By simple geometry (Fig.2.17) the new closed orbit will be

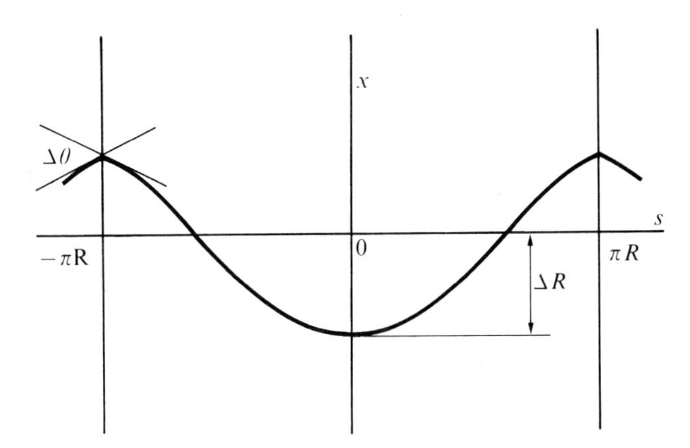

Fig.2.17. Closed orbit in uniform focusing system in the presence of an error which causes a deflection $\Delta\theta$ at an azimuth $s = \pi R$. The equation of the orbit is $x = \Delta R \cos Qs/R_0$ and the slope at $s = \pi R$ is $(Q\Delta R/R_0)\sin Q\pi = (\mu\Delta R/2\pi R_0)\sin\frac{1}{2}\mu$. Setting this equal to $\frac{1}{2}\Delta\theta$ gives equation (2.141).

displaced from the correct orbit $R = R_0$ by an amount

$$\Delta R = \frac{\pi R_0 \Delta\theta}{\mu \sin\frac{1}{2}\mu} \qquad (2.141)$$

which increases indefinitely as μ approaches 2π. For $\mu = 2\pi$ a particle starting off on the unperturbed equilibrium orbit executes oscillations of linearly increasing amplitude about it. Such a situation is termed an 'integral resonance' and may alternatively be described in terms of a resonant forcing term on the r.h.s. of equation (2.114) with frequency ω_0. When $n = 0$, $Q = 1$, so that resonance occurs. This behaviour appears at first sight to imply that there is no closed orbit under these conditions. In fact, non-linearity limits the effect, and a closed orbit does indeed exist.

A further type of imperfection occurs when the equilibrium orbit remains undistorted, but n has the wrong value at some azimuth. Such errors cause an amplitude-dependent rate of build-up. It is evident from equation (2.134) that for integral and half-integral values of Q, $\mu = N\pi$, so that $\cos\mu = \pm 1$. If a lattice with μ near $N\pi$ is perturbed by making n equal to $n + \delta n$ over a small range of s, it is found that for a small range of μ_0 the trace of the transfer matrix exceeds ± 2 so that $|\cos\mu_0| > 1$. This gives two solutions for the orbital motion with exponentially growing and decreasing oscillation amplitude respectively. The range of μ_0 over which the motion is unstable is known as a 'stop-band'. The equilibrium orbit is still circular, but oscillations about it are unstable. Resonances which arise from non-linearities in the field and from coupling between radial and vertical motion are discussed in later sections. The latter type of resonance can arise either from misalignments (such as a tilted magnet) or from non-linearities in the field gradient.

2.7.4. Momentum compaction and negative mass Before discussing the negative-mass concept we look in more detail at the momentum compaction α defined in equation (2.138). (We note at this point that sometimes, notably in Livingood's book (1961), α is defined as α^{-1}.) In a linear accelerator α is evidently zero; in a betatron field with constant value of $n = -(R_0/B)(\mathrm{d}B/\mathrm{d}r)$, $p = qBr$, $C = 2\pi R$ it is straightforward to show that

$$\alpha = (1-n)^{-1}. \tag{2.142}$$

This is always greater than unity. In strong-focusing machines, on the other hand, α is less than unity, and the orbit shape changes when p is varied. For the idealized machine described in section 2.7.2, where n alternates in sign, the equilibrium orbit of a particle with the correct momentum is a circle. For a particle with slightly different momentum, however, a circular path is not possible, and the orbit crosses a field discontinuity between focusing and defocusing sectors as illustrated in Fig.2.18. Evaluation of

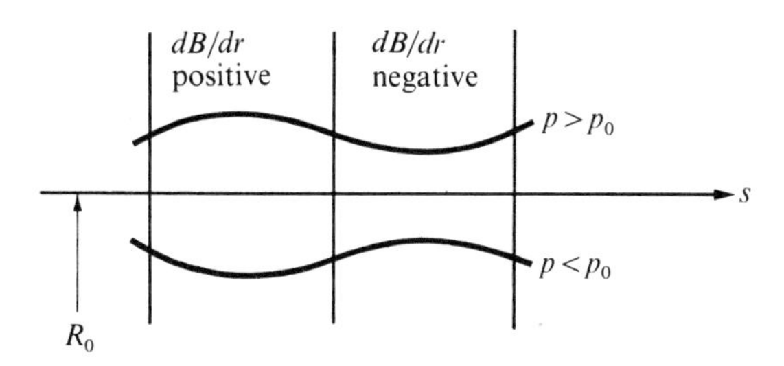

Fig.2.18. Equilibrium orbits of particles in an alternating gradient synchrotron. For momentum $p_0 = qB_0R_0$ the orbit is a circle in a uniform field; for other momenta the field varies with azimuth and the orbit consists of pieces of sine curves.

α is straightforward (see for example Livingood 1961), and yields for the particular case of an alternating-gradient field with $n \gg 1$, in which n changes sign but not magnitude

$$\alpha = 4N\left[n^{3/2}\pi\left\{\coth\left(\frac{n^{1/2}\pi}{2N}\right) - \cot\left(\frac{n^{1/2}\pi}{2N}\right)\right\}\right]^{-1} \tag{2.143}$$

where N is the number of sector magnet pairs in the machine. In the CERN 25 GeV synchrotron, which has a more complicated lattice, $\alpha = 0{\cdot}03$.

Machines with large values of α can exhibit the

'negative-mass' effect. A circumferential impulse applied to a particle increases both its velocity and the equilibrium orbit radius. The first of these tends to increase the orbital angular velocity, and the second to decrease it. If the latter effect predominates accelerating the particle decreases its angular velocity, so that in a sense it possesses a negative mass. This may be seen by defining the effective mass m^* in the following way:

$$m^* = -\frac{\Delta p}{\Delta\tau}\frac{\tau}{v} = -m_0\frac{\tau}{\Delta\tau}\frac{\Delta(\beta\gamma)}{\beta} = -\frac{m_0\tau}{\Delta\tau}\frac{\gamma^3}{\beta}\Delta\beta \qquad (2.144)$$

where τ is the time taken to complete one orbit of the machine. For a situation in which the particle moves in a straight path τ can be interpreted as the time to cover a given distance. The relation $\Delta\tau/\tau = -\Delta\beta/\beta$ applies and $m^* = \gamma^3 m_0$, the 'longitudinal mass' of the particle. The quantity $\Delta\tau/\tau$ for an orbital machine is usually expressed in terms of the parameter η, defined as

$$\eta = \frac{\Delta\tau}{\tau} \Big/ \frac{\Delta p}{p}\,. \qquad (2.145)$$

Using the relation $\Delta p/p = \Delta(\beta\gamma)/\beta\gamma = \gamma^2\Delta\beta/\beta$, it follows that

$$\frac{\Delta\tau}{\tau} = \frac{\Delta C}{C} - \frac{\Delta\beta}{\beta} = \left(\alpha - \frac{1}{\gamma^2}\right)\frac{\Delta p}{p} \qquad (2.146)$$

whence

$$\eta = \left(\alpha - \frac{1}{\gamma^2}\right)\,. \qquad (2.147)$$

The value of α for which η changes sign is known as the 'transition energy', $\gamma_t = \alpha^{-1/2}$. From equations (2.144) and (2.147) it follows that

$$m^* = \frac{m_0\gamma^3}{1-\alpha\gamma^2} = -\frac{\gamma m_0}{\eta} \qquad (2.148)$$

This is positive below transition energy, infinite at

transition energy, and negative at higher energies. In a weak focusing system γ_t is less than unity, so that the effective mass is always negative. From equation (2.142) it is evident that $\gamma_t = (1-n)^{1/2} = Q_x$. This relation is found to be approximately true in the more complicated lattices used in strong-focusing machines. If the particle is viewed in a frame of reference moving with the beam (subscript 2), then as for rectilinear motion the factor γ^3 disappears, giving

$$m_2^* = \frac{m_0}{1-\alpha\gamma^2} \ . \tag{2.149}$$

It will be seen later that there is a sense in which the energy in the moving frame, $\frac{1}{2}m_2^*\beta_2^2c^2$, is negative when $\alpha > \gamma^{-2}$.

2.7.5. Non-linear effects In this section we consider non-linear effects arising from focusing forces not proportional to particle displacement. We confine attention at present to motion in either the orbit plane or the direction perpendicular to this; coupling is considered in section 2.7.9. As in section 2.2.7 we can consider optical aberrations and calculate the appropriate coefficients; in an orbital machine, however, it is the properties associated with the periodicity which are of importance. Periodic motion in a machine in which the equations of motion for betatron oscillations are linear has been discussed earlier. For a particle with given energy a unique equilibrium orbit and Q-value can be defined; the motion is bounded unless Q has integral or half-integral values.

When non-linear terms are significant, new features appear. The Q-value now becomes amplitude dependent. In a periodic field its value can be defined as the limit as N tends to infinity of $M/2N$, where M is the number of times the orbit crosses the axis and N is the number of revolutions. The amplitude dependence of Q may be demonstrated by the following example for a focusing field which is independent of s. The equation for slightly non-linear harmonic motion in such a field is

$$x'' + \kappa x + f(x) = 0 \qquad (2.150)$$

where $f(a) \ll \kappa a$, a being the amplitude of oscillation. In a system with a rectilinear axis a quadrupole field gives rise to linear oscillations, as discussed in section 2.4.2. Sextupole and octupole components of the field give rise to $f(x)$ proportional to x^2 and x^3 respectively.

To first order sextupole field components change the amplitude of the oscillation by equal and opposite amounts in positive and negative half-cycles, but do not affect the frequency. Octupole components, on the other hand, change the amplitude symmetrically and also change the frequency. Writing $f(x)$ as $\kappa x^n/x_0^{n-1}$, where n is 2 or 3, it is readily shown (Appendix 4) that the fractional amplitude change arising from the non-linear term is $(a/x_0)^{n-1}/(n+1)$ where a is the oscillation amplitude. For $n = 2$ this alternates in sign for positive and negative half-cycles; for $n = 3$ it reduces both positive and negative half-cycles by the same amount. The fractional frequency change for an octupole field is $3a^2/8x_0$. These results are illustrative only; in an actual accelerator lattice the figures depend in a complicated way on the lattice parameters.

Although the sextupole field component does not alter to first order the oscillation frequency of a particle of given momentum oscillating about the axis $x = 0$ in equation (2.150), it does affect the chromaticity. An off-momentum particle experiences a field gradient different from that at the equilibrium orbit. To the factor $(1+n)$ in the expressions for ξ_x and ξ_y (equation 2.140) should be added an additional term $-B''R^2/Bn$. It should be noted that sextupoles aligned to give focusing in the x-plane produce no focusing in the y-plane; in this plane the force is proportional to y^2, but acts in the x-direction. At a finite value of x, however, there is a defocusing force in the y-direction proportional to xy. Sextupoles and higher multipoles (or, expressed otherwise, non-linear terms in the field) necessarily produce coupling between oscillations in the two planes, since the restoring force at a given value of x is in general a function of both x and y. A simple example illustrating this is given in section 2.7.8.

In periodic lattices the effect of non-linearities is complicated. It may be conveniently illustrated in terms of phase-space diagrams constructed in the following way. Values of x and x' for $s = s_0 + NC$ are plotted for successive values of N, where C is the circumference of the orbit. For a linear system such points lie on an ellipse, or a hyperbola if Q is complex. If $MQ = NC$, where M and N are integers, the motion repeats after N orbits have been completed and the 'phase plot' consists of a set of points spaced round an ellipse. If, on the other hand, M/N is an irrational number, the whole ellipse is ultimately traced out.

For non-linear systems, on the other hand, the topology of such plots is much more complicated and depends on the structure of the non-linear terms in the equation. Typically, there is an equilibrium orbit which corresponds to a single fixed point at the centre of the diagram. For small-amplitude oscillations the system is essentially linear, so that the point is surrounded by nested ellipses representing different initial conditions for x and x'. As an example we consider a relatively simple lattice with low Q such that $3Q$ is non-integral for small oscillations but becomes integral (but not a multiple of 3) at some finite amplitude of oscillation. As the amplitude approaches this value the ellipse becomes more triangular in shape, and when $3Q$ is integral represents three fixed points on a separatrix outside which are curves which do not encircle the origin, as sketched in Fig.2.19. The meaning of these curves at distances from the origin sufficiently great that x' is no longer small compared with s' is somewhat obscure; sometimes they are still receding in this region and the separatrix may be regarded as separating 'stable' and 'unstable' regimes. Alternatively, the curves may close while the oscillation amplitude is still relatively small, as shown by broken curves in the sketch, giving rise to further stable fixed points. These may be described as representing 'an equilibrium orbit which closes after three turns'. If $3Q$

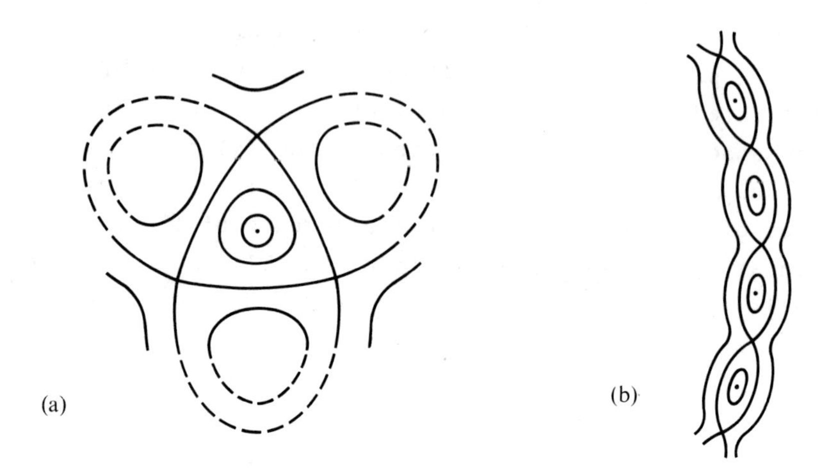

Fig.2.19. Topology of xx' plane phase plots corresponding to non-linear orbits in periodic fields. Curves consist of points corresponding to the values of x and x' of a particle at a fixed azimuth on successive revolutions. Curve A illustrates the type of behaviour exhibited by a spiral ridge cyclotron. Curve B shows part of the fine structure appropriate to a complicated system such as a large storage ring.

is integral for arbitrarily small oscillations, the central triangle disappears and the separatrix near the origin becomes three straight lines through it. This represents a 'one-third integral resonance'. In general, if the focusing parameters are slowly varied, this behaviour occurs when $3Q$ or $4Q$ are integral but not multiples of 3 or 4. For $M > 4$, however, an integral value of MQ does not in general lead to resonance. The central polygon does not vanish as the focusing parameters are varied.

In systems with a large Q value, such as alternating-gradient storage rings, the structure of the phase diagrams can be very complex. A detailed examination of the structure shows the phenomenon of 'island formation' also illustrated in Fig.2.19. This occurs when M/N is a fraction in which the denominator is not too large. Closer inspection also reveals apparent 'random' behaviour in the immediate neighbourhood of the separatrix; phase points on successive turns do not lie on a curve.

A proper description and understanding of this type of behaviour requires a far more advanced approach than that used here. This may be seen in the paper by Laslett (1974), which contains references to the considerable body of more general work on this type of phenomenon. References to the topic of non-linear behaviour of accelerator orbits, which contain justification of the descriptive statements in this section, may be found in section 2.R.6-7. A brief description of non-linear effects in coupled x and y motion is given in section 2.7.9.

2.7.6. Azimuthally varying fixed fields Fields which are constant in time but exhibit azimuthal variation are used extensively in AVF (azimuthally-varying field) cyclotrons and have been proposed for FFAG (fixed-field alternating-gradient) synchrotrons. Although several models of FFAG machines have operated successfully, no large machine has been constructed.

Azimuthally varying fields were first proposed by Thomas (1938) to provide simultaneous radial and vertical focusing in a cyclotron in which the mean magnetic field increases with radius. This type of field is required in isochronous cyclotrons in which the orbital frequency is required to be independent of radius. For constant angular velocity in the field B_y must increase with radius in proportion to γ; this implies a negative value of n, and hence, for an azimuthally uniform system, vertical defocusing. Vertical focusing may be restored by suitable azimuthal variation of the field, as illustrated by the following particular example. Suppose that the field is high and low in alternate sectors as illustrated in Fig.2.20; then the equilibrium orbit has alternately large and small curvature in the high- and low-field sectors. For this reason particles cross the sector boundary at an angle and experience vertical edge focusing between the sectors; this counteracts the defocusing which arises from the increase of field with radius. The

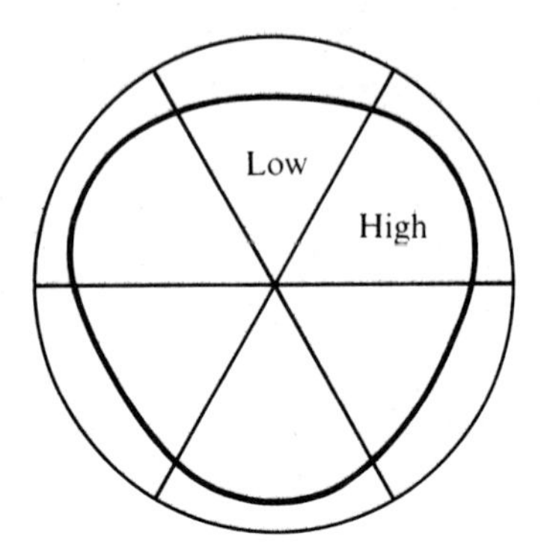

Fig.2.20. Sector-focused cyclotron. The radius of curvature of the orbit is alternately large and small in the low- and high-field sectors. Vertical focusing occurs at the edges between sectors.

degree of focusing obtainable in this way is limited, but by spiralling the sectors so that there is alternately strong focusing and (slightly weaker) defocusing the overall focusing may be strengthened. Although now used extensively in cyclotrons, spiral ridges were first suggested in connection with FFAG synchrotrons (Symon *et al.* 1956).

As indicated in section 2.7.3 the theory of orbits and oscillations in such a field is analytically complicated; in practice extensive use is made of numerical methods, which rely on computer ray tracing. The configuration of an orbit which closes after one revolution is found by iterative methods. Some special field shapes, however, give relatively simply analytical results for the Q-values; derivations of these are given for example by Livingood (1961). They are quoted here to show how the various parameters enter. For a magnetic field of form

$$\frac{B}{B_0} = (1 - \frac{nx}{R_0})\{1 + \delta \cos (N\theta - \frac{x\cot\sigma}{R_0})\} \tag{2.151}$$

it is found that

$$Q_y^2 \approx n + \delta^2(\tfrac{1}{2} + \cot^2\sigma). \tag{2.152}$$

To first order Q_x is unaffected by the ridges and is equal to $(1 - n)^{1/2}$. For an isochronous machine the orbital angular velocity ω_0 of the particles is constant, and, since $\omega_0 \propto B/\gamma$, it follows that $B(r) = \gamma(r)B(0)$. The variation of γ with r is found from a knowledge of β as a function of r, $\beta = r/R_c$, where $R_c = m_0c/qB_0$ is the hypothetical radius at which $\beta = 1$ and γ and B are infinite. Expressing γ in terms of r as $(1 - r^2/R_c^2)^{-1/2}$ enables $n = -(r/B)(dB/dr)$ to be evaluated as $1 - \gamma^2$. For large γ, n is large and negative, so that vertical focusing is difficult to achieve without excessively tight spiralling. In practice the integral resonance $Q = 2$ sets a limit at $\gamma = 2$, $n = -3$.

Modern machines make use of 'separated sectors' between which the field drops to a very low value, giving a large but far from sinusoidal modulation of the azimuthal field.

2.7.7. Linear coupling between radial and vertical motion

In the examples of orbital motion studied so far coupling between motion in the orbit plane and perpendicular to it has not been considered. In the paraxial equation, on the other hand, coupling for the special case of axial symmetry has been included. This can be dealt with conveniently by transforming to the Larmor frame or using complex variables. If, however, the fields do not have axial symmetry, the situation is more complicated, and 4 × 4 matrices are necessary to describe the motion.

Linear coupling can be introduced by longitudinal magnetic fields or by introducing an element in which the planes of symmetry are rotated with respect to those elsewhere in the orbit. The first type of coupling introduces forces in the equation for x which are proportional to $\dot{y}$, and the second introduces terms proportional to y. The 4 × 4 matrix for a uniform magnetic field has been given in section 2.2.2. For a quadrupole tilted at an angle θ the appropriate matrix may be found by rotating the axes through an angle θ and then back again:

$$M = \begin{pmatrix} I\cos\theta & -I\sin\theta \\ I\sin\theta & I\cos\theta \end{pmatrix} \begin{pmatrix} M_f & 0 \\ 0 & M_d \end{pmatrix} \begin{pmatrix} I\cos\theta & I\sin\theta \\ -I\sin\theta & I\cos\theta \end{pmatrix} \quad (2.153)$$

where I is the unit 2×2 matrix and M_{f}, M_{d} are the quadrupole 2×2 matrices given in equation (2.105). For elements containing both longitudinal fields and tilts the matrix is more complicated, and is best evaluated by following the trajectories of the four principal solutions, corresponding to initial conditions (1, 0, 0, 0), etc. For a system derivable from a Hamiltonian, as is the case when the fields are described by Maxwell's equations, there are a number of dynamical constraints on the motion which imply relations between some of the elements of the 4×4 transfer matrix. These constraints may be regarded as a consequence of the integral invariants of Poincaré (Goldstein 1950, p.247). In a system with one degree of freedom there is only one constraint, equivalent to the Wronskian relation (equation (2.36)). For coupled motion with two degrees of freedom, on the other hand, the required condition, that the transfer matrix appropriate to normalized variables be symplectic, implies that only 10 of the 16 matrix elements are independent. Details are given by Courant and Snyder (1958).

The analysis of orbital motion in the presence of weak linear coupling is of importance in orbital particle accelerator theory, since such coupling may be introduced by magnet imperfections. The appropriate basic theory has been given for example by Courant and Snyder (1958) and Laslett (1967); here we merely quote their conclusions with regard to coupling resonances in a periodic structure. This applies to coupling caused by lateral magnet displacements, magnet tilts, or longitudinal magnetic field components. Since the coupling is small, the motion in either plane viewed for a short time is unaffected. Over longer periods it is possible for partial or complete energy interchange to take place between radial and horizontal oscillations, or, alternatively, continuous resonant build-up of oscillation amplitude may occur. It may be shown that 'difference' resonances of the form $Q_x - Q_y = N$, where N is integral, produce total energy interchange between radial and vertical motion, whereas 'sum' resonances $Q_x + Q_y = N$ give rise to continuous amplitude growth. Estimating growth rates is straightforward, though sometimes tedious; the relevant theory is given in the references quoted. Further remarks on coupled motion are made in section 4.3.6 and a simple example is given in the next section.

2.7.8. Betatron with azimuthal magnetic field; an example of strong linear coupling A simple example of strong linear coupling between radial and azimuthal motion is provided by a betatron with an additional azimuthal field B_s along the orbit. This type of field configuration has been studied in connection with plasma betatrons and the electron ring accelerator. In these applications the effect of self-fields is important; nevertheless the simpler case of motion in the absence of self-fields demonstrates the essential features.

The effect of the azimuthal field is to produce a radial force proportional to the vertical velocity and *vice versa*. The equations for the orbits become

$$\begin{aligned} x'' + \frac{Q_x^2 x}{R_0^2} &= \frac{B_s y'}{B_0 R_0} \\ y'' + \frac{Q_y^2 y}{R_0^2} &= -\frac{B_s x'}{B_0 R_0} \quad . \end{aligned} \tag{2.154}$$

The azimuthal field B_s is proportional to $1/(R_0+x)$, since $\int B_s \mathrm{d}s$ round the circumference must be constant. In the paraxial approximation considered here it is sufficient initially to consider B_s as constant. The normalized ratio of fields B_s/B_0 will be written as b.

The B_s field couples the oscillations, and the normal modes are represented by ellipses in the xy plane. To find the Q values associated with these ellipses, and the ratio of their axes, we write $x = \alpha_1 \exp(\mathrm{j}Qs/R_0)$, $y = \alpha_2 \exp(\mathrm{j}Qs/R_0)$, where α_1 and α_2 are complex, and substitute in equation (2.154):

$$\begin{aligned} \alpha_1(-Q^2 + Q_x^2) &= \mathrm{j}\alpha_2 bQ \\ \alpha_2(-Q^2 + Q_y^2) &= -\mathrm{j}\alpha_1 bQ. \end{aligned} \tag{2.155}$$

From these equations

$$\frac{\alpha_2}{\alpha_1} = \frac{-j(Q_x^2-Q^2)}{bQ} = \frac{-jbQ}{Q_y^2-Q^2} . \qquad (2.156)$$

The second two terms give a biquadratic in Q which determines the Q-values for the two normal modes. These may then be inserted in equation (2.156) to find α_2/α_1. The solution of the biquadratic yields

$$Q^2 = \tfrac{1}{2}[(Q_x^2+Q_y^2+b^2) \pm \{(Q_x^2+Q_y^2+b^2)^2 - 4Q_x^2Q_y^2\}^{1/2}] \qquad (2.157)$$

Since $Q_x^2 + Q_y^2 = 1$ this may be further simplified to

$$Q^2 = \tfrac{1}{2}[(1+b^2) \pm \{(1+b^2)^2 - 4Q_x^2Q_y^2\}^{1/2}] . \qquad (2.158)$$

In general, when Q_x and Q_y are not equal, increasing b increases the larger and decreases the smaller. When b becomes large compared with unity, the two Q values become b and Q_xQ_y/b.

Since Q is real, it is evident from equation (2.156) that α_2/α_1 is imaginary; this means that the phase of the motion in the two planes differs by $\pi/2$ so that the projection on the xy plane is an upright ellipse with semi-axes $|\alpha_1|$ and $|\alpha_2|$. The value of α_2/α_1 for the two modes can be found from equation (2.156). Since $Q_x^2 - Q^2$ and $Q_y^2 - Q^2$ are of opposite sign, the sense of rotation of the two ellipses is opposite. In the limit of large b, the normal-mode ellipse associated with the high-Q mode tends to a circle, rotating at the cyclotron frequency of the particle in the field B_s; that associated with the low-Q mode has axes with ratio Q_x/Q_y.

The paraxial behaviour so far described is very simple; it is of course straightforward (though tedious) to construct a 4×4 matrix appropriate to these fields by considering four trajectories with initial conditions (1,0,0,0), etc., as described in the previous section. For the orbit with initial conditions (1,0,0,0), for example, the solution consists of the two normal modes with phases such that x' and y'

are zero for each mode and amplitudes such that the values of y for the two modes are equal and opposite.

When b is sufficiently large, the paraxial approximation breaks down because it is no longer legitimate to neglect the variation of B_s with r. Equation (2.154) becomes

$$x'' + \frac{Q_x^2 x}{R_0^2} = \frac{by'}{R_0}\left(1 - \frac{x}{R_0}\right)$$
$$y'' + \frac{Q_y^2 y}{R_0^2} = -\frac{bx'}{R_0}\left(1 - \frac{x}{R_0}\right) . \tag{2.159}$$

These equations have been discussed by Ferrari and Zucker (1971) in connection with the plasma betatron. These authors show that, when the second term on the right is large compared with the second term on the left, a drift velocity in the y direction appears, so that ultimately all particles leave the region of the equilibrium orbit. This motion is familiar in pure toroidal magnetic fields, where it may be interpreted as the 'curvature drift' associated with the curvature of the field lines.

2.7.9. Coupling arising from non-linearities; a simple example

In the last two sections linear coupling between horizontal and vertical motion was discussed, and a simple example, in which coupling was induced by a magnetic field along the orbit, was analysed. We now introduce an example of non-linear resonant coupling, and later make some general comments about this type of coupling in a periodic field.

Although in practical accelerators coupling often arises from imperfections such as stray fields and tilted magnets, essential non-linearities in an otherwise 'perfect' field can also give rise to coupling resonances. This is evident, for example, if the vertical focusing force is a function of the radial position of the particle. Radial motion produces an effective 'forcing term' in the vertical motion; if this has a component with frequency equal to the vertical oscillation frequency then resonant coupling occurs. This type of resonance is of importance, for example, in classical cyclotrons; over most of the radius Q_x is of order 1 and Q_y is small, but at large radii near the edge of the magnet pole, where the field falls off, n increases leading to a decrease in Q_x and an increase in Q_y. When $Q_x = 2Q_y$, resonant interchange between radial and vertical motion occurs.

The coupling may be demonstrated by proceeding as in section 2.5.3 but including the next term n', defined as $(R_0^2/B)(\mathrm{d}^2B/\mathrm{d}r^2)$, in the expansion of the field:

$$\begin{aligned} \frac{B_y}{B_0} &= 1 - \frac{nx}{R_0} + \frac{n'(y^2-x^2)}{2R_0^2} \\ \frac{B_x}{B_0} &= -\frac{ny}{R_0} - \frac{n'xy}{R_0^2} \end{aligned} \qquad (2.160)$$

where the form of the coefficients in the expansion is determined by setting the curl and divergence of the field to zero. The orbit equations are found to be

$$\begin{aligned} x'' + (1-n)x/R_0^2 &= \tfrac{1}{2}n'(y^2-x^2)/R_0^3 \\ y'' + ny/R_0^2 &= -n'xy/R_0^3. \end{aligned} \qquad (2.161)$$

Since n' is small, the r.h.s. can be determined to sufficient accuracy by inserting the solutions with $n' = 0$. Selecting solutions of the form $x = x_0 \exp\{\mathrm{j}(1-n)^{1/2}\theta\}$, $y = y_0 \exp\mathrm{j}n^{1/2}\theta$ and their complex conjugates, where x_0 and y_0 are complex, it is readily seen that the r.h.s. of the two equations contains terms with exponent $\pm 2\mathrm{j}n^{1/2}\theta$ and $\pm\mathrm{j}\{n^{1/2}-(1-n)^{1/2}\}$ respectively. Equating these to the respective exponents for the unperturbed equations gives rise to the condition for resonance

$$2n^{1/2} = (1-n)^{1/2} \qquad (2.162)$$

or $n = 0{\cdot}2$. So far we have merely demonstrated that resonance occurs. Rates of growth can readily be calculated in terms of n', but this simple treatment does not indicate all the features of the motion. For example a more complete treatment based on Hamiltonian methods is required to demonstrate the periodic energy interchange between radial and vertical motion (Garren *et al.* 1962). In periodic structures non-linearities can cause additional resonances to appear. If $aQ_x + bQ_y$ is integral and $a + b$ less than 4, the motion is unstable. If $a + b = 4$, then it is not possible to say without a more detailed analysis of the particular situation whether the motion is stable or not. For $a + b > 5$ the motion is stable, except under rather exceptional circumstances.

2.7.10. Trajectories, orbits, and focusing; some general observations The physical background of all the work discussed so far in this chapter has been straightforward. First, in a given configuration of static electric and magnetic fields a trajectory appropriate to a particle with particular mass, charge, and initial values of position and momentum has been identified. This special trajectory is often referred to as the 'axis' or 'closed orbit'. Sometimes, as when such a trajectory is a straight line or circle, this identification is straightforward; on other occasions it is more complicated (as in section 2.7.6). Having specified the axis or closed orbit, we have then been concerned with the properties of neighbouring trajectories, with slightly different initial conditions for the particle position, and for components of momentum transverse to or along the beam. A variety of particular situations, each with its own simplifying symmetry, has been surveyed. These represent the more important basic geometries underlying a wide range of practical devices. Optical concepts such as focusing dispersion, and aberration have been introduced and a number of subsidiary parameters, relevant sometimes only to particular applications, have been defined.

The behaviour of a single particle in static fields represents a limited but essential ingredient in the full understanding of charged-particle beams. So far the following groups of associated ideas, relevant to different contexts, have been developed.

(1) Focusing.

(2) Dispersion, momentum compaction, chromaticity.

(3) Aberrations, non-linearities.

(4) Resonant behaviour in orbital systems.

Sufficient discussion has been given to demonstrate the relevance of these ideas; all are developed further elsewhere.

The treatment so far refers to static fields only. If time variation is permitted, then further possibilities emerge. Some of these are outlined in the concluding section of this chapter.

2.R.6-7. Notes and references Except where references to specific papers are given, the material of this section is covered in the texts devoted to particle accelerators. Papers of historic interest in the field, together with comments and a discussion of priorities may be found in the volume by Livingston (1966).

Later developments are often contained in unpublished reports of the larger accelerator laboratories, though these are generally presented (often in abbreviated form) at accelerator conferences. The main series are the international meetings held every two to three years starting in 1956 at CERN, Geneva, but there are subsidiary national conferences in the USA and USSR, and specialist conferences for cyclotrons and linear accelerators.

Strong focusing and a treatment of orbit theory is naturally presented in all texts on accelerator theory. The most general of these, which cover the subject in considerably more depth and with greater thoroughness than the present treatment, are by Kolomenskij and Lebedev (1962, English version 1966) and Bruck (1966). The texts of Livingston and Blewett (1962), Livingood (1961), and Kollath (English version 1967) are more oriented to practical accelerator design and less detailed, though the latter book does contain a rather detailed section on orbit theory partly contributed by Hagedorn. The classic paper expounding in detail the linear theory of alternating-gradient focusing is by Courant and Snyder (1958). This is particularly suitable for orbits with focusing which is piecewise constant; a general treatment more appropriate to AVF and FFAG machines is given by Parzen (1961). The important topic of non-linear effects in alternating-gradient machines has been much studied; basic papers are the unpublished CERN reports of Hagedorn (1957) and Schoch (1958). The subject is treated at some length in the book by Kolomenskij and Lebedev, which contains a number of earlier references. Application to the cyclotron is covered in a general paper by Verster and Hagedoorn (1962). Many of these papers use Hamiltonian methods, not discussed in the chapter under review. (A good introduction to this approach as applied to accelerators is given in the appropriate section of the text on classical mechanics by Corben and Stehle (1960)).

Orbit theory for large accelerators and storage rings, as well as for AVF cyclotrons is now a highly developed subject, in which analytical methods are supplemented by a range of computer programmes. A number of these are described in a book by Colonias (1974).

Only single-particle motion has been considered so far. Of more relevance to the central theme of this book is the behaviour of beams, which essentially consist of ensembles of orbits. This aspect will be developed in Chapter 4, and the literature again reviewed there.

2.8. Focusing in fields which vary with time

2.8.1. Introduction So far only motion in static electric and magnetic fields has been considered. It is evident that in the presence of time-varying fields many constraints are lifted and new possibilities arise. Indeed there are so many different situations that it is difficult to cover them in a systematic and general manner. Many problems are best tackled *ab initio* using numerical techniques. An example is provided in the focusing of the first few orbits in a cyclotron, a very specialized problem for which computer programmes have been written. Another example, again tackled numerically, is the theory of large signal bunching in klystrons. A further example is found in spectroscopy which makes use of high-frequency quadrupole fields, where containment depends on mass and frequency (Blauth 1966, Dawson and Whetten 1969).

Some situations, however, can be seen as simple extensions of what has been treated already. First, there is the adiabatic variation of betatron amplitude oscillation in a guide field which is changing with time; this can be simply treated by making use of the paraxial equation, recalling that there is a field in the direction of motion, and transforming to distance along the orbit rather than time as independent variable. Alternatively, the condition for constant action $\oint p_x dx$ for an oscillation may be written down directly. For a betatron in which Q also varies with time it is readily verified that $x_0 \propto B^{1/2}/Q^{1/2}$. Second, if the field in an electrostatic lens varies in a time less than the transit time of a particle, then a wider range of focusing properties is available; this problem is relevant to cyclotrons and linear accelerators and is treated in sections 2.8.5 and 2.8.6. The third situation, which is of general interest in accelerators and also as we see later in plasmas is the motion of a particle in a moving potential well represented by a longitudinal electric field which varies harmonically with distance. Acceleration of particles in such a well, which is itself undergoing acceleration, forms the basis of all accelerators except those based on steady high-voltage generators and betatrons. The analysis of such a system may be found in all accelerator texts; we study it now in rather general form.

2.8.2. Longitudinal focusing and phase stability The focusing studied so far has been in a direction perpendicular to that in which the particles travel. In cyclic and linear particle accelerators focusing along the direction of motion is also essential to ensure synchronism with the accelerating field; such focusing introduces longitudinal structure into the beam, which forms into a series of 'bunches'. We now analyse the motion, in the limit that the beam intensity is sufficiently low that the self-fields produced by the charges can be neglected.

In the classes of accelerator listed above the particles make repeated traversals of gaps across which there is a harmonically varying field. In order that they should remain in exact synchronism with the accelerating field, it is necessary for the particles to cross the centres of the gaps at such a phase that they gain the right amount of energy. Particles crossing a gap too late or too early acquire different energy increments from that of the synchronous particle, and will take either a longer or shorter time to traverse the space between subsequent gaps. Such particles can be 'focused in phase' by ensuring that if they move faster or more slowly than the synchronous particle they will arrive at successive gaps at phases which correspond to lower or higher fields respectively.

The motion can be analysed by comparing the time of arrival at a gap of any particle with that of the special 'synchronous' particle. It is, however, possible to provide a rather more general treatment by replacing the discrete gap fields by an effective spectrum of travelling harmonic waves, and to consider the interaction of the particle with that wave which has phase velocity substantially equal to the particle velocity. This is the method used here.

The field E_s along the equilibrium orbit may be expressed as the real part of

$$\tilde{E}_s(s,t) = E_0(s)\exp(j\omega_g t) \qquad (2.163)$$

where ω_g is the frequency applied to the gap and $E_0(s)$ represents the (complex) field round the orbit. In large cyclic machines $E_0(s)$ is well represented by a δ-function or a series of δ-functions, one for each resonator. By expressing $E_0(s)$ as a Fourier series, E_s can be resolved into a spectrum of waves travelling in the s and $-s$ directions. Particles only interact significantly with that wave which has phase velocity nearly equal to the particle velocity. We select a wave component of form

$$\mathrm{Re}\tilde{E} = E_h \sin(\omega_g t - 2\pi hs/C) \quad (2.164)$$

where C/h is a distance measured along the equilibrium orbit of the particle corresponding to a change of 2π of the phase of the electric field. In a circular machine C represents the circumference of the equilibrium orbit and h the 'harmonic number' of the accelerating field. In a linear accelerator the 'circumference' has no meaning and h is taken as unity. Initial conditions are taken such that $s = 0$ at $t = 0$. The frequency ω_g is chosen such that

$$\omega_g = h\omega_s. \quad (2.165)$$

The synchronous frequency ω_s corresponds to $2\pi\dot{s}/C$ for a particle on the equilibrium orbit.

We now introduce the 'phase' of a particle at s with respect to the wave by

$$\phi = \frac{2\pi hs}{C} - \int \omega_g \mathrm{d}t. \quad (2.166)$$

For a localized field on a single gap, $-\phi$ also represents the phase of the field at the gap at the instant when the particle crosses it. It is often defined in this way rather than as in equation (2.166).

During the acceleration process the particle velocity increases; the manner in which ω_s and the mean orbit circumference vary with time depends on the type of machine. This is shown in Table 2.2. In the electron storage ring the energy does not increase; power is supplied to replenish radiation losses. The time variations implied in the table are slow, so that changes in the phase oscillation parameters occur adiabatically. The equation for phase oscillations can be found by equating the electric force to the rate of increase of momentum using the effective mass m^* of the particle defined in equation (2.144):

$$\frac{\mathrm{d}}{\mathrm{d}t}(m^*\dot{s}) = qE_h \sin\phi. \quad (2.167)$$

TABLE 2.2

Variation with time of some parameters of accelerators and storage rings

Accelerator	Energy	Gap frequency	Orbit Circumference
Linac	Increasing	Constant	—
Synchrotron	Increasing	Increasing	Constant
Relativistic synchrotron	Increasing	Constant	Increasing slightly
Synchrocyclotron	Increasing	Decreasing	Increasing
Electron storage ring	Constant	Constant	Constant

writing $\dot{\phi} = (2\pi h\dot{s}/C) - \omega_g$ (equation 2.166), $E_h C = V$ = 'volts per turn' (in an orbital machine), $h\omega_s = \omega_g$ (equation 2.165), this becomes

$$\frac{\mathrm{d}}{\mathrm{d}t}\left(\frac{m^*\dot{\phi}}{2\pi h}\right) = \frac{qV}{C}\sin\phi - \frac{1}{2\pi}\frac{\mathrm{d}}{\mathrm{d}t}(m^*\omega_s). \tag{2.168}$$

Defining $\sin\phi_s$, the 'stable phase angle', as the final term multiplied by C/qV, the equation simplifies to

$$\frac{\mathrm{d}}{\mathrm{d}t}\left(\frac{m^*\dot{\phi}}{2\pi h}\right) = \frac{qV(\sin\phi - \sin\phi_s)}{C}. \tag{2.169}$$

In general ω_s varies more slowly than ϕ; equation (2.168) can then be approximated, for small oscillations, as

$$\ddot{\phi} = \frac{\Omega_s^2}{\cos\phi_s}(\sin\phi_s - \sin\phi) \approx \Omega_s^2(\phi_s - \phi) \tag{2.170}$$

where

$$\Omega_s^2 = -\frac{2\pi hqV}{m^*C}\cos\phi_s. \tag{2.171}$$

and m^* and $\cos\phi_s$ have opposite signs for stable motion. This implies that when m^* is positive, as in linear accelerators and cyclic machines above transition energy, the stable phase is in a field which is rising in time at a given point, and thus a late particle gets an 'extra push' to help it catch up. In the negative-mass regime the point of stability is in a falling field. For a machine with a circular orbit $C = 2\pi R$, and in terms of the parameter $\eta = -\gamma m_0/m^* = \alpha - 1/\gamma^2$ (equation 2.148)

$$\Omega_s^2 = \left(\frac{hqV\eta}{2\pi\gamma m_0 R^2}\right) \cos\phi_s \quad . \tag{2.172}$$

The quantity Ω_s/ω_s is referred to as Q_s, the 'Q-value' for synchrotron oscillations. It is much less than unity.

Equation (2.170) describes the motion of a biased pendulum (Fig.2.21), or of a ball rolling in a shallow sinus-

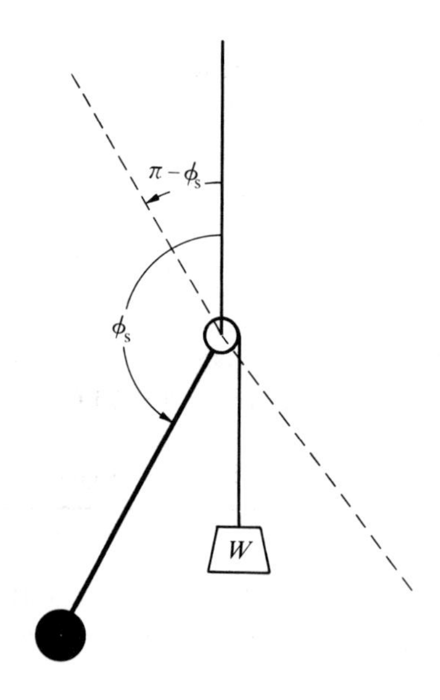

Fig.2.21. Biased pendulum, which obeys equation (2.170) and provides an analogue of phase oscillations. The broken lines represent the limits of stable motion, the upper one being a position of unstable equilibrium.

oidally modulated surface tipped at an angle $(2\pi h/\lambda)\cos\phi_s$

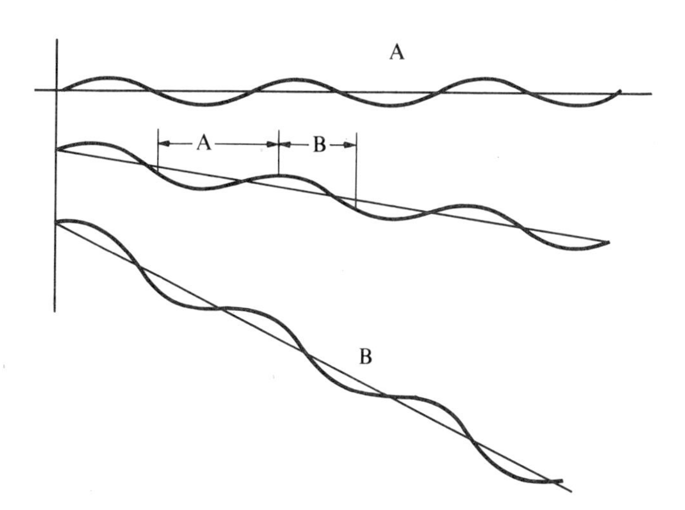

Fig.2.22. 'Wavy-slope' analogue of phase motion. A ball placed in the region A makes stable oscillations; a ball placed at B runs continuously down the slope. Balls injected with finite velocity (corresponding to non-synchronous particles) are trapped over smaller regions.

(Fig.2.22). An even better analogue is a horizontal modulated surface in a horizontal plane accelerated at a rate $(2\pi gh/\lambda)\cos\phi_s$ where g is the acceleration due to gravity. Such an analogue shows the correct damping as the parameters are varied. The characteristics of the motion can be seen immediately from these analogues. For small oscillations the motion is bounded; if the oscillation amplitude is increased the frequency decreases until the stable limit is reached, after which focusing no longer occurs and the phase increases or decreases indefinitely. The motion can be conveniently represented on a 'phase diagram', in which $d\phi/\Omega dt$ is plotted against phase (Fig.2.23). The curves can be found explicitly by writing equation (2.170) in the form

$$\frac{d}{dt}\left\{\dot{\phi}^2 - \frac{2\Omega_s^2}{\cos\phi_s}(\cos\phi + \phi\sin\phi_s)\right\} = 0 \qquad (2.173)$$

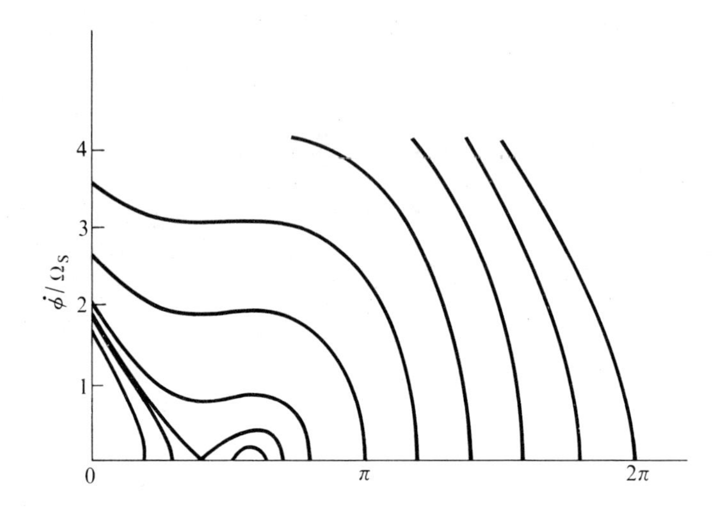

Fig.2.23. Phase motion for $\phi_s = 100^\circ$ with various values of ϕ_o, the phase when $\dot{\phi} = 0$.

and integrating to give

$$\dot{\phi}^2 - \dot{\phi}_1^2 = \frac{2\Omega_s^2}{\cos\phi_s}\{\cos\phi - \cos\phi_s - (\phi_1 - \phi)\sin\phi_s\} \tag{2.174}$$

where the subscript 1 denotes initial values. If ϕ_o is the phase when $\dot{\phi} = 0$, we can put $\phi_1 = \phi_o$ and $\dot{\phi}_1 = 0$.

Within the separatrix the motion is bounded; outside it the phase increases indefinitely. Particles represented by points within the separatrix are said to be 'trapped'; they gain energy at a mean rate equal to that of a particle at the stable phase ϕ_s:

$$\langle\frac{d\gamma}{dt}\rangle = (\Delta\gamma)_g\, \omega_g \frac{\sin\phi_s}{2\pi}\ . \tag{2.175}$$

External points 'slip' in phase and the corresponding particles experience no net gain of energy. The singularity in the separatrix represents a point of unstable phase, $\phi_u = \pi - \phi_s$; for such a point $\sin\phi_u = \sin\phi_s$ and $\cos\phi_u = -\cos\phi_s$. Setting $\phi_1 = \phi_u$ when $\dot{\phi}_1 = 0$ in equation (2.174) gives the equation for the separatrix

$$\dot{\phi}^2 = \frac{2\Omega_s^2}{\cos\phi_s}\{\cos\phi + \cos\phi_s + (\phi_s+\phi-\pi)\sin\phi_s\} \quad . \qquad (2.176)$$

This is plotted in Fig.2.24 for various values of ϕ_s. It is readily verified by differentiation that the maximum value of $\dot{\phi}$ occurs where $\phi = \phi_s$.

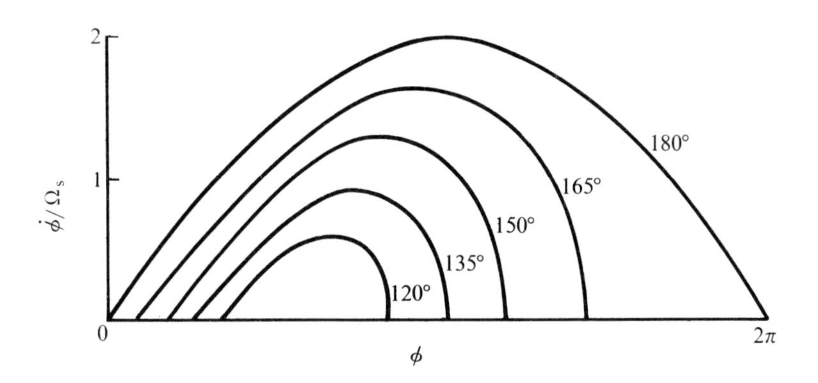

Fig.2.24. Separatrices of the phase motion for various values of ϕ_s.

It is often convenient to express the phase equation in Hamiltonian form, using canonical variables. In this form the phase-space invariants associated with the motion can be written down directly. The canonical variables are W and ϕ, where

$$W = \int_1^\gamma \frac{m_0c^2\,d\gamma}{h\omega_s(\gamma)} \qquad (2.177)$$

and the Hamiltonian is

$$H(\phi,W,t) = -\gamma(W)m_0c^2 + W\omega_g + \frac{eV}{2\pi h}\cos\phi . \qquad (2.178)$$

The area in W,ϕ space occupied by a group of particles is invariant. (This is explained in more detail in Chapter 4.) In orbital machines with circular orbits it may readily be verified that $W = 2\pi P_\theta/hc$, where P_θ is the canonical angular momentum. For linear machines $W = p_zc/\omega$.

2.8.3. Radial and energy excursions of phase oscillations; adiabatic damping It is of importance to estimate the extent of the excursions both of energy and radius associated with phase oscillations. Such information is essential for example in designing the injection system from a linear accelerator into a synchrotron; the capture efficiency depends both on the energy spread of the injected particles and the radial oscillation amplitude. The relation between $\gamma - \gamma_s$, $R - R_s$, and ϕ will now be determined with the conditions, always true in practice, that $\gamma - \gamma_s \ll \gamma$ and similarly for R. The essential relation for this calculation is obtained by integrating equation (2.169),

$$\frac{m^* C\dot{\phi}}{2\pi h m_0 c} = (\beta\gamma - \beta_s\gamma_s) \approx \frac{1}{\beta}(\gamma - \gamma_s) \tag{2.179}$$

This gives the relation between $\dot{\phi}$ and the normalized energy excursion $\gamma - \gamma_s = \Delta\gamma \ll \gamma$.

The corresponding equation for radial variation is found from the momentum compaction function (2.137) which may be written $\Delta x = \alpha_p \Delta(\beta\gamma)/\beta\gamma = \alpha_p \Delta\gamma/\beta^2\gamma$. Using this to eliminate $\gamma - \gamma_s = \Delta\gamma$,

$$\Delta x = \frac{m^* C\dot{\phi}\alpha_p}{2\pi h m_0 \beta\gamma c} = -\frac{C\dot{\phi}\alpha_p}{2\pi h \eta \beta c} . \tag{2.180}$$

In a machine with axial symmetry, $C = 2\pi R$, α_p is constant and equation (2.180) relates phase excursion to radial excursion directly. Since both $\Delta\gamma$ and $\Delta x/\alpha_p$ are proportional to $\dot{\phi}$, they are directly illustrated in Figs. 2.23 and 2.24, for which equations (2.179) and (2.180) may be regarded as scaling factors.

The adiabatic damping can be found from the action integral associated with the Hamiltonian (2.178). Alternatively, it may be found by noting that the equation of motion (2.169) for small amplitudes is of the form

$$\frac{d}{dt}(u\dot{\phi}) + v\phi = 0 \tag{2.181}$$

where u and v are slowly varying quantities. Expanding this produces a second-order equation with a term in $\dot{\phi}$, to which the solution may readily be verified to be

$$\phi = (uv)^{-1/4}\{A\cos(v/u)^{1/2}t + B\sin(v/u)^{1/2}t\} \quad (2.182)$$

giving adiabatic damping proportional to $(uv)^{-1/4}$. This method (the WKB approximation) is explained by Livingood (1961), but unfortunatley the result he obtains for damping of phase oscillations is incorrectly calculated. Applying this result to equation (2.169), with $\phi - \phi_s \ll \pi$ gives, for the invariant appropriate to the phase excursion from the stable value ϕ_s,

$$I_\phi = (\phi-\phi_s)(-m^*C^2V\cos\phi_s)^{1/4}. \quad (2.183)$$

As the factors in the second bracket vary adiabatically, $(\phi-\phi_s)$ varies so that the product of the two brackets remains constant.

A special problem which arises in alternating-gradient synchrotrons is the phase motion as the transition energy when m^* passes through infinity $(\gamma = \alpha^{-1/2})$. The adiabatic approximation is not good enough. The problem is discussed by Courant and Snyder (1958), and in much greater detail by Sørenssen (1975). A phase jump has to be provided in the accelerating field. The problems associated with this transition have been overcome successfully on large proton synchrotrons.

To obtain the corresponding invariants for the damping in energy and radial excursion from the equilibrium orbit the result expressed in equation (2.183) is combined with equations (2.179) and (2.180) and the relation $\phi_1 \propto \dot{\phi}_1/\Omega$, where ϕ_1 and $\dot{\phi}_1$ are the maximum values during one cycle of oscillation. The resulting invariants I_E and I_R are found to be

$$\frac{I_E}{\Delta E} = \frac{C^{1/2}}{\beta(-m^*V\cos\phi_s)} = \frac{\alpha C}{\beta^2\gamma}\frac{I_R}{\Delta R}. \quad (2.184)$$

These results apply equally well to synchrotrons, synchrocyclotrons, or linear accelerators, except that for the latter machine I_R is meaningless (the formula gives $I_R = \infty$).

2.8.4. Concluding remarks on phase focusing and review of some implicit assumptions The review of phase focusing in the last two sections has been brief; the aim has been to present the essential physical features rather than a detailed and rigorous treatment. Strictly, the coupling between betatron and synchrotron oscillations should be considered, but since the ratio of relative frequencies Q_s/Q is usually small, it is often a good enough approximation to treat them separately. The coupled equations are given for example by Courant and Snyder (1958).

In circular machines, with relatively slow net rate of acceleration, the radial electric fields in the gaps usually have little effect on the motion. In heavy ion linear accelerators, however, they produce significant lateral oscillations, so that phase and transverse motion need to be considered together. The lateral forces vary as γ^{-2}, and are therefore much less important for electron accelerators, except at extremely low energy where they can easily be overcome by a longitudinal magnetic field. As will be seen in the next section, simultaneous phase and transverse focusing is not possible. Transverse focusing for proton linear accelerators and large electron machines is now always provided by quadrupole lenses. The value of h, the harmonic number, is normally unity for very small machines and rises to, for example, 20 for the CERN 25 GeV proton synchrotron and 1120 for the 400 GeV machine at the Fermi National Accelerator Laboratory. It is evident that in a machine with harmonic order h the beam is 'bunched' into h bunches. An advantage of a large value of h is that the radial width of the bunches is reduced. In a synchrotron ω_g/h must be made to follow ω, the circulation frequency of the accelerated particles, as the magnetic field increases. This important problem, known as tracking, is outside the scope of the present discussion.

Finally, no mention has been made of conventional

fixed-frequency cyclotrons, though these also accelerate by means of alternating fields across a gap or series of gaps. For such machines $\alpha = 1/\gamma^2$, so that $m^* = \infty$, and consequently $\Omega_s = 0$ and there is no phase stability.

2.8.5. Transverse focusing in time-varying fields; application to cyclotron Although the main focusing force in cyclic accelerators is normally provided by a constant or slowly varying field which is independent of the high-frequency accelerating field, in some circumstances the transverse components of the accelerating field have an effect on the transverse focusing. In the cyclotron the vertical electric focusing in the first few turns is considerably stronger than the magnetic focusing, which is necessarily zero at the centre since the field is uniform there, corresponding to a field index $n = 0$. The design of the centre region is generally rather specific to a particular machine and is treated as a whole. Purely analytical methods are too complicated to apply (especially if an attempt is made to include space charge), and computer methods using ray tracing through specific fields are generally used. Nevertheless, approximate treatments have been made, valid not too near the centre, which illustrate the physical behaviour. An early discussion which identifies the effects is given by Rose (1938) and a comprehensive review has more recently been given by Reiser (1971). It is found that to first order the focusing action of the dee can be expressed in terms of two components, each with its own characteristic power. We do not calculate these in detail, but quote the results from Reiser's paper.

The first component arises from the lens action already discussed in section 2.3; this is independent of the time variation and depends on the square of the gap voltage

$$\frac{1}{f_1} = \frac{F}{2\pi b}\left(\frac{2qV_g}{m_0\beta^2c^2}\right)^2 \sin^2\phi \qquad (2.185)$$

where b is the vertical half-gap of the lens, F is a geometrical factor of order unity tabulated by Reiser, and ϕ is the phase of the field at the centre of the gap. In the papers referred to θ is used for phase angle, where $\theta = \phi - \pi/2$. Unfortunately the convention for cyclotrons and synchrotrons is different; for the sake of uniformity we use the latter.

The second component, on the other hand, depends essentially on the rate of change of field during the passage of the particle through the lens. For this component

$$\frac{1}{f_2} = -\frac{\omega q V_g \cos\phi}{m_0 \beta^3 c^3} . \tag{2.186}$$

When $\phi = \frac{1}{2}\pi$, it vanishes, as would be expected. The minus sign shows that focusing occurs when $\frac{3}{2}\pi > \phi > \frac{1}{2}\pi$, that is when the field is decreasing. This is to be expected from the argument in section 2.3.5; a particle passing through the lens in a direction such that the field produces acceleration experiences focusing and then defocusing forces as it crosses the centre of the lens. The first-order cancellation which occurs in a steady field no longer operates when the field is varying. The residual focusing is first order in V_g, whereas the other term (equation 2.185) is second order. In practical situations the effect of the second term predominates except very near the centre of the cyclotron, where the geometrical approximation is poor.

2.8.6. Transverse focusing in linear accelerators In linear accelerators the effect on the motion of the transverse components of the accelerating field is not negligible; before the introduction of strong focusing by quadrupoles it presented a considerable problem, for reasons which appear below.

In section 2.8.2 it was shown that for phase stability in a linear accelerator, the stable phase angle occurs when the field is rising. In the previous section, however, we found that for transverse focusing a falling field is necessary. Indeed it can be shown quite generally that in linear accelerators, provided that the energy change in crossing one gap is small and that there are no charges in the acceleration region, simultaneous phase and radial focusing is not possible (McMillan 1950). The proof consists essentially of making a Lorentz transformation to a frame moving with velocity equal to the phase velocity of the wave. In such a frame the magnetic field components vanish, and the particle cannot be simultaneously stable in both longitudinal and transverse directions by virtue of Earnshaw's theorem. For proton accelerators, where it is difficult to apply external focusing of adequate strength, this represented a severe limitation before before the advent of strong focusing.

As with static lenses, however, the focusing can be strengthened by the introduction of grids or foils through which the beam passes. As shown in the diagram (Fig.2.25)

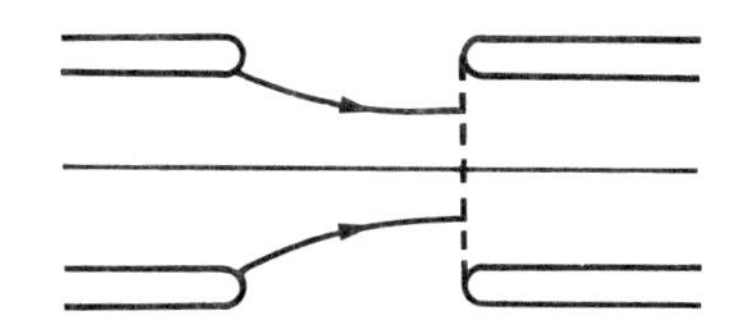

Fig.2.25. By placing a grid across the mouth of a drift tube in a linear accelerator radial focusing may be obtained at all phases of the alternating field.

the presence of a grid across the mouth of the drift tube produces a field which can be made focusing whatever the phase of the accelerating field. Grids are no longer used, however, since it is possible to incorporate magnetic quadrupoles in the drift tube structures of sufficient strength to overcome the defocusing. A full analysis of the problem, including both phase-dependent defocusing from the accelerating field and alternating-gradient quadrupole focusing, is evidently complicated, especially since the parameters vary along the machine.

In electron accelerators at low energies focusing is sometimes provided by a longitudinal magnetic field. At high energies, where the defocusing is weak, it is adequate to place quadrupoles between accelerating sections.

2.8.7. Alternating-phase focusing Shortly after the introduction of the alternating-gradient focusing concept, when it was realized that the alternation of equal and opposite focusing elements produced a net focusing effect, it was suggested that the phase of the accelerating field should be alternated along the machine by modulating the structure. Both radial and phase focusing would alternate, giving net focusing for both. In practice the focusing attainable by this method is rather weak and conditions for stability are stringent. The alternative of using magnetic quadrupoles has proved more satisfactory. It has not so far been used in an actual accelerator.

More details both of grid focusing and alternating-phase focusing (including references) are given by Boussard (1970). Also discussed are arrangements where the ends of the drift tubes do not have axial symmetry and are designed in such a way that radio-frequency quadrupole fields are generated.

2.R.8. Notes and references The theory of particle trapping, first worked out in detail by Bohm and Foldy (1946) and others soon after the synchrotron principle had been expounded by Veksler and independently by McMillan in 1944-5, is treated in the texts referred to in section 2.R.7. Kolomenskij and Lebedev (1966) and Bruck (1966) use a slightly different approach and go into more detail than the present treatment; they include more discussion of adiabatic variation and the transition-energy problem. A rather general approach to the problem, emphasizing phase-space concepts and adiabatic behaviour, is made by Lichtenberg (1969). Such detailed studies are necessary when designing injectors for accelerators; phase-space matching, discussed later in section 4.3.2 in the context of beam-transport systems, needs to be achieved.

Particle trapping and the focusing properties of the transverse components of field are essential features in linear accelerators. Here again adiabatic behaviour is important since the parameters which define the particle motion vary continuously along the machine. Indeed, because of the much greater rate of acceleration, significant changes can occur during one period of the phase oscillation. Because of the rapid variation of parameters, and also the more complicated interconnection between accelerating and focusing forces, the theory of these machines is rather special and frequently requires numerical methods. The basic principles of the single-particle dynamics have been covered in this chapter, however, and for further details the encyclopaedic compilation edited by Lapostolle and Septier (1970) should be consulted. This contains adequate accounts of, or reference to, all earlier work. Also of interest is the volume on the Stanford 'two-mile' accelerator edited by Neal (1968).

3
LAMINAR BEAMS WITH SELF-FIELDS

3.1. Introduction

The properties of the beams studied so far are determined by the external focusing fields; they do not depend on the beam intensity. For sufficiently large currents, however, both the charge in the beam and the current can play an important part in determining the beam behaviour. The disposition and motion of the charges produce fields which in turn modify the motion of the charges; this leads to a requirement of self-consistency which in all but the simplest configurations can be difficult to calculate.

In this chapter we consider laminar flow, in which the velocity distribution at a point is single valued. In such a flow the trajectories do not cross. The flow of mono-energetic particles from a point source in a paraxial system is laminar except at the source and at subsequent image points where all trajectories cross. Although pure laminar flow cannot be achieved in practice, it often represents a good approximation. In beams with self-fields it frequently permits simple models to be constructed quite easily, which illustrate many of the essential features of more realistic types of beam. The more general situation, where a distribution function is needed to specify the conditions at a point, will be discussed in the next chapter, after the relevant physical concepts for dealing with such problems have been introduced. We now consider in turn a number of simple idealized types of flow; most of these describe ideal limiting situations encountered in the design of microwave tubes and particle accelerators. For the present, steady configurations only will be considered; transient effects and wave motion are studied in Chapter 6.

Several characteristic parameters occur in the analysis. We introduce these now even though their full significance will not appear until later. Budker's parameter ν is equal to the product of N, the number of particles per unit length of the beam, and the classical particle radius r_0:

$$\nu = \frac{Nq^2}{4\pi\varepsilon_0 m_0 c^2} = Nr_0 . \tag{3.1}$$

For a uniform beam of density n particles per unit volume and radius a

$$\nu = \frac{a^2 nq^2}{4\varepsilon_0 m_0 c^2} = \frac{\gamma}{4}\frac{a^2\omega_p^2}{c^2} \tag{3.2}$$

where ω_p is the plasma frequency of the beam, defined as $\omega_p^2 = nq^2/\gamma\varepsilon_0 m_0$, and discussed further in section 6.2.4. The factor γ in the plasma frequency arises because in the frame moving with the electrons, where the frequency is defined, the density is reduced by the Lorentz factor γ. This assumes of course that all the electrons have essentially the same value of γ and β.

Closely related to γ is the characteristic Alfvén current (Alfvén 1939)

$$I_A = \left(\frac{4\pi\varepsilon_0 m_0 c^3}{q}\right)\beta\gamma = \left(\frac{cq}{r_0}\right)\beta\gamma \tag{3.3}$$

where β and γ refer to the particles which constitute the beam. For electrons the constant in brackets is equal to -17000 A. In terms of Budker's parameter we may write

$$\frac{I}{I_A} \approx \frac{\nu}{\gamma} \quad (I \ll I_A). \tag{3.4}$$

This relation is only approximate, since βc is the individual electron velocity and not the mean. Equation (3.4) is true in paraxial approximation, provided that γ still refers to the individual particles and is not defined as $(1-\bar{\beta}^2)^{-1/2}$ where $\bar{\beta}$ is the mean particle velocity. At highly relativistic energies in paraxial approximation $\beta \approx \bar{\beta}$ but $(1-\bar{\beta}^2)^{1/2}\gamma$ can be very large compared with unity. This is further discussed by Lawson (1959) in a paper in which a wide range of particle streams are discussed in terms of the parameters ν and ε, defined as $(1-\bar{\beta}^2)^{-1/2}-1$.

3.2. Characteristics of various types of flow

3.2.1. Introduction In this section we study systematically a variety of types of laminar flow in which self-fields are important. Some of these flows involve charges of one sign only; sometimes, however, neutralization by stationary charges of the opposite sign is included. For un-neutralized flows it is found that only when the particle motion is relativistic is the magnetic self-field important. When the charge is fully, or almost fully, neutralized, on the other hand, magnetic fields are of importance even in non-relativistic situations. It is sometimes found that making the calculation relativistic introduces considerable complexity, particularly when the orbits are not paraxial. Accordingly, some of the results are presented in non-relativistic approximation, especially if it is in this region that their practical interest lies.

3.3.2. Cylindrical beam in an infinite magnetic field In the approximation of infinite magnetic field, $\underline{B} = (0, 0, \infty)$, all transverse motion of the particles can be neglected, and the structure of the beam is determined by the distribution of velocity and density across the cross-section. The system would strictly be self-consistent with a completely arbitrary choice both of velocity and density; to make it interesting we introduce some other constraints.

We consider a cylindrically symmetrical beam, with radius a, containing n charges per unit volume moving with a velocity βc where n and β are both functions of radius; N denotes the number of particles per unit length of the beam and I the total current in the beam. Evidently

$$N = \int_0^a 2\pi r n(r)\,\mathrm{d}r, \qquad I = \int_0^a 2\pi q r n(r)\beta(r) c\,\mathrm{d}r. \tag{3.5}$$

The radial electric field is readily found by applying Gauss's theorem to a unit length of beam, making use of the fact that $E_z = 0$. This gives

$$E_r = \frac{q}{\varepsilon_0 r}\int_0^r rn\,dr \qquad r < a$$
$$= \frac{Nq}{2\pi\varepsilon_0 r} \qquad r > a \tag{3.6}$$

The potential of the beam with respect to an 'earth' at infinity is logarithmically infinite; in this respect it differs from a charge of finite extent in the z-direction. The zero for the potential can be chosen arbitrarily; we take the value at the beam edge multiplied by the charge to be equal to minus the kinetic energy of the charges there:

$$q\phi_a + (\gamma_a - 1)m_0c^2 = 0 \tag{3.7}$$

where the subscript a denotes the value at $r = a$. (This subscript will be used below for other variables.) Equation (3.5) implies that, if the particles are emitted from a source at zero velocity, its potential is zero. At a large distance from the emitter the potential at the beam edge differs from that at the centre, owing to the space-charge fields arising from the beam charge. This is illustrated in Fig. 3.1. The velocity of the charges is therefore a function of radius, and for a given function $n(r)$, $\beta(r)$ is determined and *vice versa*. As an example we calculate the properties of a beam for which the current density is independent of radius. For a particle at radius r in the beam

$$(\gamma-1)m_0c^2 = (\gamma_a-1)m_0c^2 - \int_r^a qE_r\,dr. \tag{3.8}$$

From equation (3.6)

$$\gamma = \gamma_a - \frac{q^2 n_a\beta_a}{\varepsilon_0 m_0 c^2}\int_0^a \frac{1}{r}\int_0^r \frac{r_1}{\beta}\,dr_1\,dr. \tag{3.9}$$

If the potential difference is small, so that the variation of β with r can be neglected, then this may be written

$$\gamma_a - \gamma \approx \nu\left(1 - \frac{r^2}{a^2}\right), \qquad \gamma_a - \gamma \ll \gamma. \tag{3.10}$$

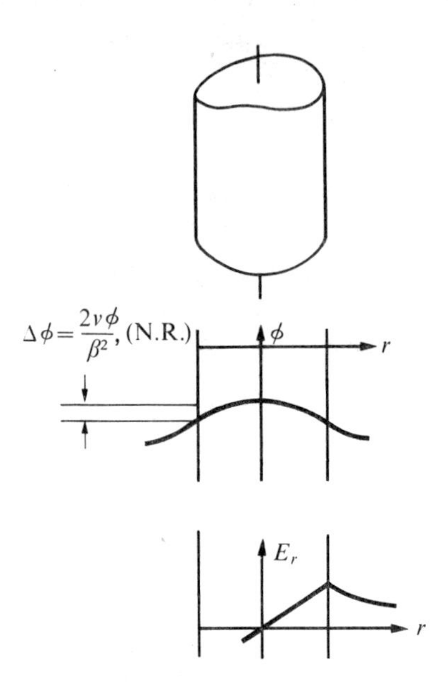

Fig.3.1. Fields and potentials near a positively charged beam at a large distance from the source.

Here we have introduced Budker's dimensionless parameter ν defined in section 3.1. It will be seen as the development proceeds that analysis in terms of this parameter, β, and γ, rather than the more traditional volts and current, often enables limitations and characteristics of different beam systems to be compared in a simple and elegant manner. In these units the current in amperes and 'voltage' of the beam are given by the following relations, where a bar denotes the average value:

$$\begin{aligned} I &= \left(\frac{4\pi\varepsilon_0 m_0 c^3}{q}\right)\nu\bar{\beta} \\ V &= \phi = -\frac{(\gamma-1)m_0c^2}{q} \\ &\approx -\frac{\beta^2 m_0 c^2}{2q} \quad \text{(N.R.)}\ . \end{aligned} \tag{3.11}$$

From equation (3.10) we see that the potential difference

between the centre and outer radius of the beam, provided that it is small, is simply $\nu m_0 c^2$. The criterion that it should be small has the simple form

$$\begin{aligned} \nu &\ll \gamma - 1 \\ \nu &\ll \tfrac{1}{2}\beta^2 \quad \text{(N.R.)} \end{aligned} , \tag{3.12}$$

It is important to note that these relations depend only on the number of particles per unit length, and not on the current density or beam radius. The second of these relations is perhaps more familiar when expressed in terms of amperes and volts. Using equation (3.7) it may be written in terms of the 'perveance' k:

$$k = \frac{|I|}{|\phi|^{3/2}} \ll \frac{4\pi\varepsilon_0 |2q|^{1/2}}{m_0^{1/2}} \tag{3.13}$$

For electrons the constant is 66×10^{-6} A $\text{V}^{-3/2}$. The perveance can be thought of as a measure of the extent to which the flow pattern of the beam is influenced by the space charge. It is the scaling factor which ensures that the flow characteristics are unchanged when the beam voltage is varied. The current in a beam is evidently limited to a value with order of magnitude given by the right-hand side of inequality (3.13), since, for some value of ν of order $\gamma_a - 1$, the potential on the axis falls to zero and the electron velocity there becomes zero. In practice a beam of radius a is often surrounded by a conducting tube of radius b. It can easily be shown that the potential difference between the tube and beam axis exceeds that between the beam edge and axis by a factor $(1 + 2\ln b/a)$.

For a beam with uniform current density, so that $n\beta$ is independent of r, the velocity βc may be determined as a function of radius from equation (3.9). Writing this in differential form and substituting for the current $I = \pi a^2 qc(n\beta)$ yields

$$\frac{1}{r}\frac{\mathrm{d}}{\mathrm{d}r}\left\{r\frac{\mathrm{d}}{\mathrm{d}r}(\gamma m_0 c^2)\right\} = \frac{q^2(n\beta)}{\varepsilon_0 \beta} \quad . \tag{3.14}$$

The solution is in general complicated; only the non-relativistic and extreme relativistic limits will be considered here. In the former equation (3.14) becomes

$$\frac{1}{r}\frac{\mathrm{d}}{\mathrm{d}r}\left(\beta r\frac{\mathrm{d}\beta}{\mathrm{d}r}\right) = \frac{q^2(n\beta)}{\varepsilon_0 m_0 c^2}\frac{1}{\beta} \quad . \tag{3.15}$$

This has a particular solution with $\beta = 0$ on the axis:

$$\beta = \left\{\frac{9q^2(n\beta)}{8\varepsilon_0 m_0 c^2}\right\}^{1/3} r^{2/3} \quad . \tag{3.16}$$

The number of particles per unit length can readily be found by integration across the beam; hence Budker's parameter ν can be found in terms of β_a, the value at the edge of the beam. This results in the simple expression

$$\nu = \frac{1}{3}\beta_a^2 \quad . \tag{3.17}$$

For electrons this represents a perveance $k = 29{\cdot}3 \times 10^{-6}$ A V$^{-3/2}$. For a beam described by this solution the particles near the axis have very low velocities but high density; they contribute little to the current but a great deal to the potential difference between the axis and beam edge. The current is not a maximum when the potential is zero on the axis. The maximum value occurs when the potential at the centre is 0·17 of that at the edge, corresponding to a particle velocity ratio of 0·42. The perveance in this case is $32{\cdot}4 \times 10^{-6}$ A V$^{-3/2}$. A more detailed analysis of this solution, with curves showing the potential variation, as well as earlier references are given by Pierce (1954a p.164).

The solution in the extreme relativistic limit $\gamma_a \gg 1$ is relatively simple. For all the charges, except those very near the axis, $\beta \approx 1$. These latter have little effect on the current, so that little error is introduced by taking $\beta = 1$ everywhere and $\gamma = 1$ on the axis. It follows immediately from equation (3.10), which holds when β is independent of r, that the limiting current is determined by the relation

$$\nu = \gamma_a \quad \text{or} \quad I = I_A. \tag{3.18}$$

If the beam is neutralized by the presence of stationary particles of opposite sign, then clearly these limitations are relaxed. If f is the 'neutralization fraction' denoting the ratio of neutralizing charge to beam charge, then, for f uniform across the beam, these current limits are increased by a factor $(1-f)^{-1}$. As we see later, however (section 6.3.2), other effects, leading to dynamic instability, can appear under these conditions. Although the infinite B_z field is an idealization, it is worth considering the problem of launching a beam of very high current to see how the limitations would set in. Before doing this we deduce Child's law for space-charge limited flow between parallel planes, neglecting thermal-emission velocities, and discuss some of its consequences.

<u>3.2.3. The planar diode</u> The equations determining the flow of charge in a planar system are

$$\frac{d^2\phi}{dz^2} = -\frac{nq}{\varepsilon_0}$$

$$nq\beta c = i \tag{3.19}$$

$$\gamma = 1 + \phi/\phi_0$$

where i is the current density and a magnetic field $(0,0,\infty)$ is assumed in order to suppress the effects of the self-magnetic field of the current. In a truly infinite system this is zero, since there is no return path for the charges. The convection current formed by the charges balances the displacement current arising from the transfer of charge, so that curl $\underline{B}$, and hence $\underline{B}$, is zero. Eliminating n and ϕ in equation (3.19) we find

$$\frac{d^2\gamma}{dz^2} = \frac{nq^2}{\varepsilon_0 m_0 c^2} = \frac{qZ_0}{m_0 c^2}\frac{i}{\beta} \tag{3.20}$$

where Z_0 is the impedance of free space. Only in the non-relativistic limit, $\gamma = 1+\frac{1}{2}\beta^2$, can this equation be solved in closed form. Under these conditions the l.h.s. can be replaced by $\beta(d^2\beta/dz^2)$. After multiplying both sides by $2\beta(d\beta/dz)dz$ this may be integrated to give

$$\beta^2\left(\frac{d\beta}{dz}\right)^2 = \frac{4qZ_0 i}{m_0 c^2}(\beta-\beta_0) \qquad (3.21)$$

where β_0 is a constant of integration. Now $\beta = 0$ at the cathode, where $z = 0$. For space-charge limited flow the electric field is also zero, so that $d\beta/dz$ and hence β_0 are zero. Equation (3.22) becomes

$$\frac{d\beta}{dz} = \left(\frac{qZ_0 i}{m_0 c^2}\right)^{1/2} \qquad (3.22)$$

from which, by integration,

$$\beta^3 = \frac{9qZ_0 i}{4m_0 c^2} z^2 \quad . \qquad (3.23)$$

This is Child's law. It is usually written in terms of the potential†

$$\frac{iz^2}{\phi^{3/2}} = 2{\cdot}34 \times 10^{-6}\ \mathrm{A\ V^{-3/2}}. \qquad (3.24)$$

If ϕ is fixed at some value of $z = d$ by placing a conducting plane there, then i is determined. The time taken for a charge to cross the diode can readily be found by evaluating $\int(dz/\beta c)$; if the result is expressed in terms of the plasma frequency corresponding to the value of n at $x = d$, then it has the very simple form $t = 1/2\omega_p$.

An exact relativistic solution of equation (3.19) has been given by Jory and Trivelpiece (1969). This is in terms of elliptic integrals, and is in general somewhat involved.

---

†Modulus signs, as in equation (3.13), are to be understood in this expression and in similar expressions which occur hereafter.

An approximation which is in error by 5 per cent at $\gamma = 2$ but decreases rapidly with increasing γ may be written

$$\left(\frac{iZ_0 q}{m_0 c^2}\right)^{1/2} z = (\gamma+1)^2 - 0{\cdot}85 \qquad (3.24)$$

Curves showing behaviour under current-limited conditions are also presented; practical systems with high γ are more commonly of this type.

In order to launch a beam, the conducting plane can in principle be replaced by a grid, through which the particles pass with velocity βc. We now enquire into what happens if the space beyond the grid is at the same potential as the grid in the absence of the beam. Beyond the grid a space charge builds up as the particles penetrate through into this region; this space-charge produces a retarding field which slows the electrons down, and by symmetry they come to rest at a distance d beyond the grid. If an 'absorbing' cathode at zero potential were placed at this point ($z = 2d$), the system would be symmetrical with regard to the fields but not to the flow of charge. In the absence of such an absorber the electrons are reflected back towards the cathode, the region beyond the reflection plane being at zero potential. If now a further conducting plane at positive potential is placed at $z > 2d$, some charges are reflected, but others pass beyond the zero potential plane and are collected on the anode. This plane is known as a 'virtual cathode'. This at least is the behaviour predicted by equation (3.19); the problem is discussed in more detail in the book by Birdsall and Bridges (1966), where it is shown that this is not the only solution for flow between the grid and second plane; there exists a second solution with a potential minimum, which does not, however, reach zero, in which no electrons are reflected. What happens in practice is complicated; the results are modified by the existence of thermal velocities and made more difficult to calculate by the finite geometry. When time variation is permitted, it is found that the flow is often unstable, so that oscillations are produced.

This is an important and extensive field, bordering on topics treated here. A coherent account of the subject, containing extensive references, is given in the book by Birdsall and Bridges.

So far we have assumed that the flow is planar. In the non-relativistic region the self-magnetic field is unimportant, and in the absence of an applied field the flow is determined by the boundary conditions, being essentially planar when the cathode-to-anode distance is small compared with the transverse width of the system. Closely allied to the planar flow is cylindrical and spherical flow between appropriate electrodes, studied by Langmuir. An account of this type of flow, and its application to the design of guns, is described by Pierce (1954a). Although of considerable interest, it is outside the scope of the present work. A very detailed account of the theory of guns of this and related types, together with references to earlier work is given in the book by Kirstein *et al.* (1967).

In the relativistic region, in the absence of an external magnetic field, the self-field produces a force towards the axis so that strong pinching can occur, and the flow departs very considerably from being planar. In a drifting un-neutralized cylindrical beam of infinite axial extent, as we see in section 3.2.5, this magnetic force is weaker than the outward radial electrostatic force and the beam spreads out. In the diode, however, where the 'ends' are important and the paraxial equation is very far from being valid, the radial component of electric field is relatively small, and consequently the beam pinches strongly towards the axis. A simple analytical treatment of what happens is not in general possible, and the problem has been treated by numerical methods. A discussion of this subject, which includes a general theoretical formulation and its relation to the paraxial approximation, has been given by Poukey and Toepfer (1974). Their paper includes also the results of a number of numerical simulations for various values of ν/γ. The application of pinched beams of this sort to the problem of fusion has been reviewed by Yonas *et al.*

(1974). A general treatment which includes other geometries has been given by Creedon (1975). All these papers contains references to earlier work and to a great deal of recent experimentation.

Because of the intense power carried by relativistic beams with ν/γ of order unity, such beams are available only in short pulses, of order 10^{-8}s; the behaviour in experiments is transient and complicated. The geometry associated with the field-emission cathodes is necessarily considerably more elaborate than the simple planar system so far discussed; furthermore, during the pulse plasma is generated from the electrodes, and this influences the flow in a transient manner.

At the time of writing this is a rapidly developing field, and we do not attempt a comprehensive survey of the present position.

3.2.4. Launching a cylindrical beam We return now to a qualitative discussion of the problem of launching a cylindrical beam. Suppose that the plane at $x = d$ consists of a plate containing a gridded hole of radius a centred on the z-axis. The potential beyond the hole is now a function of r; the slowing-down process which occurs in the infinite system still occurs, but the charges on the axis come to rest first. The point of reflection is a function of r, the outer electrons travelling furthest. The field distribution, and hence the flow, is of course influenced by any conductors placed near the beam.

The beam described in section 3.2.2. was analysed with the assumption that its outer edge was at zero potential. This could be provided by surrounding it with a close-fitting tube joined on to the plate as shown in Fig.3.2. Charges which pass through the grid at $r = a$ see no retarding field; at other radii, however, there is a retarding field near the grid. The injected current is independent of radius, and at distances greater than several times a from the grid the model described in section 3.2.2 should be a fair description. If the injected current is continuously increased (by decreasing d, or using an intermediate grid between the cathode

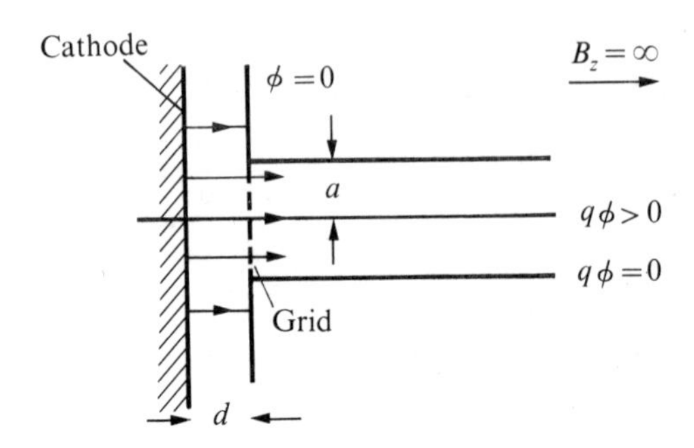

Fig.3.2. Beam launched from a cathode through a grid into a tube of radius a. At the grid the potential throughout the beam is constant. Further along the tube the beam at the beam centre is depressed (for electrons) and the charges there are retarded.

and $z = d$ plane), then it is difficult to see what happens when the axial potential falls below the critical value for maximum perveance. One might expect a discontinuous jump to a situation in which a virtual cathode is formed somewhere in the system and electrons are returned to the cathode. This would imply a sudden drop in current, perhaps to a value determined by the solution with $\phi = 0$ on the axis, described by equation (3.16).

A more 'practical' arrangement (in the sense that it is somewhat closer to actual devices) would be to remove the grid and allow the charges to pass through a hole. In such a system the retarding field would penetrate into the region $z < d$ leading to a decrease in current density, especially near the axis. This would lead to a beam at large z with higher current density at the edge than on the axis. As before, a virtual-cathode condition would ultimately be reached, though it is not evident that there would always be a discontinuity. It may easily be shown, for example, that, in a beam model where β rather than $n\beta$ is independent of r, the maximum current occurs when $\phi = 0$ on the axis and $\nu = \frac{1}{2}\beta_a^2$.

Discussion of these somewhat academic examples does

help in the understanding of the phenomena which occur in very much more complex practical systems. Jumps between one operating condition and another and oscillations of frequency of order of the plasma frequency are frequently observed. As exemplified in the discussion after equation (3.22), the plasma frequency is of the order of the reciprocal of the time taken for a particle to make a transit between the cathode and virtual cathode. Hysteresis effects showing clearly the existence of two stable configurations have been studied by Haeff (1939), and relaxation oscillations between two states have been described by Sutherland (1960).

Some practical electron guns with 'immersed' cathodes have much in common with the systems already discussed. The differences can be listed as follows:

(1) The field is not infinite.

(2) The transverse extent of the cathode is not infinite. Electrodes may, however, be placed so that the fields at the beam edge simulate the fields in an infinite system, and the flow is therefore the same.

(3) Thermal emission velocities smear out the beam edge and virtual cathode.

(4) The beam does not completely fill the tube. (This changes its potential and decreases the maximum current.)

(5) Very often, in order to reduce the current density the flow is arranged to converge from a cathode which forms part of a sphere, through an anode hole which acts as an aperture lens.

In the last section brief mention was made of non-relativistic guns based on spherical or cylindrical flow in the absence of a magnetic field. A beam from such a gun can be guided into a confining magnetic field with the aid of a suitably defined transition region. Intermediate between 'shielded' and 'immersed' guns are those in which some of the flux contained within the beam passes through the cathode. We return to this topic in section 3.3, after analysing flow in a finite magnetic field.

The design of the many types of electron gun, and the necessarily more complex sources of positive and negative ions is not discussed here, but some references will be found in section 3.R. In the relativistic regime, in practical situations, the self-magnetic field has an important effect on the flow. As indicated at the end of section 3.2.3 extensive experimental and theoretical studies have been made on beams in the range $\gamma = 2\text{-}10$, with ν/γ of order unity .

Theoretical work necessitates numerical computer studies, and at the time of writing the main processes involved are basically understood, but many details are obscure. We return to a consideration of this type of beam after a discussion of pinched beams in section 3.2.6.

3.2.5. Strip beams It is straightforward to repeat the analysis of the preceding sections for strip beams. There are however two significant differences; the first is that the external electric field is uniform rather than decreasing as $1/r$. This means that, if the earth planes are not close to the beam, then the difference of potential between planes and beam can easily have a more important effect on the flow than the potential variation in the beam. The second important difference is that the perveance per unit length increases as the beam is made narrower. For a uniform beam of width $2b$ between planes of fixed potential the perveance per length equal to $2b$ is independent of size, as might be expected by analogy with a cylindrical system, where the perveance is independent of diameter for a given current. The properties of such a square section of strip are comparable with those of a circular beam. For example, the following formulae per square of a strip beam can be compared with equations (3.6) and (3.10):

$$\begin{aligned} E_x &= \frac{q}{\varepsilon_0}\int n\,\mathrm{d}x \qquad x < b \\ &= Nq/4\varepsilon_0 b \qquad x > b \end{aligned} \tag{3.26}$$

$$\gamma_b - \gamma = \pi\nu(1 - x^2/b^2) \tag{3.27}$$

where N and ν refer to a square of side $2b$ and in the second equation n is assumed uniform.

3.2.6. Properties of a beam with no externally applied fields

An important type of motion is the spreading (or self-constriction) of a charged-particle beam under the influence of its own self-fields. Perhaps the simplest configuration is one in which a uniform parallel beam of particles is injected into a field-free space. For vanishingly small currents the beam continues as a uniform cylinder; if the source of particles were a planar diode with gridded hole of radius a, then the beam radius would remain constant and equal to a.

We consider first the situation where the potential difference between the centre and edge of the beam, and hence any fields in the z-direction, are small and can be neglected. The possibility of partial neutralization of the radial space-charge force arising from a uniform distribution of charges of opposite sign in the beam, which have no motion in the z-direction, is allowed for. The effect of such neutralization is to reduce the electric force arising from the space charge by a factor $1-f$, where f is the ratio of the neutralizing charge to the beam charge, assumed uniform across the beam cross-section. The neutralizing charges are in a transverse parabolic potential well and will consequently have transverse velocities; a velocity distribution which gives rise to the uniform spatial distribution required here is described in section 4.3.3. We make also the paraxial assumption that z' is small; the implication of the assumptions will be evident after the calculation has been made.

By virtue of these assumptions a particle in the beam experiences only radial forces. There is an electrostatic outward force equal to $(1-f)Nq^2r/2\pi a^2\varepsilon_0$ and an inward 'pinch' force arising from the motion of the charge in the self-magnetic field in the beam; here r refers to the radial position of the electron under consideration and a to the radius of the beam. For a given line density N both the field and the charge are proportional to β, and it is readily verified that the force is equal to

$$\frac{\beta^2 N q^2 c^2 r}{2\pi a^2 \mu_0} = \frac{\beta^2 N q^2 r}{2\pi a^2 \varepsilon_0} .$$

Equating the sum of the electric and magnetic forces to the transverse acceleration, we find

$$\gamma m_0 \ddot{r} = \frac{N q^2 r}{2\pi a^2 \varepsilon_0} (1 - f - \beta^2). \tag{3.28}$$

Since the force is proportional to r, the orbits of all the particles are identical apart from a factor r/a. Consequently there is no loss of generality in considering the orbit of a particle starting at $r = a$. Putting $a = r$ and writing $r'' = \ddot{r}/\beta^2 c^2$, equation (3.28) becomes, for the beam profile,

$$r r'' = \frac{2\nu}{\beta^2 \gamma} \left(\frac{1}{\gamma^2} - f\right) = K.$$

For $f < \gamma^{-2}$ the electric force predominates and the beam spreads; for $f > \gamma^{-2}$, on the other hand, the beam pinches. We define the r.h.s. of the equation as the generalized perveance K:

$$\begin{aligned} K &= \frac{2\nu}{\beta^2 \gamma} \left(\frac{1}{\gamma^2} - f\right) \\ &= \frac{2\nu}{\beta^2 \gamma^3} \quad \text{when } f = 0 \\ &= \frac{-2\nu}{\gamma} \quad \text{when } f = 1 \\ &\approx \frac{2\nu}{\beta^2} = \frac{I}{\phi^{3/2}} \left(\frac{m_0^{1/2}}{4\sqrt{2}\pi\varepsilon_0 q^{1/2}} \right) \quad \text{(NR) when } f = 0 \end{aligned} \tag{3.30}$$

For non-relativistic un-neutralized beams K differs from the conventional perveance k by the factor in parentheses; numerically this is $15 \times 10^3\ \mathrm{V}^{3/2}\,\mathrm{A}^{-1}$.

Before solving equation (3.30) we note a convenient asymptotic form due to Harrison (1958), which is a good approximation when $r - a_0$ is less than a_0, the value of a when $r' = 0$. Under these conditions the quantity r'^2 is small compared with rr'' so that it can be added to the equation without causing much error:

$$rr'' + r'^2 \approx K. \tag{3.31}$$

For positive K this represents the hyperbola $r^2 - Kz^2 = a_0^2$ with asymptotes at an angle $K^{1/2}$; for negative K it is an ellipse. This suggests that the condition for the paraxial approximation to be good is that K should be much less than unity. The condition that the potential difference between the axis and beam edge should be small is, from equation (3.12) with the factor $1 - f$ included, $\nu \ll (\gamma-1)(1-f)$. In the limit of a non-relativistic un-neutralized beam these conditions become the same, namely $\nu \ll \frac{1}{2}\beta^2$ or $k \ll 66 \times 10^{-6}\text{A V}^{-3/2}$.

The solution of equation (3.29) may be obtained by integrating twice. The beam-profile equation is found to be

$$\begin{aligned} \frac{z}{a_0} &= \left(\frac{2}{K}\right)^{1/2} \int_0^{\{\ln(r/a_0)\}^{1/2}} \exp u^2 \mathrm{d}u\,, \quad (K > 0) \\ \frac{z}{a_0} &= \left(\frac{-2}{K}\right)^{1/2} \int_0^{\{\ln(a_0/r)\}^{1/2}} \exp(-u^2)\mathrm{d}u\,, \quad (K < 0). \end{aligned} \tag{3.32}$$

Curves of r against z are shown in Fig.3.3. The profile is symmetrical about the minimum or maximum at $z = 0$, so that the behaviour of initially converging or diverging beams can readily be found from them.

The equations (3.32) for a beam with positive K can be written in terms of the function

$$F(x) = \exp(-x^2) \int_0^x \exp u^2 \mathrm{d}u\,. \tag{3.33}$$

In terms of it the first equation (3.32) can be written

$$z = \left(\frac{2}{K}\right)^{1/2} r\,F\left\{\left(\ln \frac{r}{a}\right)^{1/2}\right\}. \tag{3.34}$$

For small x, $F(x) \approx x$, and equation (3.34) becomes

$$\frac{a_0}{r} = \exp -\left(\frac{Kz^2}{2r^2}\right) \approx 1 - \frac{Kz^2}{2r^2} . \tag{3.35}$$

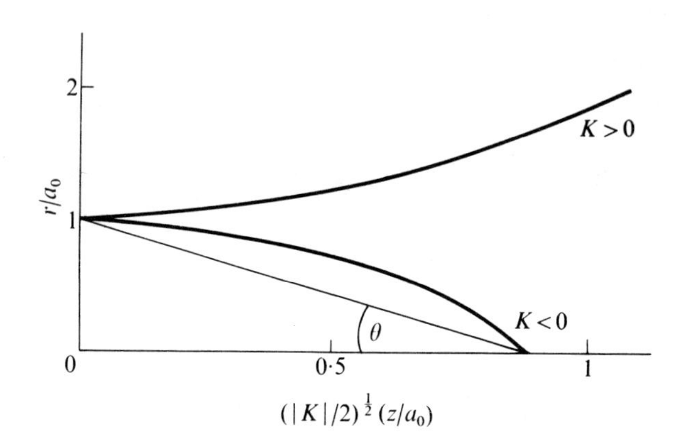

Fig.3.3. Profiles for paraxial beams spreading or being 'pinched' under the influence of self-fields. These are plotted from equations (3.32) in terms of the normalized variables r/a_0 and z/a_0.

For $Kz^2/2r^2 \ll 1$ this is equivalent to the hyperbolic solution represented by equation (3.31), as may be seen by squaring left- and right-hand sides of equation (3.35). In the other limit, x large, $F(x) \approx \frac{1}{2}x$, so that equation (3.34) becomes

$$\ln \frac{r}{a_0} = \frac{r^2}{2Kz^2} . \tag{3.36}$$

For a converging beam, with initial conditions such that $r/a_0 \gg 1$, r/z can be approximated by $dr/dz = \alpha$. Equation (3.36) then becomes

$$a_0 = r \exp\left(\frac{-\alpha^2}{2K}\right) . \tag{3.37}$$

If the waist is formed from an initially parallel beam by a lens of focal length f, then $r = \alpha f$, so that for a given perveance and angle of convergence the minimum spot size which can be produced is proportional to f.

For a strip beam the analysis is much simpler. The

transverse force (in the x direction) is independent of beam width. As was shown earlier in section 3.2.5, the perveance for a given length of strip is inversely proportional to its width, and the significant parameter as before is the perveance per square. It is easily shown from equation (3.26) that the profile equation is a parabola defined by

$$x''b = \tfrac{1}{2}\pi K_s \tag{3.38}$$

where b is the half-height of the beam and K_s refers to a square of side $2b$.

Further details on the subject of space-charge spreading of a beam in free space, including a graph of $F(x)$, are given in Pierce's (1954a) book. Also included is an estimate of the maximum current which can be passed through a cylindrical tunnel without interception. References to the original papers on this topic may be found there too.

As with the beam in an infinite field, the behaviour when ν is no longer small compared with $\frac{1}{2}\beta^2$ is complicated and depends on the boundary conditions. The way in which the beam is launched and the disposition of conducting walls around the beam exert a strong influence on the flow. The potential depression (for electrons) at the beam centre slows down the charges there, which further increases the field. For this reason the current transmitted through a tube or a pair of spaced apertures is less than that which might be expected from applying the formulae for small K. Practical beam systems always operate well below the critical perveance of 66×10^{-6}A V$^{-3/2}$. For a typical electron beam of 1 A and energy 10 keV the perveance is 10^{-6}A V$^{-3/2}$ and $K = 0{\cdot}015$, so that the potential depression is 150 V and $\theta \approx 1°$. Calculated curves of the decrease in axial velocity of a cylindrical beam in a cylindrical box of radius and length large compared with the beam radius are given in the review by Amboss (1969).

Since 1965 there has been a considerable amount of theoretical and experimental work with relativistic electron beams having $|K|$ of order unity and above. The first experiments in this class, with beams obtained by the discharge of a large coaxial line charged up to 2 - 3 MeV had $|K|$ somewhat less than unity, and showed behaviour in general accordance with the analysis presented here (Graybill and Nablo 1966). Later work, at higher currents, revealed a number of additional effects which sometimes make the behaviour very much more complicated. An important feature of beams injected into plasma or neutral gas where the electron density greatly exceeds that of the beam is the appearance of a reversed current which tends to cancel the self-magnetic field and thus reduce the effective pinching (Cox and Bennett 1970). Thus, beams with $\nu/\gamma \approx 2{\cdot}5$ are described by Levine *et al.* (1971). This is essentially an inductive effect which therefore brings in time-dependent phenomena associated with the setting up of the beam. Details and references to earlier work may be found in the papers quoted. Since most beams in this class are essentially non-laminar, we defer further discussion until the next chapter.

3.2.7. Uniform laminar pinch In the previous section an equation was derived for the pinching of an initially parallel beam which is partly neutralized, so that the inward magnetic self-force overcomes the outward electric repulsive force. It is possible in principle to balance this inward force by an outward centrifugal force by imparting an angular velocity to the beam. This could be done by launching it from a cathode immersed in a magnetic field, which imparts a component of angular velocity to the particles when they leave the field.

Although such a beam is unlikely to be achieved in practice, it is basically very simple, and exhibits properties of interest in discussing more complicated configurations later on. We therefore calculate its basic features. The paraxial approximation will again be made, and the circumferential velocity at the beam edge assumed small compared with $\beta_z c$. The condition for a uniform beam is found by replacing the left-hand side of equation (3.28) by the centrifugal force $\gamma m_0 \beta_\theta^2 c^2/r$. Assuming constant angular velocity $\beta_\theta c/r$ we find

$$\beta_{\theta a}^2 = 2\langle\beta_\theta^2\rangle = -K\beta_z^2 \tag{3.39}$$

where the bracket denotes the average over radius and K is the perveance defined in equation (3.30). The condition that β_θ should be much less than β_z is $(-K)^{1/2} \ll 1$. For the special case of $f = 1$, corresponding to a fully neutralized beam,

$$\frac{2\langle\beta_\theta^2\rangle}{\beta_z^2} = \frac{2\nu}{\gamma}\,. \tag{3.40}$$

It may readily be verified that $2\nu/\gamma$ is also the ratio of the beam radius to the Larmor radius of an electron in the self-field at edge of the beam. Equation (3.40) is a form of the 'Bennett pinch relation' discussed in section 4.4.5.

The restriction to paraxial approximation produced a simple result, but it is evident that by relaxing it, and also the condition that $\beta_\theta \propto r$, a wide variety of equilibria with varying velocity and density profiles can be constructed. By making $\beta_\theta \gg \beta_z$ self-consistent flows with $2\nu/\gamma$ arbitrarily large can be constructed analytically, though it is difficult to see how they could be realized in practice.

3.2.8. Equilibrium of a uniform cylindrical beam in a uniform magnetic field The properties of a beam in an infinite magnetic field have already been described; we consider now a system which is more realistic in the sense that the field is finite, but which is still idealized in that the flow is laminar and there is no variation with z. Laminar flow and the assumption of azimuthal symmetry together imply that individual charges move in helices of constant radius. Under these conditions the following radial forces act on the charges

(1) Centrifugal force.
(2) Lorentz force arising from electron motion in external

field, proportional to $\beta_\theta B_z$.

(3) Lorentz force arising from azimuthal self-magnetic field of beam proportional to $\beta_z B_\theta$.

(4) Correction to (2) arising from self-magnetic field in the z direction from the azimuthal current.

(5) Electric force from space charge in beam.

It is simple (though perhaps not profitable) to write down a general integral equation for the forces on a charge at radius r in terms of n, β_θ, β_z as functions of radius. There are obviously many solutions to such an equation; we attempt to select the more interesting ones by imposing relevant restrictions.

First, we treat a beam with no angular shear ($\dot{\theta}$ = constant) so that the beam forms a 'rigid rotator'. To simplify the algebra and bring out the essential physical features we initially take $f = 0$ and consider only non-relativistic velocities; n is assumed uniform with radius though β is allowed to vary. Only forces described under (1), (2), and (5) above need to be considered, and the radial force balance gives

$$m_0 r\dot{\theta}^2 = -\frac{nq^2 r}{2\varepsilon_0} - qB_z r\dot{\theta} \tag{3.41}$$

in terms of the plasma and Larmor frequencies ω_p and Ω_L; this may be written

$$\dot{\theta}^2 + \tfrac{1}{2}\omega_p^2 - 2\dot{\theta}\Omega_L = 0 \tag{3.42}$$

whence

$$\dot{\theta} = \Omega_L \pm (\Omega_L^2 - \tfrac{1}{2}\omega_p^2)^{1/2} . \tag{3.43}$$

For $\Omega_L^2 > \tfrac{1}{2}\omega_p^2$ this has two solutions, one higher and one lower than the Larmor frequency. When $\Omega_L^2 < \tfrac{1}{2}\omega_p^2$ no equilibrium is possible; the limit $\Omega_L = \omega_p/\sqrt{2}$ represents the special case known as Brillouin flow (Brillouin 1945), which we discuss below in more detail. In terms of other parameters the condition for Brillouin flow is

$$\Omega_L^2 = \tfrac{1}{4}\omega_c^2 = \tfrac{1}{2}\omega_p^2 = 2\nu c^2/a^2. \qquad (3.44)$$

So far the component of velocity in the z direction has not been specified; we now assume that all the charges have such a velocity, and furthermore that they originate from a cathode at constant potential. With this assumption, and a specified potential at the beam edge, we can calculate the variation of potential with radius and also the beam perveance. We first, however, discuss conditions at the cathode implied by Busch's theorem (equation (2.5)). This may be written as

$$\dot{\theta} = \Omega_L(1 - \Psi_0/\Psi) \qquad (3.45)$$

where Ψ is the flux linking a circle of radius r and Ψ_0 is the flux linking the cathode. Comparing with equation (3.43) we find

$$\Psi_0 = \pm\,\Psi\left(1 - \frac{\omega_p^2}{2\Omega_L^2}\right)^{1/2} = \frac{2\pi P_\theta}{q} \qquad (3.46)$$

A schematic diagram illustrating possible flux distributions at the cathode for these two solutions is shown in Fig.3.4. Some limiting cases are of interest; in the low-density limit $\omega_p = 0$ so that Ψ_0 is equal either to $+\Psi$, in which case the charges move along the line of force, or $-\Psi$, when they receive an impulse in the θ direction sufficient to project them into an orbit centred on the axis. This is consistent with fact that, when Ψ_0 is equal to zero, the particles receive an impulse which is half as great, so that the orbit passes through the axis. The high-density limit, where the two solutions coincide, represents Brillouin flow.

We now find the distribution of z-velocity across the beam. Since the density is assumed uniform, the potential variation is parabolic (equation (3.10)) and the equation relating velocity components is

$$\tfrac{1}{2}(\beta_z^2+\beta_\theta^2-\beta_{z0}^2) = \nu r^2/a^2 \qquad (3.47)$$

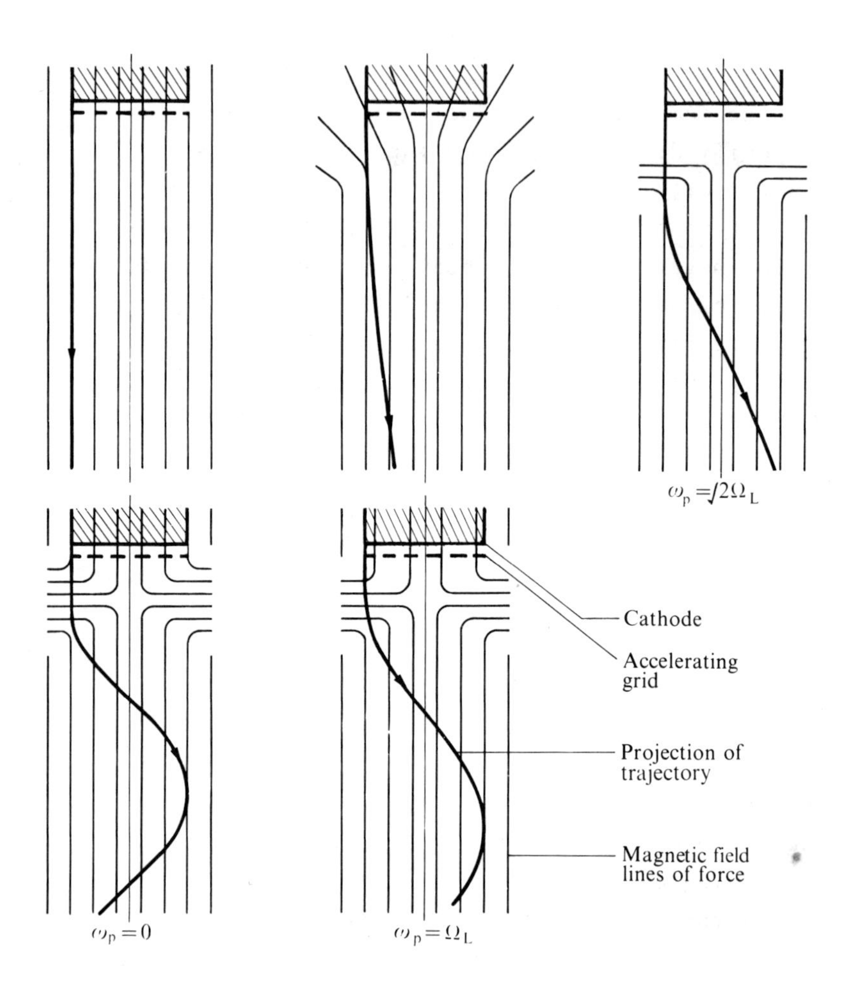

Fig.3.4. Trajectories of electrons launched into a magnetic field, illustrating schematically the two solutions of equation (3.43) which merge when $\omega_p = \sqrt{2}\Omega_L$ to give Brillouin flow. The values of ω/Ω_L corresponding to the three values of ω_p are 0 and 2, $\frac{1}{2}(\sqrt{2} \pm 1)$ and 1.

where $\beta_{z0}c$ is the velocity on the axis. Substituting $\beta_\theta = r\dot{\theta}$ and $\nu = a^2\omega_p^2/4c^2$ this becomes

$$\beta_z^2 - \beta_{z0}^2 = \frac{r^2}{2c^2}(\tfrac{1}{2}\omega_p^2 - \dot{\theta}^2). \qquad (3.48)$$

There is therefore a variation of z-velocity across the beam, unless $\omega_p^2 = 2\dot{\theta}^2$. This condition implies Brillouin flow. It may be argued that this calculation is somewhat artificial and that it is more realistic to assume a constant current distribution across the beam. We could of course have done this, but the consequent non-uniformity of n with radius would have given rise to shear in the θ direction.

In non-relativistic Brillouin flow we have seen that the beam moves as if it were a rigid body. As n increases the pitch of the spirals decreases, so that β_z decreases. It is of interest to calculate the maximum current for fixed potential, allowing the magnetic field to increase to maintain the equilibrium condition. Defining ϕ_a as the potential at the beam edge and $\phi_0 = -m_0c^2/q$, we have the following relation:

$$\frac{\phi_a}{\phi_0} = \tfrac{1}{2}(\beta_z^2+\beta_{\theta a}^2) = \tfrac{1}{2}(\beta_z^2+\frac{a^2\Omega_L^2}{c^2}) = \tfrac{1}{2}(\beta_z^2+2\nu). \qquad (3.49)$$

For constant ϕ_a we find by differentiating the last part of this equation that $d(\nu\beta_z)/d\beta_z = 0$ when

$$\nu = \beta_z^2 = \frac{1}{3}(\beta_z^2+\beta_{\theta a}^2) . \qquad (3.50)$$

The maximum current occurs therefore when the ratio of the θ-velocity at the edge of the beam to the z-velocity is $\sqrt{2}$. Furthermore $\phi_{axis}/\phi_a = \frac{1}{3}$.

In practical units this becomes

$$k_{max} = \frac{16\pi\varepsilon_0 q^{1/2}}{3\sqrt{6}m_0^{1/2}}$$

$$= 25{\cdot}4 \times 10^{-6}\ \mathrm{A\ V^{-3/2}} \quad \text{for electrons.} \qquad (3.51)$$

The first and third parts of equation (3.49) can be used to find a general relation between current, magnetic field, and potential. Multiplying through by $a^2\Omega_L^2 = 2\nu c^2$ (equation 3.44) gives

$$\frac{\phi_a}{\phi_0}\; a^4\Omega_L^4 = \beta_z^2\nu^2c^4 + \frac{a^6\Omega_L^6}{2c^2}\;. \tag{3.52}$$

Expressing Ω_L in terms of B and $\nu\beta_z$ in terms of I gives

$$\begin{aligned} I &= 2\pi\varepsilon_0 B^2a^2\left(\frac{q}{2m_0}\right)^{3/2}\left(-\phi_a - \frac{qB^2a^2}{8m_0}\right)^{1/2} \\ &= 1{\cdot}45\times10^6B^2a^2\,(\phi_a - 2{\cdot}2\times10^{10}B^2a^2)^{1/2} \\ &\qquad\text{for electrons.} \end{aligned} \tag{3.53}$$

When the potential variation is small, the second term in the bracket can be neglected. For $a = 10^{-3}$m, $\phi_a = 10$ kV, $B = 10^{-1}$T, we find $I = 1{\cdot}43$ A.

It is interesting to look at the structure of a beam defined by equations (3.41) and (3.42) in a frame of reference rotating with frequency ω and drifting with velocity β_z, assumed constant across the beam. There is no shear, so that the particles could not have originated from a unipotential cathode (except in the special case of Brillouin flow). From equation (2.21) we see that in the rotating frame a compensating electric field appears which for suitable choice of ω_f, can be arranged just to balance the space-charge force so that the particles remain at rest. The magnetic field is reduced by a factor $1 - \omega_f/\Omega_L$ and thus vanishes for Brillouin flow. The fictitions field which appears in the rotating frame is precisely equivalent to a field which would be provided by a neutralizing background if the system were in the laboratory frame. This suggests a close relationship between the behaviour of neutral stationary plasma in a magnetic field, and un-neutralized confined beams, a topic to which we return in Chapter 6.

A fully relativistic treatment of Brillouin flow has been given by de Packh and Ulrich (1961). This includes all the five force components listed earlier. The beam has no shear, but the current density is no longer uniform, being greater at the beam edge than at the centre. The value of β also increases with radius, remaining proportional to γ to maintain ω constant. Defining γ_0 as the value on the axis, equal to $(1 - \beta_z^2)^{-1/2}$, and γ_a that at the beam edge they obtain the result

$$\nu = \gamma_0\left(\frac{\gamma_a^2}{\gamma_0^2}-1\right) \tag{3.54}$$

as the relativistic equivalent of equation (3.49) (second and fourth parts). For the maximum current condition they find

$$\gamma_0^6 + (2\gamma_a^2 + \gamma_a^4)\gamma_0^2 + 2\gamma_a^4 = 0 \tag{3.55}$$

and the equation corresponding to (3.50) is obtained by eliminating γ_a or γ_0 from these two equations. This is clearly impossible in closed form, but in the extreme relativistic limit $\gamma_0 = \surd 2$, $\beta_0 = 1/\surd 2$, and

$$\nu = \frac{\gamma_a^2}{\surd 2} - \surd 2 \tag{3.56}$$

Since $\dot{\theta}$ is independent of radius, the radial variation of field under these conditions varies by a factor of γ_a/γ_0 between the axis and the beam edge. This is equal to $\gamma_a/\surd 2$.†

Such a beam is of somewhat academic interest, since it is hardly possible to make high-γ guns of suitable form. More practical, though more complicated, treatments have been given by Yadavalli (1958) and Neugebauer (1967). The latter papers also considers partially immersed flows for which the flux linking the cathode is not zero.

3.2.9. Practical neutralized beams Several situations involving the neutralization of beams by particles of the opposite sign have been encountered in this chapter. Simple assumptions have been made, though in practice the behaviour is often rather complicated.

Neutralization is commonly produced from the residual gas in the apparatus; an electron beam, for example, ionizes

---

†Later calculations by Reiser (1977) lead to results somewhat different from those in equations (3.54) and (3.55).

residual air atoms, the secondary electrons are then repelled to the wall, and the ions remain trapped in the potential well in the beam. This process continues until equilibrium is reached and the rate at which electrons and ions leave the beam is equal to the rate of formation. The form that this equilibrium takes depends on such factors as the beam energy, gas pressure, ionization cross sections, the energy of the secondary electrons, and whether the ions can escape from the end of the beam. If this is possible, then the degree of neutralization varies along the beam. This process has been described in detail, with experimental comparison, by Hines, Hoffman, and Saloom (1955). Earlier work is also reviewed, including experiments by Field, Spangenberg, and Helm (1947) where space-charge spreading of a long beam was prevented.

Drainage of ions out of the ends of the beam can be prevented by making use of an ion trap as shown in Fig.3.5.

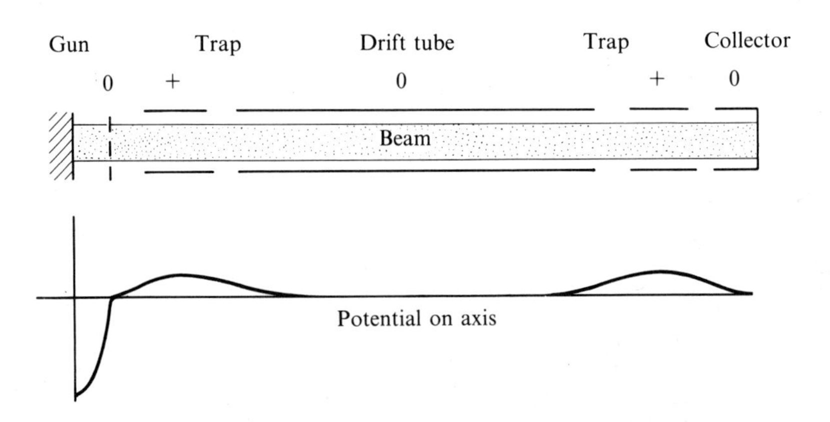

Fig.3.5. Schematic representation of ion traps at the ends of a drift tube through which an electron beam passes.

A small region is provided at each end of the beam in which a longitudinal potential difference of a few volts is applied, by grids or apertures, in such a direction as to reflect the drifting ions. At the gun end of an electron beam a trap can be used to prevent ions from being accelerated back through the gun and damaging the cathode by bombardment.

If the ions are constrained from flowing along the beam, they necessarily leave radially. This situation has been studied by Dunn and Self (1964). The electrons formed with an energy of a few eV, are assumed to have a Maxwellian distribution, whereas the ions fall freely to the walls under the influence of the radial electric field. As in a gas discharge, it is assumed that the ion and electron currents must be equal at the wall. The variation of potential is found from the simultaneous solution of Poisson's equation, the Boltzmann relation (equation (4.55)) and the equation of motion of the ions, together with a source term representing the production of ions and electrons by the beam.

In some circumstances a net positive core appears in the beam which causes 'electrostatic pinching.' This is essentially the process of 'gas focusing' used in early cathode ray tubes. An analysis of this problem, together with a description of experiments to test the theory, is given by Halsted and Dunn (1966). This paper contains references to earlier work, including a number of papers before 1935. More recent work with beams of much higher intensity, is described by McCorkle and Bennett (1971).

In the absence of a magnetic field, neutralization can substantially alter the flow characteristics; in particular, space charge spreading can be reduced. Although this method can be effective in thin low perveance beams, it is difficult to apply it in long beams at higher perveance. Experimental studies by Ginzton and Wadia (1954), on a beam with $I/V^{3/2} = 10^{-6}$ A $V^{-3/2}$ show that the behaviour is strongly influenced by the detailed shape of the fields in the ion trap which separates the neutralized beam from the unneutralized gun region. Considerable aberrations are necessarily introduced which give rise to divergence between rays near the axis and those at the beam edge. It is difficult to give a universal theory for effects associated with beam neutralization. Ionization rates depend very much on such factors as the energy of the beam and the composition and pressure of the residual gases. For this reason equilibria take a finite time to build up, and it is often found that a pulsed beam has very different characteristics from those of a steady beam. Relaxation oscillations between two configurations are sometimes observed. An experimental study of 'spurious modulation' of electron beams, in which many of these effects were found, has been made by Cutler (1956). It is evident from his results that the behaviour depends in a complicated way on the particular parameters of the apparatus.

An interesting application of the variation with time of the properties of a beam has been proposed by Atkinson (1963), for use as a high-vacuum gauge. A high-perveance electron beam of about 100 V passes into a 'field-free' drift space. Initially, the space-charge causes the beam to diverge radially. As ions become trapped in the beam the divergence decreases. The rate at which the divergence decreases is proportional to gas pressure. This is easily measured as an increase of current on a target placed behind the anode in which there is a hole.

A more recent application for neutralized electron beams is in the field of trapped-ion spectroscopy. The ions to be studied are generated in an electron beam with ion traps. They are then extracted by applying a local electric field. By analysing the ions in a mass spectrometer and measuring the current at varying electron beam and extraction voltages, it is possible to find various ionization and excitation cross-sections. Ions can be retained long enough for multiple ionization to be observed and measured. The technique is explained in papers by Baker and Hasted (1966) and Redhead (1967). A number of ingenious configurations of beams and electrodes have been devised, details of which may be found in the recent review by Redhead (1974) and the references therein.

Slightly neutralized ring beams are an essential ingredient of the electron ring accelerator proposed by Veksler (1967). Such a ring, typically containing 10^{11}-10^{12} electrons with major and minor radii of order a few centimetres and 1 mm respectively, represents an object with charge-to-mass ratio considerably higher than that of the trapped ions. If it is accelerated in a direction perpendicular to the plane of the ring, either in an electric field or a magnetic-field gradient, the ions are accelerated in the self-field of the ring, which is itself much higher than the external field accelerating the electrons. The motion in such rings, however, is essentially non-laminar, both for electrons and ions; their characteristics will be discussed further in the next chapter.

Discussion so far has been of the neutralization of electron beams by ions. The reverse problem is also of importance in some contexts. It is essential, for example, to the successful operation of high-current isotope separators; in the absence of neutralization the space-charge spreading destroys the resolution required for separation. This problem is discussed in detail by Walcher (1958) and Nezlin (1968). Neutralization is also essential in intense deuteron beams currently being studied for heating plasmas for the purpose of producing fusion reactions. This topic is reviewed by Green (1974).

A further field in which neutralized beams are important is that of ion thrusters for satellites. In these systems not only must a beam of ions be ejected, but an equal number of electrons must leave too in order to preserve charge neutrality. An extensive technology, mainly using beams of singly charged mercury ions, has evolved in recent years. The fundamentals are presented in a book by Brewer (1970).

In this section a large number of important fields of work and their associated techniques, all of which involve neutralized beams, have been touched on briefly. The physical ideas involved, although at times ingeniously deployed are on the whole straightforward, and reasonably well covered by the more general situations studied in this chapter and elsewhere in the book.

3.3. The paraxial equation for beams with space-charge in axial magnetic fields

3.3.1. Solid beams In previous sections the effect of a large variation of potential across the diameter of the beam has been studied; in many practical systems, however, this is small. If this assumption is made, then the paraxial equation can be used to describe the motion in which the beam radius varies with z in an external field which can also vary with z. The term to be added to equation (2.13) representing the effects of space charge and self-magnetic fields in a beam of uniform cross-section is $-Kr/a^2$. In the presence of accelerating fields E_z, K is a function of z. For laminar flow the radial position of any particular particle will always be such that r/a is constant. In this case we can write $r = a$, and the equation for the edge electron in the beam is the equation for the beam profile. (This is not true when the trajectories cross, as we shall see later.) The appropriate paraxial equation required in this section, including the space-charge term but excluding the effect of external electric fields, is accordingly

$$a'' + \frac{\Omega_L^2}{\beta^2 c^2} a - \left(\frac{P_\theta}{\beta\gamma m_0 c}\right)^2 \frac{1}{a^3} - \frac{K}{a} = 0. \qquad (3.57)$$

This equation is subject to the normal paraxial approximations, but it is more general than the equation used in section 3.2.7 describing Brillouin flow in that it is relativistic and includes the effect of space-charge neutralization. The Brillouin-flow condition under these circumstances is found by putting $a'' = 0$ and $P_\theta = 0$; it is simply

$$K = \left(\frac{a\Omega_L}{\beta c}\right)^2. \qquad (3.58)$$

The r.h.s. is the square of the ratio of transverse to longitudinal velocities of a particle at the beam edge. The paraxial approximation demands that this ratio be small, or $K \ll 1$.

It is evident that in a uniform field the solution of

equation (3.57) is in general oscillatory. It would apply for example to a cylindrical beam launched into a magnetic field from a gun, which could either be shielded ($\Psi = 0$) or immersed ($\Psi > 0$). In general, unless conditions are carefully matched in the transition region, the profile in the uniform-field region is 'scalloped'. Proper design of this transition region is a specialized art, described in detail elsewhere (for example, by Brewer (1967a) and in the book by Kirstein *et al.* (1967)). In many practical guns the cathode is partially immersed; the flux through the cathode is finite, but less than that linking the beam in the uniform-field region.

The wavelength of small oscillations in a constant field can be found by linearizing equation (3.57), which may be accomplished by setting $r = a + \rho$ where $\rho \ll a$. The result is

$$\frac{1}{\lambda\!\!\!^{-2}} = \frac{\Omega_L^2}{\beta^2 c^2} + 3\left(\frac{P_\theta}{\beta\gamma m_0 c a^2}\right) + \frac{K}{a^2} \ . \tag{3.59}$$

Setting $P_\theta = 0$ and making use of equation (3.58) this becomes for Brillouin flow,

$$\lambda\!\!\!^{-} = \frac{\beta c}{\sqrt{2}\Omega_L} \ . \tag{3.60}$$

For non-relativistic flow $\omega_c = \sqrt{2}\omega_p$ so that an individual electron oscillates radially at $\sqrt{2}$ times the plasma frequency. The quantity $\lambda\!\!\!^{-\,-1}$ is sometimes known as the 'stiffness' of the beam. For large oscillations where the situation is non-linear the profile cannot be expressed in simple analytical form.

Equation (3.57) can also be used to determine the equilibrium beam radius in a uniform field B_z of a beam launched from a cathode which links flux $\Psi_0 = \pi B_z R_c^2$ where R_c is a constant with dimensions of length. If the cathode is also in the uniform field containing the beam, then R_c is the actual cathode radius. Substituting $P_\theta = q\Psi_0/2\pi$ into equation (3.57) with $a'' = 0$ (to determine the equilibrium beam radius) gives a biquadratic equation for a which has for the only acceptable root

$$a = \{2R_e^2K+(2R_e^4K^2+R_c^4)^{1/2}\}^{1/2} \tag{3.61}$$

where $R_e = \beta c/2\Omega_L$, the radius of curvature of a particle in the field B_z. For $R_e = 0$ (infinite magnetic field) or $K = 0$ this gives $a = R_c$, as expected. It is interesting to compare the radius of such a beam with that of a Brillouin beam with radius R_c. Setting $a = R_c$ in equation (3.58) gives $R_c^2 = 4KR_e^2$; substituting for KR_e^2 in equation (3.61) we find

$$a = \{\tfrac{1}{2}+(\tfrac{1}{8}+1)^{1/2}\}^{1/2} R_c = 1{\cdot}27R_c\,. \tag{3.62}$$

If the cathode were in the same uniform field as the beam, then the beam profile would in fact oscillate about the value of r given above unless some suitable matching lens system were introduced beyond the cathode.

The modulation on cylindrical beams which has been described can be considered as a 'standing-wave' deformation. The more general case which includes travelling-wave motion will be studied in section 6.4.7. These examples have indicated some of the characteristics of laminar cylindrical beams in magnetic fields; in the next section some other geometrical configurations will be examined.

3.4. Some further types of space-charge flow

3.4.1. Hollow and planar Brillouin beams Many of the essential physical features of laminar space-charge flow have been illustrated in the examples of the previous sections. It is evident that further types of flow are possible; many have indeed been studied, some in connection with particular devices and others as academic exercises. Most of the work is restricted to un-neutralized non-relativistic beams, since these restrictions apply in the majority of applications, which are to microwave tubes. Since the derivation of the properties of hollow beams follows closely the methods already used, and furthermore are set out in the book by

Kirstein *et al.* (1967), we quote only the results here.

The generalization of Brillouin flow to a hollow beam with inner and outer radii of r_a and r_b, sometimes known as 'Samuel flow' (Samuel 1949), results in a beam which has uniform velocity in the z-direction but exhibits circumferential shear. The current density is a function of radius, and decreases between r_a and r_b by a factor of between 2 (for a beam with a small hole) and 1 (for a thin shell); as r_a reaches the limit zero and the beam becomes solid the region of high current density, which becomes increasingly localized near the centre, suddenly disappears. The characteristic frequencies are related to the radius by the expressions

$$\omega = \Omega_{\mathrm{L}}\{1-(r_a/r)^2\} \tag{3.63}$$

$$\omega_{\mathrm{p}}^2 = 2\Omega_{\mathrm{L}}^2\{1+(r_a/r)^4\} \; . \tag{3.64}$$

The relation between current, field, and radius may be obtained from equation (3.53) by putting $a = r_b$ and multiplying the whole expression by $\{1 - (r_a/r_b)^4\}$ and the second term in the bracket by $\{1 - (r_a/r_b)^2\}^2$. The maximum perveance (equation (3.51)) is increased by the factor $(r_b^2 + r_a^2)/(r_b^2 - r_a^2)$. The variation of ω_{p} and ω with respect to r for a typical beam is shown in Fig.3.6, together with a schematic view of the magnetic field and cathode topology.

By placing a cylinder at a suitable potential (higher than that of the anode for electrons) inside the beam and arranging the cathode flux at a radius less than r_1, where $\omega = 0$, to pass by the cathode in the region $r < r_a$, the current can be increased. If the potential at r_a is made equal to that at r_b, then Samuel shows that the current is approximately doubled and $r_1^2 = r_a r_b$. The sign of ω changes at $r = r_1$.

By making $r_a - r_b \ll r$ it is easy to proceed to the limit of a strip beam; such a beam has a perveance per square of $2/\pi$ times that of a circular beam. As with a circular beam the outer particles again have a velocity equal to $\surd 3$ times

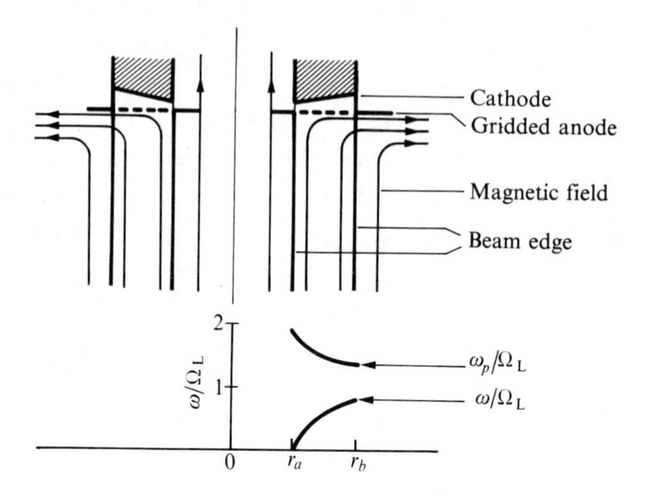

Fig.3.6. Brillouin flow in a hollow beam, sometimes called Samuel flow. The cathode and flux distribution are schematic only. The variation of ω and ω_p with radius are given by equations (3.63) and (3.64).

that of those at the centre. Further properties of such beams, and also of immersed strip-beams have been given by Lawson (1954). They have the property that the space-charge force is independent of the distance x of a particle from the symmetry plane. If the current is too small to sustain Brillouin flow beam scalloping occurs, and if the current is less than half the Brillouin current the trajectories pass through the symmetry plane ($x = 0$). When this occurs the direction of the space-charge force suddenly reverses, giving a beam envelope consisting of discontinuously joined pieces of sine curves. In practice such analysis cannot be taken too seriously. The characteristics would quickly be smeared out owing to non-laminarity, even if the beam were to survive the lateral instability described in section 6.4.13.

For hollow Brillouin beams the scalloping wavelength for small perturbations can be estimated by the methods described in the last section. It is found that the wave-length on the outer radius exceeds that on the inner radius.

For a thin beam, the wavelength ratio is about r_a/r_b.

The very important form of planar Brillouin flow in crossed fields can also be seen as a limiting case of a hollow beam with $\beta_z = 0$ and $r_a \to \infty$. Here, however, we calculate it directly in relativistic form taking into account the self-magnetic field. Neutralization is not allowed for, since it would be almost impossible to achieve in a realistic configuration. Flow is assumed laminar in the z-direction, extending from $y = 0$ to $y = w$, and there is a plane at zero potential at $y = 0$. The applied magnetic field is $(B_x, 0, 0)$. Both n and β_z vary with y. Equating the net force on a charge in the y direction to zero:

$$E_y + \beta_z c\left\{B_x - \mu_0 q\left(\int_0^y - \int_y^w\right) n\beta_z c\,dy\right\} = 0 \tag{3.65}$$

where, from Poisson's equation,

$$\varepsilon_0 E_y = \int_0^y qn\,dy\,. \tag{3.66}$$

Eliminating E_y between these two equations gives an integral equation relating $\beta_z(y)$ and $n(y)$. This is in general complicated, but an interesting particular solution may be found for a non-relativistic system in which n is constant. The self-magnetic field may be neglected, and equation (3.65) reduces to

$$E_y + \beta_z c B_x = 0. \tag{3.67}$$

From equation (3.66), $E_y = nqy/\varepsilon_0$, so that

$$\varepsilon_0 \beta_z c B_x + nqy = 0. \tag{3.68}$$

This may be written

$$\beta_z c = \frac{\omega_p^2}{\omega_c}\, y. \tag{3.69}$$

If the potential is zero at $y = 0$, then at $y = w$

$$-q\phi = \tfrac{1}{2}nq^2w^2. \tag{3.70}$$

It may be simply shown that this is equal to the kinetic energy at $y = w$ and that $\omega_p = \omega_c$ throughout the beam; this latter result may be compared with equation (3.44), which shows that for cylindrical flow $\omega_p = \omega_c/\sqrt{2}$. It is not necessary of course that β_z should be zero at $y = 0$; part of the beam can be removed provided that a plane at suitable potential is inserted to give the appropriate value of E_y at the new beam edge.

This type of flow is very important as an idealization applicable to crossed-field microwave tubes such as the linear magnetron. The corresponding flow in cylindrical geometry is as for a Brillouin hollow beam but with $\beta_z = 0$. In practical situations the flow is more complicated; it is always non-laminar.

Further types of Brillouin flow, including relativistic and self-field effects, are described by Creedon (1975).

3.4.2. Elliptical beams and elliptical Brillouin flow So far all the beams which have been considered have either been circular or planar. Beams with elliptical cross-section are sometimes of interest (especially in cyclic particle accelerators), and we introduce some of the properties of elliptical charge distributions before proceeding to the (somewhat academic) topic of elliptical Brillouin flow.

The problem of calculating the potential in a circular beam with known variation of density with radius is trivial. Also, the equipotentials are circles, so that there is no difficulty in choosing an external boundary which does not upset the flow. For an ellipse, however, the shapes of the equipotentials outside the beam are not represented by simple curves; they vary with radius, approximating to a circle at infinity.

A method of analysing this problem is presented by Kellogg (1930). We first consider the fields inside an elliptical distribution with semi-axes a and b, assuming that no conducting walls are present. We work with a dimensionless co-ordinate ρ such that $x = \rho a \cos\theta$ and

$y = \rho b\cos\theta$ and assume a radial charge distribution which is a function of ρ only. The x-component of field at a point $(x_0,0)$ on the x-axis from a charge $nq\,\mathrm{d}x\mathrm{d}y$ at (x,y) is

$$\mathrm{d}E_x = \frac{nq}{2\pi\varepsilon_0}\frac{x_0-x}{y^2+(x-x_0)^2}\,\mathrm{d}x\mathrm{d}y\ . \tag{3.71}$$

We now change the variable to ρ defined above and integrate

$$E_x = \frac{q}{2\pi\varepsilon_0}\int_0^1\int_{-\pi}^{\pi}\frac{n(x_0-\rho a\cos\theta)}{b^2\rho^2\sin^2\theta+(x_0-a\rho\cos\theta)^2}\,\mathrm{d}\theta\,\mathrm{d}\rho\,. \tag{3.72}$$

The θ integration may be performed by setting $t = \tan\frac{1}{2}\theta$. This gives

$$E_z = \frac{q}{2\pi\varepsilon_0}\int_0^1 n\rho\left[\int_{-\infty}^{\infty}\frac{(1+t^2)x_0-2a\rho(1-t^2)}{4b^2\rho^2t^2+\{x_0(1+t^2)-a\rho(1-t^2)\}^2}\,\mathrm{d}t\right]\mathrm{d}\rho\,. \tag{3.73}$$

The inner integral, which has poles at values of t

$$t = \frac{i\,[\pm ab\pm\{x_0^2+(b^2-a^2)\rho^2\}^{1/2}\,]}{x_0+a\rho} \tag{3.74}$$

can be integrated by the method of residues. It is found that the behaviour depends on whether x_0 is greater or less than a; furthermore, for $x_0 < a$ the integral is zero for $\rho > x_0/a$. This means that the contribution to the field at a point comes only from charges inside an elliptical cylinder passing through the point. It is not necessary to consider what happens when $x_0 > a$, since a can be defined to be arbitrarily large. With the conditions $\rho < x/a < 1$ the integral (3.73) becomes (dropping the subscript)

$$E_x(x,0) = \frac{q}{\varepsilon_0}\int_0^{x/a}\frac{n\rho\mathrm{d}\rho}{\{(b^2-a^2)\rho^2+x^2\}^{1/2}}\ . \tag{3.75}$$

We consider now a beam with charge density n uniform up to $x = a$ and zero for $x > a$. For such a beam

$$E_x = \frac{nqbx}{\varepsilon_0(a+b)}$$
$$E_y = \frac{nqay}{\varepsilon_0(a+b)} \quad . \qquad (3.76)$$

This may be compared with the field $E_r = nqr/2\varepsilon_0$ for a circular beam of the same area. Distorting the circle into an ellipse leaves the mean radial field unchanged, but introduces a quadrupole component whose strength compared with the radial field is $\pm(a-b)/(a+b)$.

The potential distribution inside the uniformly charged elliptical beam may be readily found by writing down the general solution of Poisson's equation for a uniform charge density and equating coefficients to those obtained by differentiating equation (3.76). The result is

$$\phi = -\frac{qn}{4\pi\varepsilon_0}\left\{x^2+y^2 - \frac{a-b}{a+b}(x^2-y^2)\right\}. \qquad (3.77)$$

By rearranging this expression it is simple to show that the equipotentials are ellipses with semi-axes proportional to $a^{1/2}$ and $b^{1/2}$. These are illustrated in Fig.3.7. Fields outside the beam may be found from equation (3.75) by taking a in that equation to be larger than the beam radius. At large distances from the beam only the $1/r$ term and quadrupole term are significant.

As previously indicated, if the beam is surrounded by a tube which is not an equipotential in the 'free-space' configuration analysed above, additional fields will be induced in the beam. Even for relatively simple situations such as elliptical tubes co-axial with the beam the analysis becomes complicated. There is one special case which is simple but not of great practical interest. If the beam is surrounded by a close-fitting conducting tube, this forces the surface to be an equipotential. Under these circumstances it is easily verified that the additional field arising from the presence of the tube is quadrupolar, giving rise to a potential

$$\phi = -\frac{qn}{4\pi\varepsilon_0}\left\{x^2+y^2 - \frac{a^2-b^2}{a^2+b^2}(x^2-y^2)\right\}. \tag{3.78}$$

Returning now to Brillouin flow, the behaviour of elliptical beams has been analysed by Walker (1955) and

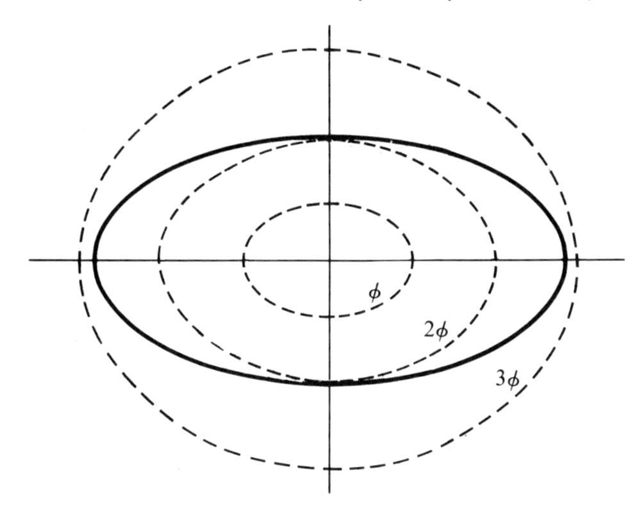

Fig.3.7. Equipotentials (broken curves) associated with a uniform elliptical charge distribution with axes of ratio 2 : 1. The internal equipotentials are ellipses with axes of ratio √2 : 1; the external equipotentials have a more complicated shape.

Pöschl and Veith (1962). Ellipticity is induced by superposing an external quadrupole field. The behaviour of the elliptical beam is characterized by a frequency ω_0 which vanishes for a circular beam. This frequency is a function of the external field and the parameters of the flow; for small ellipticity it corresponds to the ratio of the externally applied quadrupole field to the radial field of a beam of circular cross-section having the same area as the elliptical flow. The formal definition of ω_0 is through the equation

$$\omega_0^2 = \tfrac{1}{2}\omega_p^2 - \Omega_L^2 \tag{3.79}$$

Walker shows that the particle orbits are given

$$(\Omega_L - \omega_0)x^2 + (\Omega_L + \omega_0)y^2 = \text{constant}. \qquad (3.80)$$

These are ellipses if $\omega_0 < \Omega_L$; when $\omega_0 = \Omega_L$ (and is also equal to ω_p from equation (3.70)) the flow becomes planar, as described in the last section (continued for negative values of x); when $\omega_0 > \Omega_L$ it becomes hyperbolic.

Pöschl and Veith calculate the external field using elliptical co-ordinates; they point out also that the quadrupole component of the externally applied field does not increase monotonically as the beam ellipticity increases, but attains a maximum value and then decreases. For planar flow, for example, it has completely disappeared, and the field varies linearly between plane electrodes on either side of the beam.

3.4.3. Hollow laminar flows with zero axial velocity It is possible to construct relativistic self-consistent laminar flows carrying circumferential currents in a thin cylindrical shell. The forces acting on a charge are the centrifugal force, the Lorentz force arising from the external field, the outward radial forces arising from the space charge, and the self-magnetic field, which is in the z-direction but of opposite sign to the external field. The electric force increases from zero at $r = r_a$ to a maximum $r = r_b$, whereas the magnetic force is a maximum at $r = r_a$ and decreases to zero at the outer radius. For a balanced system the force on a charge at radius r is given by

$$\frac{\gamma m_0 \beta_\theta^2 c^2}{r} = -qB_z\beta_\theta c - \frac{1}{\varepsilon_0 r}\int_{r_a}^{r} nq^2 r\,\mathrm{d}r - \mu_0 \int_{r}^{r_b} n\beta_\theta q^2\,\mathrm{d}r\,. \qquad (3.81)$$

By adjusting γ (and hence β_θ) and n as functions of r, it is evident that a variety of solutions may be found. These are all somewhat artificial, however, since practical flows in thin shells of this type are invariably non-laminar. For this reason the topic will not be pursued further; non-laminar systems are analysed in section 4.5.4.

3.4.4. Laminar flow in a betatron focusing field Although laminar flow in a betatron field is unlikely to be obtained in practice, it does represent a limiting condition for the maximum current which can be held in a field of this type.

Here we present a very much simplified discussion, neglecting both the effects of beam curvature and image forces in the walls. These are considered in section 4.5. Anticipating the results of this later discussion we can say that the present treatment is a good approximation if the beam is confined in a tube of radius large compared with the beam radius but small compared with the orbit radius, which shields the electric and magnetic fields which arise from the beam. We assume a beam of particles with energy $\gamma m_0 c^2$ and neutralization fraction f. For a beam of elliptical cross-section, with a field index n, the laminar flow condition may be obtained by equating the betatron focusing force to the defocusing force arising from the self-field. From equation (2.17) for the betatron force and equations (3.28) and (3.76) for the space-charge force

$$\begin{aligned} \gamma m_0(1-n)\omega_0^2 x &= \frac{Nq^2 x}{\pi\varepsilon_0 a(a+b)}(1-f-\beta^2) \\ \gamma m_0 n\omega_0^2 y &= \frac{Nq^2 y}{\pi\varepsilon_0 b(a+b)}(1-f-\beta^2). \end{aligned} \tag{3.82}$$

From these equations it follows immediately that $a/b = n/(1-n)$. Eliminating n by addition and introducting the perveance K defined in equation (3.30) gives

$$\begin{aligned} \frac{ab}{2R_0^2} &= \frac{2\nu}{\beta^2\gamma}(1-f-\beta^2) = K \\ &= \frac{2\nu}{\beta^2\gamma^3} \quad \text{when } f = 0. \end{aligned} \tag{3.83}$$

where R_0 is the radius of the equilibrium orbit. It is clear from the form of the expression for K that if $f = 0$ the magnetic and electric forces very nearly balance for large β, giving rise to the small factor $1/\gamma^2$, this implies a large value of ν and hence that a large current can be held. Substituting for ν in equation (3.83) from equation (3.2), with $f = 0$ yields the very simple result $\omega_p = \beta\gamma c/R_0 = \gamma\omega_0$, where ω_p is measured in the frame of reference moving with the beam.

3.4.5. Electrostatically confined flow In section 3.4.1 Brillouin flow in hollow beams was studied. It was found that the current could be increased by inserting a cylinder inside the beam at a different potential from the external cylinder surrounding the beam. In the absence of a magnetic field, it is evident that a hollow beam requires an internal cylinder, with potential such that the force on the charges are inward, if equilibrium is to be achieved. An equilibrium of this type was studied by Harris (1952) in connection with travelling-wave tube devices. Such a flow, in which all particles have the same axial velocity, is capable of producing a maximum perveance equal to that obtainable from Brillouin flow. Under these conditions the minimum potential (for electrons) is at the outer radius, and the helix angle at the inner radius is $\tan^{-1}(1/\sqrt{3})$. This is the same as that for Brillouin flow at the outer radius (section 3.2.8).

In practice it is difficult to arrange an efficient launching system for such flow; this problem is discussed by Kirstein *et al.* (1967). Since the angular velocity of the electrons is not zero, P_θ is finite and the beam must be launched from a cathode immersed in a magnetic field.

Analysis of ripples on such beams, and on hollow Brillouin beams, can be made after the manner of section 3.3; it is found that the rippling frequency on the inner surface is higher than that on the outer. The same observation applies to rippling in the circumferential direction.

3.5. Space charge in periodic systems

As explained in Chapter 2, alternating-gradient focusing is used in particle accelerators and beam lines in order to provide strong focusing and contain the beam within a relatively small aperture. Such focusing is essentially different in the x and y planes, so that a laminar beam of elliptical cross-section changes its eccentricity as it proceeds. If self-field forces are present, their magnitude in either plane depends on the dimensions in both planes, so that the beam profile is governed by two coupled equations for x and y. Practical beams in particle accelerators are never laminar; the more general situation is analysed in section 4.3.6. The behaviour of a laminar beam in a quadrupole lens has been analysed by Bick (1965); an interesting difference between such a system and an axially symmetrical system is that a cross-over of orbits can occur; since the space-charge force does not tend to infinity as the elliptical cross-section of the beam degenerates to a line.

Periodic focusing is used also in microwave tubes, but in a different way. A uniform beam in a longitudinal magnetic field requires a solenoid or a permanent magnet with large pole area. The former consumes power, and the latter is clumsy and heavy. A more elegant practical solution is to provide a succession of annular permanent magnets of alternating polarity around the beam. These concentrate the field near the beam where it is needed and provide a more economical system. Such an arrangement has the slight disadvantage that the equilibrium beam is of necessity rippled though it is not difficult to ensure that the rippling is acceptably small.

Similar focusing can be obtained with an alternating electric field, provided, for example, by a series of rings on which the potential varies between ϕ and $\phi+\Delta\phi$. In practice this normally turns out to be rather more complicated, and consequently periodic electrostatic focusing is not widely used.

The type of focusing to be studied here differs from that studied in section 2.4 in several ways. We consider first the focusing of individual particles in the absence of space charge. Here the difference is that, even though the field reverses, every lens element produces focusing. In the unrealistic limit of a 'square-wave' field in which

B_z periodically reverses sign but does not change in amplitude, the radial variation of particle position with z is the same as if it were in a uniform magnetic field. A sudden change in B_z, however, implies a δ-function B_r in the reversal plane, proportional to r, so that the θ variation would be very different in the two systems; for a particle at constant radius, $\dot{\theta}$ reverses sign at each field reversal.

In practice of course the variation of B^2 is smooth and the system is essentially a chain of focusing lenses. Even though the lenses are all focusing, it is possible for the motion to be unstable if the lenses are too far apart. 'Overfocusing', described in section 3.2.9, occurs. To demonstrate this we calculate the matrix element for one period consisting of a thin lens of focal length f and a drift space D:

$$M = \begin{pmatrix} 1 & D \\ 0 & 1 \end{pmatrix} \begin{pmatrix} 1 & 0 \\ -\frac{1}{f} & 1 \end{pmatrix} = \begin{pmatrix} 1 - \frac{D}{f} & D \\ -\frac{1}{f} & 1 \end{pmatrix}. \tag{3.84}$$

If now $D > 4f$, the trace of the matrix numerically exceeds 2 and the particle makes oscillations of increasing amplitude about the axis.

When the field varies sinusoidally with periodicity d the paraxial equation takes the form

$$r'' + \frac{1}{2}\left(\frac{\Omega_{L0}}{\beta c}\right)^2 \sin^2\left(\frac{2\pi z}{d}\right) r - \left(\frac{P_\theta}{\beta\gamma m_0 c}\right)^2 \frac{1}{r^3} = 0. \tag{3.85}$$

When $P_\theta = 0$ it simplifies to

$$r'' + \left(\frac{\Omega_{L0}}{\beta c}\right)^2 \left\{1 - \cos\left(\frac{4\pi z}{d}\right)\right\} r = 0. \tag{3.86}$$

This represents a Mathieu equation, usually written in the form

$$\frac{d^2y}{dx^2} + (a - 2q\cos 2x)y = 0 . \tag{3.87}$$

The (a,q) plane may be divided into bands which represent regions of bounded and unbounded solutions (McLachlan 1947). The growing solutions are oscillatory, but have an envelope which for large amplitudes is approximately exponential. Equations (3.86) and (3.87) correspond when

$$a = 2q = \frac{1}{2}\left(\frac{\Omega_{L0}d}{2\pi\beta c}\right)^2 . \tag{3.88}$$

This represents the square of d measured in revolutions at the Larmor frequency. If the line $a = 2q$ is drawn on the (a,q) plane, the first two bands it crosses are as follows:

Ranges of a for stable motion : 0-0·66, 1·75-3·7 etc.
Ranges of a for unstable motion : 0·66-1·75, 3·7-6·2 etc.

In the presence of space-charge the paraxial equation for laminar flow, equation (3.57), is non-linear. For small oscillations about the mean this may be linearized using the method employed in section 3.3. Again, stable and unstable regions may be found. For practical use, however, computer studies of the non-linear equation have been made over ranges of practical interest, and sets of design curves are available in the literature.

A more detailed study of this topic, in which some of these curves are reproduced and references to the literature are given, may be found in the book by Kirstein *et al.* (1967).

3.6. Hydrodynamic approach to space-charge flow

The geometrical configurations of the flows so far considered in this chapter have been simple. In many instances they represent idealizations which cannot be achieved in practice, because no way can be found of devising suitable gun configurations or (in the case of ring beams) injection arrangements. This difficulty arises partly from such factors as the inevitable thermal spread of energies from the source, and partly from the conflicting requirements of the gun or injector design for beams which are both precisely shaped and intense.

Although all practical flows are to some extent non-laminar, the assumption of laminarity (defined in section 3.1) is often a reasonable approximation and provides a basis for design in more complicated geometrical configurations than those considered so far. This remark applies to steady-state electron beams generated from extended cathodes. Closed annular configurations such as that described in section 3.4.4, on the other hand, require time-dependent fields during their formation. Such beams are in practice highly non-laminar and laminar models are not often appropriate; they do, however, usefully indicate ultimate theoretical limitations.

Many of the studies of laminar flows have been made in connection with the design of microwave devices. Although practical designs are very often partly empirical, considerable insight has been gained in a general hydrodynamic approach to laminar space-charge flows, and guns of the 'crossed field' and 'magnetron injection' type have been designed making use of hydrodynamic concepts. For other than the simplest problems analysis becomes complicated and difficult. Applications to gun design are described by Kirstein *et al*. (1967) and by Amboss (1969). The magnetron problem is covered by papers in the book edited by Okress (1961).

Here we introduce some of the physical properties of the flow and take a more general look at some results obtained earlier in this chapter. Relativistic laminar flows are difficult to consider in a general way; we confine attention therefore to the non-relativistic situation, where momentum is proportional to velocity and, in the absence of space-charge neutralization, self-magnetic fields are unimportant.

The following discussion then applies to flow which is laminar, un-neutralized, non-relativistic, and steady in time. It is convenient to work in terms of the potentials

$\underline{A}$ and ϕ rather than the fields; furthermore, since the treatment is non-relativistic, it is convenient to write the velocity as v rather than βc. The equation of motion for a particular electron may be written

$$\frac{m_0 \mathrm{d}\underline{v}}{\mathrm{d}t} = q(-\mathrm{grad}\,\phi + \underline{v} \times \mathrm{curl}\,\underline{A}). \tag{3.89}$$

If, instead of considering a particular electron, we consider a fixed point in the medium and observe the velocity and velocity gradient of the particles at that point the variable $\mathrm{d}/\mathrm{d}t$ is replaced by $\partial/\partial t + \underline{v}.\mathrm{grad}\,\underline{v}$. (This standard hydrodynamical transformation, from a 'Lagrangian' to a 'Eulerian' description, may readily be proved; a more detailed discussion may be found in texts on hydrodynamics.) Under steady-state conditions $\partial/\partial t = 0$, so that

$$\frac{\mathrm{d}\underline{v}}{\mathrm{d}t} = \underline{v}.\mathrm{grad}\,\underline{v} \ . \tag{3.90}$$

Making use of the vector identity

$$\underline{v}.\mathrm{grad}\,\underline{v} = \mathrm{grad}(\tfrac{1}{2}v^2) - \underline{v} \times \mathrm{curl}\,\underline{v} \tag{3.91}$$

and substituting for $\mathrm{d}\underline{v}/\mathrm{d}t$ in equation (3.90) yields

$$\mathrm{grad}(\tfrac{1}{2}m_0 v^2 + q\phi) = \underline{v} \times \mathrm{curl}\,\underline{P} \tag{3.92}$$

where $\underline{P} = (m_0\underline{v} + q\underline{A})$, the non-relativistic canonical momentum. The quantity curl $\underline{P}$ is known as the 'vorticity', denoted by $\underline{w}$. Writing the total energy as

$$H = \tfrac{1}{2}m_0 v^2 + q\phi \tag{3.93}$$

equation (3.92) becomes

$$\mathrm{grad}\,H = \underline{v} \times \underline{w} \ . \tag{3.94}$$

For a flow which has originated at a cathode, where electrons are emitted with zero velocity, a further simplification follows. The value H is a constant of the motion, equal to zero, so that

$$\underline{v} \times \underline{w} = 0. \tag{3.95}$$

This implies that $\underline{w}$ is either zero or parallel to $\underline{v}$. Since $\underline{w}$ is the curl of $\underline{P}$, $\operatorname{div} \underline{w} = 0$. Furthermore, since $\underline{w}$ is parallel to $\underline{v}$, the total flux of $\underline{w}$ in a tube-shaped bundle of orbits is conserved. Across any surface

$$\int_S \underline{w} . d\underline{s} = \int_S (m_0 \operatorname{curl} \underline{v} + q\underline{B}) d\underline{s} = \text{const.} \tag{3.96}$$

By Stokes's theorem, which states that the integral of the curl of a vector over a surface is equal to its line integral round the circumference,

$$\oint_L (m_0 d\underline{v} + q\underline{A}) dL = q\Psi_0 \tag{3.97}$$

where Ψ_0 is the flux through the cathode, at which $\underline{v} = 0$. This equation will be recognized as a more general form of Busch's theorem. By considering a system with axial symmetry it becomes

$$2\pi m_0 r^2 \dot{\theta} + q\Psi = q\Psi_0 \tag{3.98}$$

which is the same as equation (2.5).

If the vorticity is zero, the flow is irrotational and $\Psi_0 = 0$. For systems with axial symmetry this implies that $P_\theta = 0$. This property has already been discussed in section 3.2.8. To establish more clearly the link with this earlier discussion, it is convenient to write $\underline{w} = m_0 \underline{\omega}_v$, where $\underline{\omega}_v$ is the 'vortex frequency'. This will be used again in Chapter 6. For the moment we note its relation to quantities discussed in section 3.2.8 in connection with flow in a uniform axial magnetic field. In such a situation $\underline{\omega}_v$ has only a z component;

$$\omega_v = 2(\Omega_L - \dot{\theta}) = \omega_c - 2\dot{\theta} = \frac{2\Omega_L \Psi_0}{\Psi} = \frac{-2P_\theta}{m_0 r^2} \quad \text{(N.R.)}. \qquad (3.99)$$

When the flow is paraxial, it is permissible to replace m_0 by γm_0, leading to

$$\omega_v = \frac{-2P_\theta}{\gamma m_0 r^2} \qquad (3.100)$$

If the vorticity $\underline{w} = \text{curl}\,\underline{P} = 0$, then $\underline{P}$ can be expressed as the gradient of a scalar function W

$$\text{grad}\,W = m_0\underline{v} + q\underline{A}. \qquad (3.101)$$

In the absence of magnetic fields this is essentially the hydrodynamic 'velocity potential' or action function.

The action function only exists when the vorticity is zero. Under these circumstances surfaces can be found which are orthogonal to the canonical momentum $\underline{P}$; the velocity is not, however, parallel to the canonical momentum, and the flow is called 'skew congruent'. Such flow occurs in a magnetic lens when the beam is launched from a finite cathode outside the field (or from a point cathode in the field). In the absence of a magnetic field $\underline{P}$ and $\underline{v}$ are parallel, and the flow is called 'congruent'. Such flow occurs in electrostatic lenses. In paraxial situations, in the absence of space charge, congruent flows can always be focused to a point. If the flow is not congruent, there is a term in $1/r^3$ in the equation and the trajectories cannot be made to pass through a point (section 2.2.2). The action function W clearly forms a useful compact description of the flow; we now find a general expression for it. From equations (3.93) and (3.101)

$$q\phi + \frac{1}{2m_0}(\text{grad}\,W - q\underline{A})^2 = 0 \qquad (3.102)$$

Combining this with Poisson's equation, $\nabla^2\phi = nq/\varepsilon_0$ and the continuity equation $\text{div}(\underline{nv}) = 0$ yields the required equation

$$\mathrm{div}[(\mathrm{grad}\, W - q\underline{A})\, \nabla^2\{(\mathrm{grad}\, W - q\underline{A})^2\}] = 0 \tag{3.103}$$

with boundary conditions that $\underline{v}$, Ψ, and $\underline{B}$ are zero at the cathode.

This equation is somewhat unwieldy; nevertheless when $\underline{B} = 0$, by taking W as a separable function with certain special properties, a number of flow patterns can be found. If $W(r,z,\theta)$ is of the form $f_1(r)f_2(z)$ for example, and the potential is chosen such that $\phi = \phi_1(r)\phi_2(z)$, an ordinary rather than a partial differential equation can be found which specifies $f(x)$. Using this technique a number of flows with axial symmetry from conical and cylindrical cathodes have been found. Another flow found using this technique by Meltzer (1956) has circular equipotentials and hyperbolic trajectories. Details of these and other flows, and their application to the design of crossed-field and magnetron-injection guns, may be found in the book by Kirstein *et al*. (1967).

3.R. Notes and references

Much of the material in this chapter has been developed in connection with the design of guns and beams for microwave tubes, where laminar-flow theory provides a good guide. The simpler types of flow are well covered in the book by Pierce (1954a). The longer book by Kirstein *et al*. (1967) treats the material in a more advanced way, and includes more topics and more detail. These books contain references to the original papers describing early work in this field. A more recent treatise by Nagy and Szilágyi (1974) contains a very detailed discussion of many space-charge and electron optics problems. The hydrodynamical aspects of electron flow are treated in a paper by Gabor (1945), which may be read as an introduction to section 3.6.

Lengthy reviews have been written by Dow (1958) on magnetically focused beams and Brewer (1967a,b) on guns and high-density beams. A review by Amboss (1969) on dense electron beams contains a particularly comprehensive list of early papers. In the field of particle accelerators the approximation of laminar flow is not a good one, and has only been used to illustrate limiting cases. The limit to the current which can be accelerated in a betatron. for example, (see section 3.4.4) was calculated by Blewett (1946).

Laminar flow is seldom a useful approximation in ion beams; the material in the next chapter is more relevant to them.

4

NON-LAMINAR BEAMS WITHOUT COLLISIONS

4.1. Introduction

In Chapter 2 the motion of beams of non-interacting particles through various types of focusing system was studied. For a linear focusing system and monoenergetic particles which make a small angle to the axis, the motion is particularly simple. Orbits originating from a point intersect at image points, but not elsewhere. Between the object and image points the flow is laminar. When the source consists of an extended object, each point of which emits particles over a range of directions, then the flow is essentially non-laminar. Even when the object is a point source, non-laminarity can develop in the presence of non-linearity, represented by aberrations. Both classes of non-laminar flow will be considered in this chapter.

The way in which non-laminar beams have been described and characterized has been somewhat different in the fields of electron microscopy, cathode-ray tube design, particle accelerators, and plasma physics. An attempt is made to unify the viewpoints and compare some of the work in the different fields. All make use of Liouville's theorem, and the remainder of this section is devoted to a brief summary of the basic dynamics leading to this theorem.

4.2. Hamiltonian formalism and Liouville's theorem

So far, Hamiltonian methods have not been used for analysing the properties of beams; it is convenient to look briefly at this formalism before introducing Liouville's theorem. A more formal and general treatment is given for example by Goldstein (1950). A good basic discussion with examples from accelerator theory is given by Corben and Stehle (1960). A much more detailed discussion, with particular reference to particle-accelerator problems, may be found in the book by Lichtenberg (1969). The treatment here is not the most general possible, and is partially at least particularized to situations encountered in charged-particle beams. The Hamiltonian formalism itself does not of course introduce any new 'physics', but it does provide a framework for formulation in terms of variables which simplify complicated problems and leads to rules which can be followed without constantly referring back to the physical situation. It is needed for a full appreciation of the concept of phase space.

We start from the Lagrangian for an assembly of particles in an external potential defined by ϕ and $\underline{A}$, which are in general functions of time:

$$L = -\frac{m_0c^2}{\gamma} - q(\phi - \underline{\beta}.\underline{A}c). \tag{4.1}$$

Next, a co-ordinate system is introduced in which x_i denotes the 'position' of the ith particle. The canonical momenta P_i, are defined as

$$P_i = \frac{\partial L}{\partial \dot{x}_i} . \tag{4.2}$$

For cartesian co-ordinates and $\underline{A} = 0$, $\underline{P} = \underline{\beta}\gamma m_0 c$; when $\underline{A} \neq 0$, $\underline{P} = \underline{\beta}\gamma m_0 c + q\underline{A}$, and in polar co-ordinates $P_\theta = r(\beta_\theta \gamma m_0 c + qA_\theta)$.

The Hamiltonian is defined as

$$H = \Sigma P_i x_i - L ; \tag{4.3}$$

using this definition, equation (4.2) and the expression for finding the equations of motion from L,

$$\frac{\mathrm{d}}{\mathrm{d}t}\left(\frac{\partial L}{\partial \dot{x}_i}\right) - \frac{\partial L}{\partial x_i} = 0 ,$$

equation (4.1) yields

$$-\dot{p}_i = \frac{\partial H}{\partial x_i} , \qquad \dot{x}_i = \frac{\partial H}{\partial P_i} . \tag{4.4}$$

The set of equations (4.4) are known as Hamilton's canonical equations, and the variables P_i and x_i are said to be canonically conjugate. The product Px has the dimensions of action, ML^2T^{-1}, but x does not necessarily have dimensions of length. (For example, angle and angular momentum are canonical.) There are twice as many equations as degrees of freedom, f, and the motion may be represented as the trajectory of a point in $2f$-dimensional space. Such a space is known as 'phase space'. With momentum and position as canonical variables the phase-space trajectory of a particle moving in a one-dimensional potential well is an ellipse, or, with suitable normalization, a circle. The radius of this circle and the angle swept out by the phase point can be used to define canonical 'action-angle' variables. These are often useful in discussing motion in systems where the Hamiltonian is periodic, such as sector-focused cyclotrons. When N particles are present, phase space contains $6N$ dimensions; this is referred to as 'Γ-space'. It is more convenient for our purposes to represent the motion as that of an assembly of points in six-dimensional phase space, or 'μ-space'. For paraxial situations, in which all particles have the same velocity component along the axis, the four dimensions of xyp_xp_y space are sufficient. Sometimes $xy\dot{x}\dot{y}$ and $xyx'y'$ space are useful, as we see later; the pairs of variables are not canonical conjugates, however, and they do not possess some of the general attributes of variables which are conjugates. The actual construction of the Hamiltonian function, and best choice of co-ordinates for a particular situation, is not always obvious or straightforward. Two cases for which this is straightforward, however, have been treated in Chapter 2. These are betatron oscillations and phase oscillations (section 2.8.2). In the former, the canonically conjugate variables are p_y and y, p_x and x, and the Hamiltonian is

$$H = \tfrac{1}{2}\left\{\frac{p_x^2}{\gamma m_0} + \gamma m_0\omega_0^2(1-n)x^2\right\} + \tfrac{1}{2}\left\{\frac{p_y^2}{\gamma m_0} + \gamma m_0\omega_0^2 y^2\right\}. \quad (4.5)$$

In this geometry A_x and A_y are zero, so that $p_x = P_x$ and $p_y = P_y$. The Hamiltonian for the phase oscillations represented by equation (2.173) has already been quoted (equation (2.178)).

If H is independent of time, then it corresponds to the Jacobian integral of the motion and represents the mechanical energy of the system. In applications to accelerator theory both adiabatically varying and periodic Hamiltonians are encountered. In these situations the invariant no longer corresponds to H; for example, in an adiabatically varying oscillatory system it is the action integral $\oint p\,dq$ which is invariant. For a harmonic oscillator of amplitude a this is

$$J = \pi\omega a^2 = 2\pi H/\omega. \tag{4.6}$$

The Hamiltonian approach has been extensively used to study non-linear effects in alternating-gradient systems and orbits in azimuthally varying cyclotrons, both of which involve periodically varying Hamiltonians. These topics are outside the scope of the present work, but examples of this approach may be found in the references to non-linear motion given in section 2.R.6-7.

The concepts necessary for stating Liouville's theorem, in particular the definition of phase space, have now been introduced. The theorem may be stated in many ways; for our purpose the following is convenient. *The density in six-dimensional phase space of non-interacting particles, in a conservative dynamical system, measured along the trajectory of a particle, is invariant.* This result follows immediately from considerations of continuity, plus the Hamiltonian property of conservative systems. The continuity equation for a distribution function of form $f(P_i, x_i, t)$, which specifies the time dependence of the density in phase space, is

$$\frac{\partial f}{\partial t} + \frac{\partial}{\partial x_i}(f\dot{x}_i) + \frac{\partial}{\partial P_i}(f\dot{P}_i) = 0 \tag{4.7}$$

where summation over i = 1 to 3 is assumed. Expanding the second and third terms

$$\frac{\partial f}{\partial t} + f\frac{\partial \dot{x}_i}{\partial x_i} + \dot{x}_i\frac{\partial f}{\partial x_i} + f\frac{\partial \dot{P}_i}{\partial P_i} + \dot{P}_i\frac{\partial f}{\partial P_i} = 0 \qquad (4.8)$$

From the Hamiltonian condition, equation (4.3),

$$\frac{\partial \dot{x}_i}{\partial x_i} + \frac{\partial \dot{P}_i}{\partial P_i} = 0 \qquad (4.9)$$

so that equation (4.7) reduces to

$$\frac{\partial f}{\partial t} + \dot{x}_i\frac{\partial f}{\partial x_i} + \dot{P}_i\frac{\partial f}{\partial P_i} = 0 \qquad (4.10a)$$

or

$$\frac{Df}{Dt} = 0 \qquad (4.10b)$$

where the total derivative D/Dt denotes the rate of change at the position of a given particle as it moves in phase space. The flow of particles in phase space is analogous to the flow in real space of an incompressible gas. (This analogy is expressed in the term 'waterbag' model; a sharp-edged volume of uniform density can change its shape and position but not its volume, like an elastic bag filled with water.) If the density of particles is a function of position, it follows that the volume enclosed by a surface of given density is conserved.

For particles which interact the Hamiltonian contains terms which arise from the mutual interactions; Liouville's theorem remains true in $6N$-dimensional Γ-space, but is not true in six-dimensional space. (That this is so may be seen from a similar argument to that used in deriving equation (4.10), bearing in mind that the Hamiltonian now contains terms containing co-ordinates of all the particles. The function f now represents a probability density of finding a particle at a given point, not an actual density of particles in a particular system.) There are, however, conditions under which, even with interacting particles, the theorem remains a good approximation. Formally, if

$F_N(p_{i1} \ldots p_{iN} \ldots x_{i1} \ldots x_{iN}, t)$ represents the distribution function in $6N$-dimensional phase space, the distribution function in six-dimensional space can be found by integrating over N sets of six variables to yield F_1, the distribution denoted by f in equation (4.10). When, however, this integration is performed, it yields equation (4.10) together with an additional term which involves integration over F_2, a function containing information about the correlation of the positions of pairs of particles. If no such correlation exists, the additional term is zero, and equation (4.7) applies; if there is correlation, then F_2 may be determined by integrating the original equation over only N-1 variables. This, however, contains a term in F_3, determination of F_3 involves F_4, and so on until the original equation is reached. This chain of equations is known as the 'BBGKY hierarchy', after the initials of those who originally formulated the theory (N.N. Bogoliubov, M. Born, H.S. Green, J.G. Kirkwood, and J. Yvon). Details may be found in texts on plasma kinetic theory, such as those of Montgomery and Tidman (1964) or Wu (1966).

This open set of equations can only be closed by introducing approximations. The simplest approximation can be made when a single particle interacts more strongly with the collective fields of the other particles than with its nearest neighbour. Quantitatively this condition may be satisfied by requiring that the Debye length (section 4.4) be large compared with the interparticle distance. Under these circumstances the collective space-charge fields and magnetic fields can be treated on the same basis as the external fields and the collisions between individual particles ignored. Liouville's theorem is thus retained, with the collective forces represented by a smooth 'external' potential which appears the same to all particles. This approximation forms the basis of the Vlasov equation, introduced in section 4.6.

In many practical situations scattering can be neglected, and Liouville's theorem is a good approximation even in the presence of collective fields. In other circumstances (for example, in particle-storage rings) the scattering causes steady attentuation or diffusion of the beam; nevertheless on a small enough time scale the approximation may still be good. In non-Hamiltonian situations of course Liouville's theorem does not apply. (These arise when simplifications in the detailed dynamics, for example by using the concept of dissipation, are introduced.) Examples, such as gas scattering in accelerators and radiation damping in electron synchrotrons, will be studied in Chapter 5.

4.3. The emittance concept

4.3.1. Definition of emittance; brightness The paraxial and laminar beams which have been considered in Chapters 2 and 3 represent idealizations of what can be achieved in practice. In actual beams thermal velocities at the source, aberrations, and other imperfections always give rise to non-laminar behaviour. To provide a quantitative basis for describing beam quality, a figure of merit known as the 'emittance' may be defined. This is closely related to the projection on a plane of the volume in phase space occupied by the particles which constitute the beam, as will be demonstrated later. Most practical beams have two planes of symmetry, or else are axially symmetrical. We consider a beam symmetrical about the xz and yz planes, and define the x-plane emittance ε_x as $1/\pi$ times the area in xx' space occupied by the points represented by the particles of the beam at a given value of z, the distance along the beam.[†] (Instantaneously, there may of course be few or no particles precisely at some particular value of z. For a steady beam one can include the particles which pass the appropriate value of z during a fixed time.) Clearly this is a somewhat incomplete definition; in practice the density of points is non-uniform and decreases gradually at the edges.

[†]Sometimes the factor $1/\pi$ is omitted in the definition; often the factor π is then included explicitly in the quoted figure.

A convenient convention is to specify the area within which say 90 per cent of the points lie; alternatively the areas within equi-density contours containing different fractions of the beam can be presented as a curve.

Full information about the beam requires that not only the area but also the shape of the distribution should be specified. This is often presented as an 'emittance plot', an example of which is shown in Fig.4.1. For reasons which

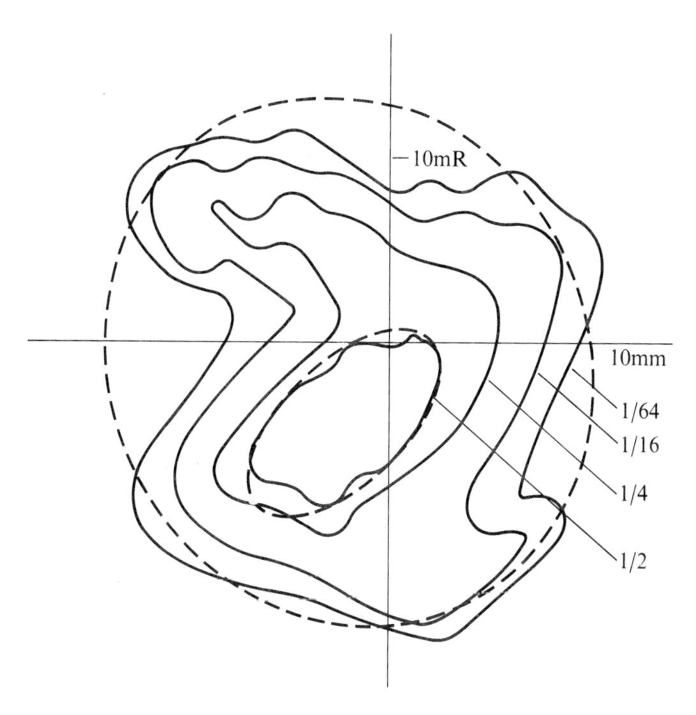

Fig.4.1. Emittance plot in the xx' plane of an axially symmetrical beam from a helium ion source, showing contours of constant brightness, (equation 4.11). The numbers represent the brightness relative to the maximum, and the broken curves are ellipses. (After Greenway and Hyder 1971.)

will be apparent later, a plot of elliptical shape is generally to be preferred to a more irregular one.

Such plots can be determined experimentally with suitable arrangements of slits and screens, as indicated in Fig.4.2. A slit is scanned across the beam, and the density

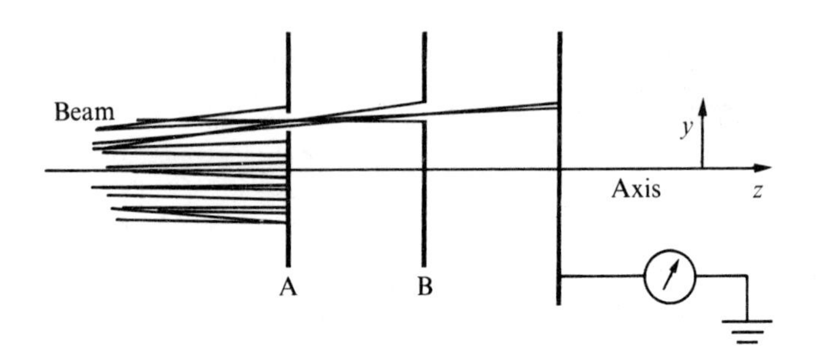

Fig.4.2. Apparatus for measuring beam emittance in yy' plane. For various positions of the slotted screen A, the screen B is moved, and the current on the third screen measured. The position of the first slit determines y, and that of the second determines for the chosen value of y.

distribution of that part of the beam which passes through is measured with another slit. A detailed evaluation of this method has been given by van Steenbergen (1967). Although simple in principle, the method is not always easy to apply in practice; for example, if the particle range in matter is too long the slits need to be in a thick material, and this gives rise to 'slit scattering'. Emittance measurements of internal beams in circular accelerators are somewhat more difficult; several methods have been devised to meet particular situations.

The emittance plot has a very important property, which can be deduced from Liouville's theorem. In the paraxial approximation, with $A_x = A_y = 0$ so that $P_x = p_x$, x' is proportional to p_x/p_z. Consequently, if there is no coupling between x and y motion, the density of points in xx' space in the neighbourhood of a particular point remains proportional to p_z as z varies. The area enclosing a given set of points is therefore inversely proportional to p_z. If p_z does not vary with z the density of points in xx' space in the neighbourhood of a particular point remains constant; consequently the area enclosing a given set of points, and therefore the emittance, are invariant quantities. If p_z does vary with z, then the emittance varies in inverse proportion, since $p_x \propto \beta\gamma x'$. The invariant 'normalized emittance' $\varepsilon_n = \beta\gamma\varepsilon$ is therefore often used. This is convenient when considering beams in which there is acceleration, and also for comparing beams at different energies and with different particles. In beams without axial symmetry ε_x and ε_y are only independent and invariant if there is no coupling. If there is coupling then it is possible to define an invariant four-dimensional volume, though this is seldom useful.

In the presence of an axial magnetic field there is always coupling between the x and y motion. For an axially symmetrical beam the emittance remains constant and equal in the two planes. If measured in a rotating frame, the emittance of an axially symmetrical beam is different from its value in a stationary frame. For a beam consisting of particles moving in helical trajectories with angular velocity ω about the axis, the emittance is obviously zero in a frame also moving with angular velocity ω. For reasons which emerge later it is convenient to measure the emittance in the Larmor frame when there is an axial magnetic field present. The emittance of a Brillouin beam is therefore zero, though in the stationary frame it would be finite. Working in the Larmor frame removes the B_z field and ensures that A_θ is zero.

For monoenergetic particles and $\dot{z} \gg \dot{x}, \dot{y}$, the momentum p_z is the same for particles with the same value of z, so that a four-dimensional phase space is adequate to describe the beam. In practice, however, there is often a significant energy spread, and for a full description the six dimensions of phase space must be considered. Instead of z and p_z it is convenient in travelling-field accelerators such as synchrotrons and linear accelerators to use ϕ and W (section 2.8.2) as the canonical variables describing the longitudinal degrees of freedom. The transverse and longitudinal components are referred to as 'betatron' and 'synchrotron' phase space.

The emittance is closely related to the 'brightness' concept used in electron optics, defined as the current per unit area per unit solid angle

$$B = \mathrm{d}I/\mathrm{d}A\mathrm{d}\Omega \tag{4.11}$$

in A m^{-2} s-rad^{-1}. In general B varies across the beam, but average values are often used. For small angles $\mathrm{d}\Omega\mathrm{d}A$ may be written $4\pi^2 rr'\mathrm{d}r'\mathrm{d}r$. For a beam confined uniformly within the limits $r<r_1$ and $r'<r_1'$ the emittance diagrams are rectangles of non-uniform density in the xx' and yy' planes with area $\pi\varepsilon = 4xx' = 4rr'$, and $\mathrm{d}\Omega\mathrm{d}A = \pi^2 r^2 r'^2 = \pi^4\varepsilon^2/16$. For such a beam, carrying current I, $B = 16I/\pi^4\varepsilon^2$. In general there is some correlation between r and r' and the beam is not uniformly distributed; under these circumstances an average must be taken, and

$$B = \eta I/\pi^2\varepsilon^2 \tag{4.12}$$

where η is a form factor of order unity. The invariant corresponding to normalized emittance is $\beta^2\gamma^2/B$, which in the non-relativistic limit is proportional to ϕ/B.

Conservation of brightness divided by momentum is implied in equation (2.62) relating magnification to the ratio of the angles between a ray and the axis. If x_o and x_i are corresponding points on object and image, and θ_o and θ_i are the angles between a pair of rays from x_o to x_i, then it follows from equation (2.62) that

$$\frac{x_i\theta_i}{x_o\theta_o} = \frac{x_i x_i'}{x_o x_o'} = \frac{\beta_o\gamma_o}{\beta_i\gamma_i} . \tag{4.13}$$

The corresponding principle in light optics is that the image brightness is equal to the object brightness multiplied by the square of the ratio of the refractive indices of the image and object spaces. The relation expressed in terms of ray angles and image sizes is known as the Helmholtz-Lagrange relation; in light optics $\beta\gamma$ is replaced by the refractive index n.

The relation between light optics and charged-particle optics is developed in most texts on electron optics. The correspondence can most readily be seen by comparing Fermat's principle with the principle of least action; the refractive index and canonical momentum are corresponding quantities. The permitted spatial variation of refractive index in charge- and current-free regions in particle optics is restricted by the need for the fields to satisfy Poisson's equation as has been noted earlier in the discussion of lenses and their aberrations. No such restriction exists in light optics; on the other hand, it is more convenient in practice to use regions of constant refractive index bounded by curved or plane surfaces. Because of these differences the value of detailed comparison of the two subjects is limited. Nevertheless, an interesting approach to transmission optics using matrix techniques has been described by Marcuse (1972).

The discussion in this section has not been very detailed. In particular, it has been restricted to the paraxial approximation, which is hardly adequate when discussing the formation of beams from an emitting surface. Furthermore, the correspondence between brightness and emittance has been made in a rather approximate manner. More careful discussions of these points have been given by van Steenbergen (1965), Walcher (1972) and Fink and Schumacher (1975). The emittance concept has been critically reviewed by Lapostolle (1969a, 1972).

4.3.2. Phase-amplitude variables and beam matching So far the paraxial ray equation has been developed for laminar beams including both space-charge and self-magnetic fields in a beam with uniform density across the cross-section. As explained in section 3.3.1 the envelope equation may then be found from the ray equation by writing a, the beam radius, in place of r. When the beam is non-laminar, or the ray equation is given in cartesian co-ordinates, this can no longer be done. In this section we calculate the appropriate equation using cartesian co-ordinates and reduced variables. This applies to elliptical beams with different diameters in the x and y directions.

We consider motion in the x direction and use the reduced variables $X = (\beta\gamma)^{1/2} x$, and write the paraxial equation in the form

$$X'' + \kappa(s)X = 0 \tag{4.14}$$

where $\kappa(s)$ includes both the external focusing force and the linear self-force in a beam of uniform transverse density. It is a function of s and is in general continuous; in some applications, such as the alternating-gradient synchrotron, it is piecewise constant and periodic. Laminar flow is not assumed, so that an individual particle can move from the centre to the outside of the beam. We leave for the moment the question of the form of distribution function which gives a uniform beam with non-laminar flow and merely postulate that such a distribution exists.

Equation (4.14) is transformed by changing to phase amplitude variables A, Φ. These are defined by

$$X = Aw(s)\cos\{\psi(s)+\Phi\} \tag{4.15}$$

$$X' + A[w'(s)\cos\{\psi(s)+\Phi\} - \frac{1}{w}\sin\{\psi(s)+\Phi\}] \tag{4.16}$$

where a restriction has been placed on the choice of w and ψ by writing

$$\psi' = 1/w^2. \tag{4.17}$$

The reason for this seemingly arbitrary choice will become evident later. Four quantities, w, ψ, A, and Φ, have been introduced; of these w and ψ are functions of s and the same for all trajectories, whereas A and Φ are independent of s and replace the values of X and X' at $s = 0$ as specifying a particular trajectory. Substituting equation (4.15) into equation (4.14) it is found that w satisfies the equation

$$w'' + \kappa w - \frac{1}{w^3} = 0. \tag{4.18}$$

Equation (4.18) has many solutions, and it remains to choose an appropriate one. To do this we determine the relation between A, X, X', w, and w' found by eliminating ψ from equations (4.15) and (4.16) (using the fact that $\cos^2\psi + \sin^2\psi = 1$). The required relation may be written

$$A^2 = \frac{X^2}{w^2} + (wX' - w'X)^2 \qquad (4.19)$$

or

$$A^2 = \gamma_0 X^2 + 2\alpha_0 XX' + \beta_0 X'^2 \qquad (4.20)$$

where

$$\begin{aligned} \alpha_0 &= -ww' \\ \beta_0 &= w^2 \\ \gamma_0 &= w^{-2} + w'^2 = \frac{1+\alpha_0^2}{\beta_0} \end{aligned} \qquad (4.21)$$

Equation (4.20) represents an ellipse in the XX' plane. The size, eccentricity, and orientation of this ellipse are determined by A and the coefficients α_0 and β_0. These two coefficients are determined in turn by w and w'. To follow the motion as s varies of an ensemble of points lying on a given ellipse, therefore, we need to know how w varies. This can be found by inserting appropriate initial values in equation (4.18). Points lying on a particular ellipse have a unique value of A, and a range of Φ from 0 to 2π. Points corresponding to trajectories with a different value of A lie on an ellipse scaled in size but otherwise geometrically similar. As s varies, the shape and orientation of the ellipse change in accordance with the variation of w and w', and points corresponding to a given value of Φ move round the ellipse, as indicated in Fig.4.3. The area of the ellipse defined by equation (4.20) is

$$S = \pi A^2 (\beta_0\gamma_0 - \alpha_0^2)^{-1/2} . \qquad (4.22)$$

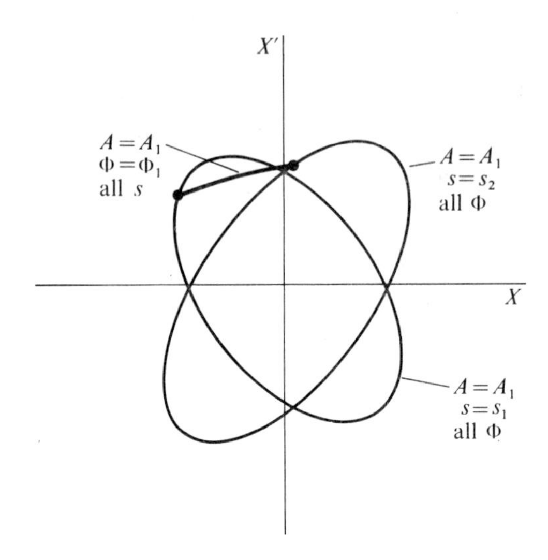

Fig.4.3. Transformation of ellipse defined by equation (4.21) in XX' plane. All points with given A at a given axial position s lie on an ellipse; as s varies the points move round the ellipse, which changes in eccentricity and orientation but not in area.

From equation (4.21) the quantity in brackets is equal to unity, so that the area is invariant and equal to πA^2.

Consider now an ensemble of trajectories corresponding to points uniformly distributed inside the ellipse; as s varies these all move on geometrically similar ellipses with the appropriate value of A. Since the area of the ellipse is invariant, the density of points remains constant, as expected from Liouville's theorem; furthermore, since its area is just $\pi\times$ the normalized emittance of the beam, $A_0 = \varepsilon_n^{1/2}$, where A_0 is the maximum value of A.

The envelope of a beam consisting of trajectories constituting a uniformly filled ellipse in phase space (all values of Φ, $A < A_0$) can readily be calculated. At any value of s, the edge particle in the beam corresponds to $A = A_0 = \varepsilon_n^{1/2}$, $\psi+\Phi = 0$. From equation (4.15) $X = A_0 w$ for such a particle, so that from equation (4.18) the envelope equation is

$$X'' + \kappa X - \varepsilon_n^2 X^{-3} = 0 \tag{4.23}$$

with initial conditions determined from equations (4.15-4.17) and (4.21).

An interesting particular solution arises when $\kappa = 0$, so that there is no focusing force. Equation (5.19) reduces to $X''X^3 = \varepsilon_n^2$ with solution

$$X = \left(X_1^2 + \frac{\varepsilon_n^2 s^2}{X_1^2} \right)^{1/2} . \tag{4.24}$$

Initial conditions have been chosen such that $X = X_1$, $X' = 0$ at $s = 0$. The profile is hyperbolic, with a 'waist' at $s = 0$; individual trajectories are all straight lines.

We now study the behaviour of the ellipse in the XX' plane for a number of typical situations encountered in the design of beam lines. The simplest of these is a uniformly focusing channel where κ is constant. Fig.4.4 shows the transformation of an ellipse along such a channel. For a beam emerging from a point source the ellipse degenerates into a line; for a beam containing a uniform distribution in Φ for each value of A present the ellipse remains upright and is independent of s. Such a beam is said to be 'matched'. It is represented by a solution of equation (4.23) for which κ is constant and $X'' = 0$. The dimension of the beam is given by

$$X = \varepsilon_n^{1/2} \kappa^{-1/4} \tag{4.25}$$

For a channel of finite width represented by a given value of $X = X_1$, the 'acceptance' is defined as the value of emittance given by equation (4.25), $\varepsilon_n = \kappa^{1/2} X_1^2$.

To transfer a beam efficiently from a wide weak-focusing channel to a narrower channel with stronger focusing requires a 'quarter-wave' matching section of appropriate focusing strength $\kappa_m^2 = \kappa_1 \kappa_2$. Trajectories and phase-space diagrams appropriate to this situation are shown in Fig.4.5. The beam is matched in both channels. Fig.4.6 illustrates the passage of a beam through a thin lens.

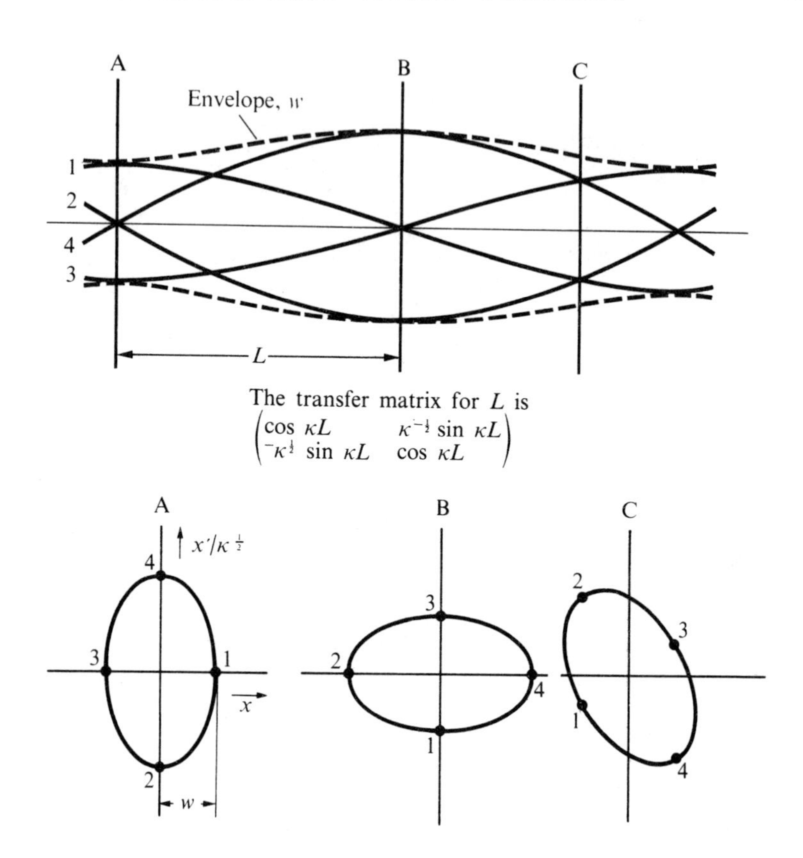

Fig.4.4. Trajectories, beam envelope, transfer matrices, and phase-space ellipses for a typical beam in a uniform focusing channel. The scale is chosen such that individual phase points move in circles.

More generally the 'acceptance diagram' of a piece of apparatus is a contour in xx' space showing the limiting co-ordinates of particles which can pass through the system without striking the walls. In general it will not be elliptical. If a beam of known emittance is available, say from a source, then, provided that the emittance is less than the acceptance, a matching section to transform the shape of the emittance can in principle be designed. Only if the two diagrams are elliptical can this be done in a straightforward way with linear elements. For non-linear systems xx' and yy' are likely to be coupled, making the problem even more complicated. In practice it is often con-

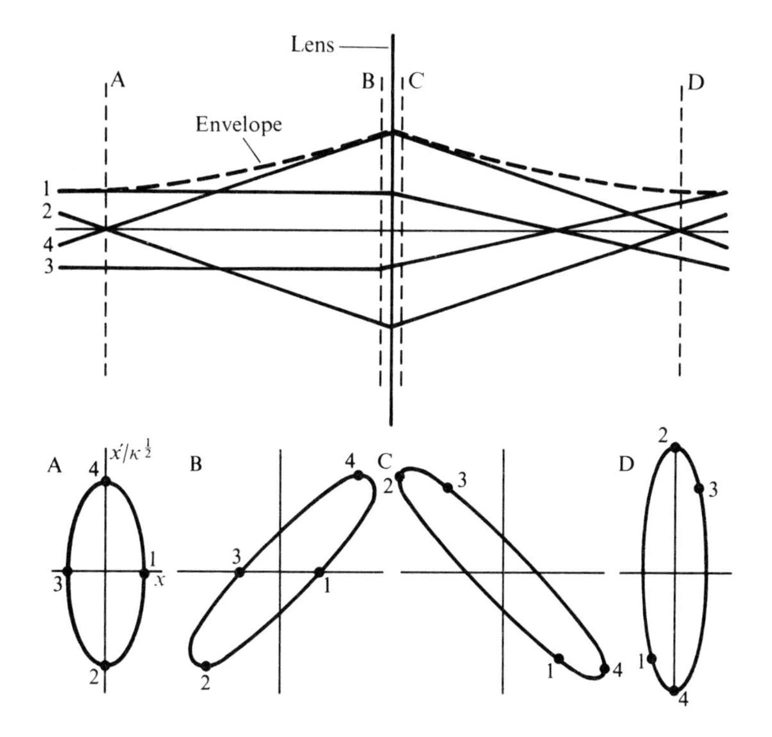

Fig.4.5. Trajectories, beam envelope, and phase-space ellipses for a beam matched from a channel with $\kappa = \kappa_1$ via a transformer with $\kappa = \sqrt{2}\kappa_1$ to a channel with $\kappa = 2\kappa_1$. The transfer matrix is as in Fig.4.4, with appropriate values of κ.

venient to work with ellipses circumscribing the emittance diagram of the beam, or inscribed in the acceptance diagram of the apparatus.

4.3.3. The distribution of Kapchinskij and Vladimirskij A distribution function in $xx'yy'$ space which projects to a uniform elliptical distribution on the xx', yy', xy, $x'y'$, xy', and $x'y$ planes was introduced by Kapchinskij and Vladmirskij (1959). It will be referred to as the 'K-V distribution'. (The less satisfactory term 'normal distribution' was introduced by Walsh (1963).) It represents a uniformly filled hyper-ellipsoidal three-dimensional shell in four-

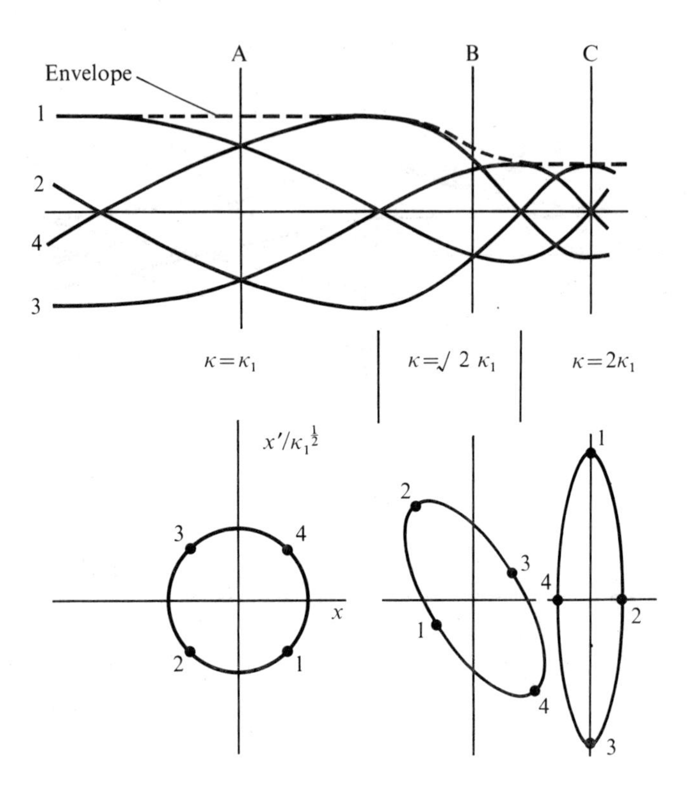

Fig.4.6. Trajectories, beam envelope, and phase-space ellipses showing passage of a drifting beam through a thin lens; AB = CD = $2F$, where F is the focal length of the lens. The transfer matrix for the system $M_{\text{DA}} = M_{\text{BA}}\, M_{\text{CB}}\, M_{\text{DC}}$ is given by

$$M_{\text{DA}} = \begin{pmatrix} 1 & 2F \\ 0 & 1 \end{pmatrix} \begin{pmatrix} 1 & 0 \\ -1/F & 1 \end{pmatrix} \begin{pmatrix} 1 & 2F \\ 0 & 1 \end{pmatrix} = \begin{pmatrix} -1 & 0 \\ -1/F & -1 \end{pmatrix} .$$

dimensional phase space. In a uniform matched focusing channel, where the envelope is parallel to the axis, the axes of the hyper-ellipsoid are parallel to the co-ordinate axes and the distribution takes the simple form

$$f(x,y,x',y') = \delta\left(\frac{x^2}{a^2} + \frac{y^2}{b^2} + \frac{a^2 x'^2}{\varepsilon_x^2} + \frac{b^2 y'^2}{\varepsilon_y^2} - 1\right). \qquad (4.26)$$

By an extension of the theorem of Archimedes all two-dimensional projections of such a three-dimensional hyper-ellipsoidal shell are uniform, so that the distribution has the convenient property that the charge density across the beam is constant and the forces associated with the self-fields vary linearly with radius. If $a/b = \varepsilon_x/\varepsilon_y$ the transverse energy (kinetic and potential) of all the particles in the distribution is the same, and it may be termed 'microcanonical'.

In an axially symmetrical beam, with $a = b$, $\varepsilon_x = \varepsilon_y$, the projections of particle trajectories of a matched beam on the xy plane are circular (Fig.4.7). All orbits extend to a

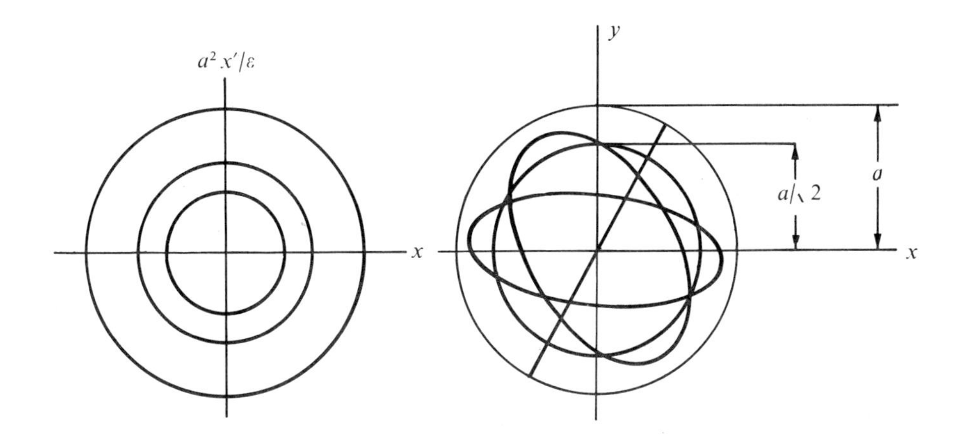

Fig.4.7. Orbits in matched uniform circular beam with Kapchinskij-Vladimirskij distribution, projected on the xx' and xy planes. In the latter the sum of the squares of the axes for all ellipses is constant.

a radius between $a/\sqrt{2}$ and a, and the transverse energy of all particles is the same. Orbits in the xx' projection are concentric circles, whereas in the xy' projection they consist of ellipses of all eccentricities and sizes such that the major axis is less than a. For the K-V distribution it may readily be shown that the form factor in equation (4.12) connecting brightness and emittance is 2 (Walsh 1963).

The uniformity of density in the various projections

makes this a very convenient distribution to handle analytically. It is physically unrealistic, however, since particles with small oscillations in both planes are excluded. Practical distributions tend to have a four-dimensional phase-space density which decreases with radius rather than being hollow. Nevertheless, for the reason stated the K-V distribution provides a convenient model which will be used in section 4.3.4 to discuss linear self-field effects. It is important to note that the K-V distribution has zero four-dimensional volume, even though it represents crossing orbits in an actual beam. The four-dimensional phase-space volume corresponding to an actual beam is not a very useful physical parameter.

4.3.4. Phase-amplitude variables in periodic systems Phase-amplitude variables, and in particular the parameter β_0 in equation (4.22) are very convenient for studying the properties of a matched beam in a periodic lattice, such as is used in particle accelerators and storage rings.[†]

In a system in which κ varies periodically an important solution of equation (4.14) (or equation (4.18)) is that for which the beam profile varies periodically with the lattice period. This represents a matched beam, for which the distribution in Φ for a given A is uniform. At corresponding points in different periods of the lattice the emittance ellipse is identical, though points corresponding to a particular particle are at different points on the ellipse. This is evident from equation (4.15); since the beam profile is periodic, w is periodic, Φ is constant, but ψ changes by an amount not equal to a multiple of 2π in one lattice period.

In order to find that solution of equation (4.20) which is periodic, it is necessary first to find the transfer matrix in terms of w, w', and ψ. This can be done by finding values of A and Φ for the principal solutions of equation (4.14), with initial conditions $(X,X') = (1,0)$ and $(0,1)$ at

---

[†]The subscript used here to avoid confusion with $\beta = v/c$ is not used in other treatments.

s_1, and writing the matrix in terms of these solutions, as in equation (2.32). The parameters A and Φ can be expressed in terms of w, w', ψ at $s = s_1$, from equations (4.15) and (4.16), and the matrix as a function of s_2 through the values of w, w', and ψ at $s = s_2$. If w is to be periodic with period L, then $w(s_1) = w(s_1+L)$. Setting $w(s_1) = w(s_1+L) = w_1$, $w'(s_1+L) = w_1'$ and $\psi(s_1+L)-\psi(s_1) = \psi_L$, the matrix for one period is found to be

$$M_{21} = \begin{pmatrix} \cos\psi_L - w_1 w_1' \sin\psi_L & w_1^2 \sin\psi_L \\ -(w_1^{-2}+w_1'^2)\sin\psi_L & \cos\psi_L + ww'\sin\psi_L \end{pmatrix} \tag{4.27}$$

Using equation (4.14) this may be written

$$M_{21} = \cos\psi_L \begin{pmatrix} 1 & 0 \\ 0 & 1 \end{pmatrix} + \sin\psi_L \begin{pmatrix} \alpha_0 & \beta_0 \\ -\gamma_0 & -\alpha_0 \end{pmatrix} \tag{4.28}$$

which is of the same form as equation (2.132), with $\psi_L = \mu$. Since $\beta_0 = w^2 = 1/\psi'$, the important relation follows that

$$\psi_L = \int_0^L ds/\beta_0 . \tag{4.29}$$

If NL represents the total circumference of a periodic machine, then $Q = N\psi_L/2\pi$.

To summarize, w or β_0 is a periodic function with the periodicity of the lattice, which, through equations (4.20) or (4.23), specifies the profile of a matched beam; ψ is a phase function, with rate of change periodic with the lattice frequency. The displacement of a given particle is proportional to $w\cos(\psi+\Phi)$, which is *not* periodic with the lattice frequency.

This outline discussion indicates only some of the features of motion in periodic systems. For example, it has been assumed that a periodic solution for w exists; this will not be so if the transfer matrix is such that $\cos\psi_L$ exceeds unity. Under these circumstances resonant build-up of oscillations occurs, as described earlier (section 2.7.3).

4.3.5. Linear self-field effects; self-constricted beam In section 4.3.2 the envelope equation for a beam with a K-V phase-space distribution was obtained; this consists of the paraxial ray equation for a single particle plus an extra term ε_n^2/X^3. Longitudinal magnetic fields were specifically excluded, and the analysis confined to one plane of the beam.

For an axially symmetrical beam the x-plane can be taken at any orientation, and the same equation applies with R written for X. The envelope equation in terms of a, the beam radius, may accordingly be written

$$a'' + \frac{\gamma' a'}{\beta^2\gamma} + \left(\frac{\gamma''}{2\beta^2\gamma} + \frac{\Omega_L^2}{\beta^2 c^2}\right)a - \frac{K}{a} - \frac{\varepsilon_n^2}{\beta^2\gamma^2 a^3} = 0. \qquad (4.30)$$

The canonical angular momentum P_θ was taken as zero in section 4.3.2, and it does not appear therefore in equation (4.30). It will be noted however that the angular momentum term in equation (2.14) has the same form as the emittance term in equation (4.30). Indeed, the projections on the xx' plane of the orbits associated with a K-V distribution, or a helical laminar distribution, are the same when the corresponding coefficients are equal:

$$\varepsilon_n = \beta\gamma\varepsilon = \frac{p_{\theta a}}{m_0 c} = \frac{a\beta_{\theta a}}{\beta_z} \qquad (4.31)$$

where, as before, the suffix a denotes values at the edge of the beam. Equation (4.31) defines the emittance associated with the rotation characterized by $p_{\theta a}$ or $\beta_{\perp a}$.

It is readily verified that for such laminar helical motion the projections of the orbits on the xx' and yy' planes are upright ellipses with emittance $a\theta$, where θ is the helix angle. This laminar distribution can be represented by a two-dimensional shell in four-dimensional phase space, defined by the intersection of the three-dimensional volumes

$$\begin{aligned} x &= y'a^2/\varepsilon \\ y &= -x'a^2/\varepsilon \end{aligned} \qquad (4.32)$$

where $\varepsilon = a\theta$.

The equivalence of the emittances associated with the K-V distribution, in which crossing orbits occur, and the laminar rotating flow emphasizes that the emittance is a property of projections and does not take into account correlations between, for example, x' and y. For this reason there is no direct correspondence between emittance and quantities such as pressure and temperature which involve the distribution function in both x' and y' at a single point in the beam. The K-V and laminar distributions are compared again in section 4.4.2 where the relation between pressure, temperature, and emittance is explored.

For a beam in which the tendency to spread arising from the finite emittance just balances the self-constriction or 'pinching' arising from a negative value of K, only the last two terms of equation (4.30) are non zero. The equilibrium beam radius is therefore given by

$$a^2 = \frac{-\varepsilon_n^2}{\beta^2\gamma^2 K} = \frac{-\varepsilon^2}{K} . \tag{4.33}$$

This corresponds to the value for a fully neutralized laminar pinch obtained in section 3.2.7;

$$-K = \left(\frac{\beta_\theta a}{\beta_z}\right)^2 = 2\,\frac{\langle \beta_\theta^2 \rangle}{\beta_z^2} . \tag{4.34}$$

The first and last two terms of equation (4.30) taken together define the beam-spreading curve modified by finite emittance. This has been evaluated for some cases by Walsh (1963) and Weber (1964). A criterion for whether space charge or emittance dominates can be found by comparing the last two terms; at a waist their effect is roughly equal when r' for a ray on the axis is equal to $K^{1/2}$.

4.3.6. Equations of Kapchinskij and Vladimirskij If the beam does not have axial symmetry the situation is more complicated, especially in the presence of a longitudinal field B_s. Under these circumstances the envelope is given by two coupled equations. In the absence of B_s the equations are relatively simple. They are sometimes known as the K-V equations, having been first derived by Kapchinskij and Vladimirskij (1959).

They can be obtained in a straightforward way by specifying a focusing force different in the x- and y-planes, using equation (3.76) to give the self-field associated with an elliptical beam. As originally presented, in the absence of E_s, they are

$$\begin{aligned} a'' + \kappa_x a - \frac{2K}{a+b} - \frac{\varepsilon_x^2}{a^3} &= 0 \\ b'' + \kappa_y b - \frac{2K}{a+b} - \frac{\varepsilon_y^2}{b^3} &= 0 \end{aligned} \qquad (4.35)$$

where κ accounts for the external focusing, K the self-fields, and ε the finite emittance.

Equation (4.35) is the *envelope* equation for a beam in which κ_x and κ_y are functions of s. For a matched beam, with κ_x and κ_y independent of s, the dimensions $x = a$ and $y = b$ can be found by setting $x'', y'' = 0$ and solving for x and y. The *trajectory* equation for particles in such a uniform elliptical beam is then given by

$$\begin{aligned} x'' + \left\{\kappa_x - \frac{2K}{a(a+b)}\right\}x &= 0 \\ y'' + \left\{\kappa_y - \frac{2K}{b(a+b)}\right\}y &= 0 \quad . \end{aligned} \qquad (4.36)$$

The special case of laminar flow, when $x'' = y'' = 0$, for $\kappa_x = (1-n)/R^2$ and $\kappa_y = n/R^2$ was discussed in section 3.4.4.

<u>4.3.7. Complete paraxial envelope equation in reduced variables</u> It is convenient at this point to summarize previous work by writing the paraxial ray envelope equation in normalized variables, including all effects so far studied. In terms of the normalized radius, $A = (\beta\gamma)^{\frac{1}{2}}a$ it may be written (cf. equation 2.40)

$$A'' + \left\{\frac{\gamma'(\gamma^2+2)}{4\beta^4\gamma^4} + \frac{\Omega_L^2}{\beta^2c^2}\right\}A - \frac{\beta\gamma K}{A} - \frac{\varepsilon_n^2}{A^3} = 0. \qquad (4.37)$$

1 2 3 4 5

Many situations have been discussed so far in which only two or three terms are present. These are shown in Table 4.1. The term 5a indicates a beam with finite emittance but laminar flow so that $\varepsilon_n = p_{\theta a}/m_0 c$.

TABLE 4.1

Situations described by particularizations of equation (4.37)

Combination of terms	Physical situation	Section where studied
1, 2	Electrostatic lens	2.2.5
1, 4	Magnetic lens	2.2.6
1, 4	Space charge spreading	3.2.6
1, 5 or 5a	Beam waist	4.3.2
1, 2, 5a	Electrostatic lens system	4.3.2
3, 4	Brillouin flow	3.2.8
3, 4, 5a	Equilibrium flow in axial magnetic field	3.2.8
3, 5	Matched beam in B_z field	4.3.2
4, 5	Uniform pinch	4.3.5
4, 5a	Laminar pinch	3.2.7

4.3.8. Non-linear optical systems; emittance growth So far only systems where the external and self-fields provide linear focusing have been studied. It is evident that non-linearities and aberrations in the optical systems distort the phase-space ellipses. The area of the distorted curves remains constant of course, but for practical purposes (for example, matching a beam into a focusing channel) it is the area of the circumscribing ellipse which is the significant quantity. This, and hence the effective emittance of the beam, is increased by non-linearities.

As an example, we consider a point source in a uniform channel with non-linear focusing such that the wavelength at large amplitudes is decreased. The source emits uniformly over an angle $\pm x_0'$, and its 'emittance plot' at the source is a straight line. As s increases this line curls up as illustrated in Fig.4.8. (This type of behaviour is also evident to a small extent in Fig.4.1.) The actual area

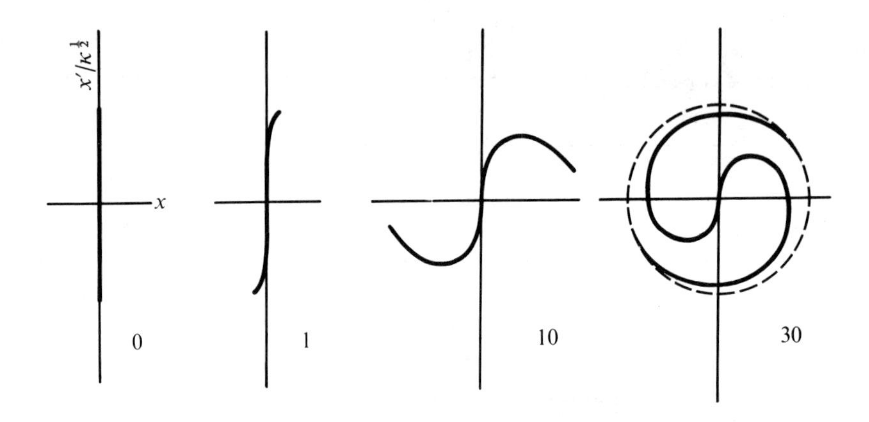

Fig.4.8. Phase-space diagram for a beam launched from a point in a uniform focusing channel with a small non-linearity. The number denotes the number of oscillations made by a particle with small amplitude. Although the line has zero area, after sufficient time the dotted circle is effectively 'filled'.

occupied by the line remains zero, as required by Liouville's theorem, but ultimately the phase points effectively occupy the whole of the dotted circle. The microscopic emittance is still zero, but the macroscopic value has increased to slightly more than $x_0' \kappa^{-1/2}$.

The dilution of phase space in this way, sometimes referred to as 'filamentation', is a feature of non-linear systems; it occurs particularly in phase oscillations (section 2.8.2) where the frequency becomes zero on the separatrix. Phase-space concepts have been profitably applied to the study of phase oscillations, especially in calculating capture conditions into synchrotrons and the stacking of accelerated beams in storage rings. This aspect is treated comprehensively in the book by Lichtenberg (1969). This type of filamentation is familiar also in the description of the development of star clusters moving in their gravitational self-field (see for example Lynden-Bell 1967). In Fig.4.9 the progressive filamentation of the emittance ellipse associated with an array of lenses with spherical aberration is illustrated.

Returning to the example of the beam in the non-linear focusing channel, it is clear that the emittance is zero at $t = 0$ and has a definite finite (macroscopic) value at $t = \infty$; the question arises of how one can best represent this quantitatively as a steady deterioration. One method, as indicated earlier, is to specify the emittance of the circumscribing ellipse. A further alternative is to use the 'r.m.s. emittance', defined by Lapostolle (1971) as

$$\bar{\varepsilon} = 4(\langle x^2\rangle\langle x'^2\rangle - \langle xx'\rangle^2)^{1/2} . \tag{4.38}$$

This form gives a result which is independent of the orientation of the axes and which agrees with the usual definition for a two-dimensional projection of the K-V distribution. For a straight line through the origin $\bar{\varepsilon} = 0$. For a curved line on the other hand, even though the phase area is zero, $\bar{\varepsilon}$ is finite. It is invariant in a linear focusing system, but not otherwise.

To illustrate this we now calculate the way that $\bar{\varepsilon}$ varies after a parallel beam passes through a thin lens with spherical aberration coefficient C_s defined in equation (2.89). If f is the paraxial focal length of the lens, then the equation of a ray passing through the lens at a distance x_0 from the axis is given by

$$\begin{aligned} x &= x_0 + x'z \\ &= x_0 - z\left(\frac{x_0}{f} + \frac{C_s x_0^3}{f^4}\right) . \end{aligned} \tag{4.39}$$

Setting $z = f$ and $x_0 = a$, the radius of the lens, gives

$$x = C_s\left(\frac{a^3}{f^3}\right) = C_s \alpha_i^3$$

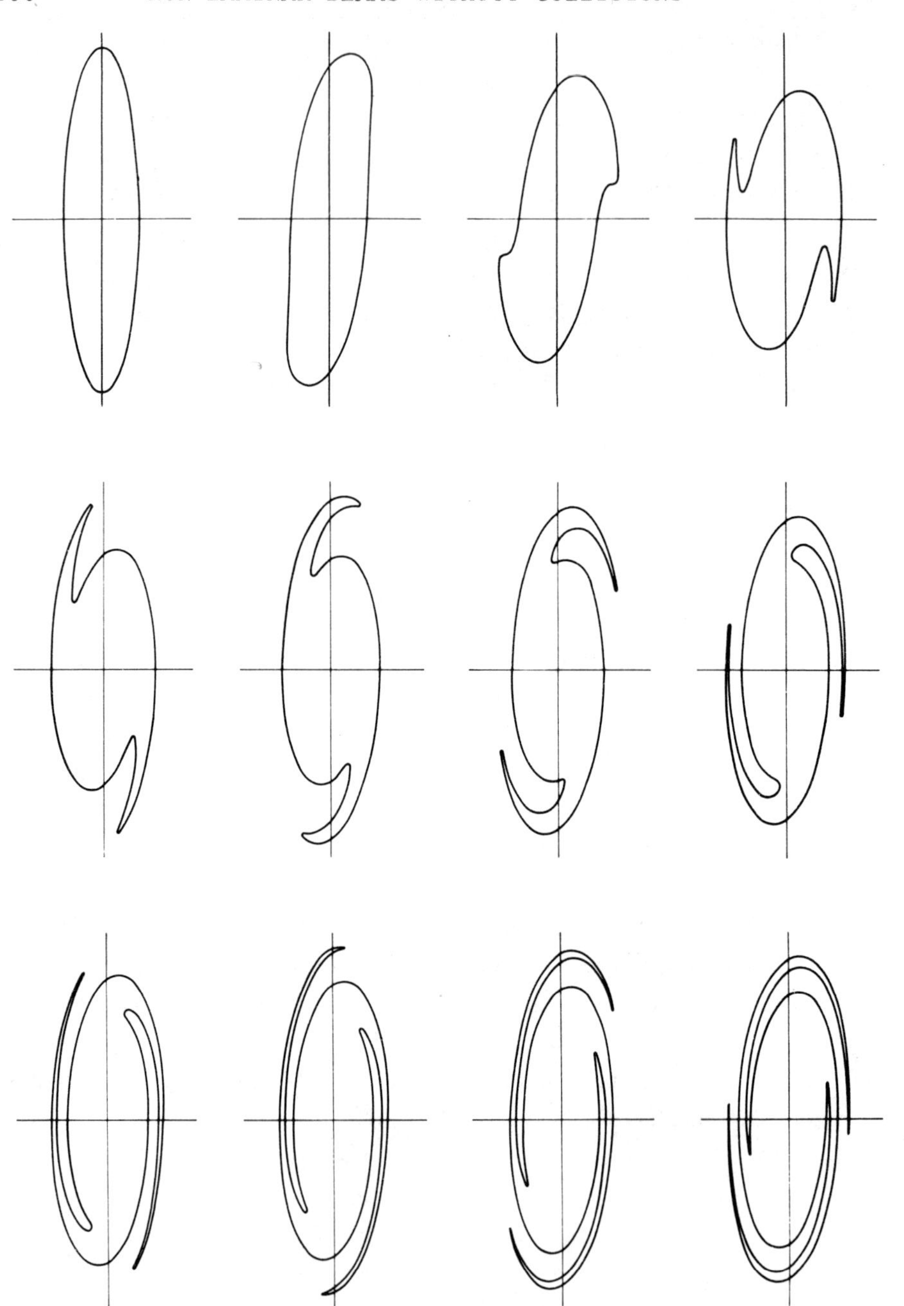

Fig.4.9. Successive distortion of the phase-plane ellipse corresponding to a beam focused by an array of lenses with small spherical aberration. The diagrams are for points along the beam exactly 10 wavelengths apart, so that in the absence of aberrations the diagrams would all be identical ellipses.

as the radius of the aberration disc, in accordance with the definition. This is illustrated in Fig.2.10.

Substitution into equation (4.38) is straightforward for a beam uniformly distributed in x from $x = 0$ to $x = \pm a$ at the plane of the lens, $z = 0$. For such a beam, from equation (4.39),

$$\begin{aligned} \langle x^2 \rangle &= \frac{1}{2a}\int_{-a}^{a}\left\{x_0 - z\left(\frac{x_0}{f} + C_s\frac{x_0^3}{f^4}\right)\right\}^2 \mathrm{d}x_0 \\ \langle x'^2 \rangle &= \frac{1}{2a}\int_{-a}^{a}\left(\frac{x_0}{f} + \frac{C_s x_0^3}{f^4}\right)^2 \mathrm{d}x_0 \end{aligned} \tag{4.40}$$

with a corresponding expression for $\langle xx' \rangle$. Evaluating these expressions to find $\bar{\varepsilon}$ leads to the result

$$\bar{\varepsilon} = \frac{4}{5\surd(21)}\frac{C_s a^4}{f^4} . \tag{4.41}$$

This is independent of z and has the form which might be expected. The radius of the circle of confusion (section 2.2.8) is $r_c = \frac{1}{4}C_s\alpha^3 = \frac{1}{4}C_s a^3/f^3$. Multiplying this by $\alpha = a/f$ gives a quantity of the same order as the emittance $\alpha r_c = \frac{1}{4}C_s a^3 f^3$; comparison with equation (4.38) shows that $\alpha r_c/\bar{\varepsilon} = 5\surd(21)/16 = 1{\cdot}4$.

A practical distribution for which the r.m.s. emittance will be required later (section 4.4.7) is that of a beam with radius a and uniform transverse temperature kT. It may readily be shown from equation (4.38) that

$$\bar{\varepsilon}_{\mathrm{th}} = 2a\left(\frac{kT}{m_0\beta^2c^2}\right)^{1/2} . \tag{4.42}$$

It is interesting to note both the formal and physical analogy between emittance and entropy; both are a measure of disorder. Also, for a beam capable of being focused to form an image, aberrations increase the disorder and give rise to loss of information. This connection has been further explored by Lawson, Lapostolle, and Gluckstern (1973). They show that for a uniform distribution, containing N

particles, the entropy S is given by

$$S = kN(\ln\pi\varepsilon - \ln A) \tag{4.43}$$

where k is Boltzmann's constant and A is the area of the cell in phase space. The entropy is defined in the usual way as $k \log W$, where W is the number of ways that the phase points can be assigned to the cells to produce the given distribution. If the cells are sufficiently small that the variation of phase-space density across a cell is small, then S remains constant by Liouville's theorem. If, however, the filamentation proceeds to a degree such that this is no longer true, then information about the detailed structure is lost and the calculated entropy is higher. (This is illustrated in Fig. 4.10.) In practice the cell size is conveniently determined

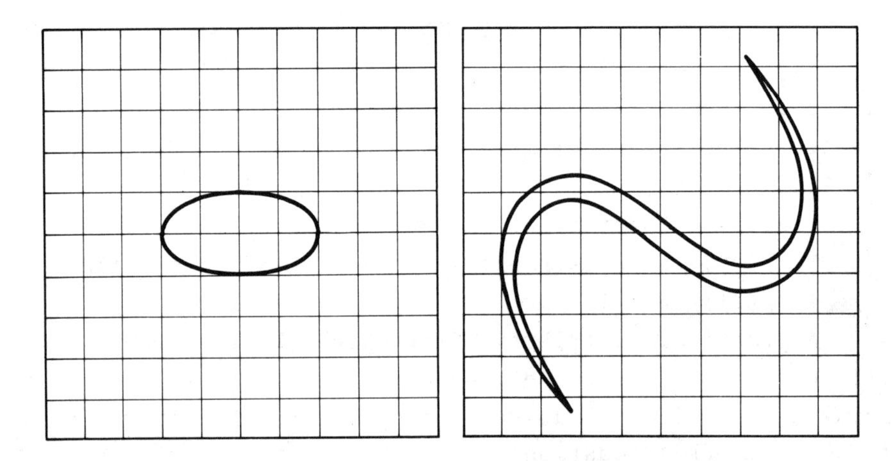

Fig.4.10. Two phase-space distributions with the same area, but, for the cell size shown, different entropy. In the left-hand diagram only 8 cells are occupied; in the right-hand diagram, on the other hand, 25 cells contain some particles. As the cell size tends to zero, the same number of cells is occupied in both cases, and the entropy is the same.

by the resolution of the apparatus used to measure the emittance.

There is further discussion in the paper cited, including a thermodynamic description of a beam as a two-dimensional gas, where the connection between the entropy and adiabatic behaviour of individual particle oscillations is noted.

4.3.9. Systems with non-linear self-field forces Practical beams have non-uniform transverse density profiles. The transverse oscillations of particles in such beams are non-linear, even if the external focusing field is linear, unless the self-field effects are negligible. For matched beams, where the distribution is independent of position along the beam, self-consistent distributions may readily be constructed, for example by the methods to be discussed in section 4.6. A number of such distributions have been found for example by Kapchinskij (1966), Sacherer (1971), and Lapostolle (1971).

The properties of mismatched beams, for which the radius varies along the length, are difficult to calculate whether or not the focusing is uniform. It has, however, been shown by Lapostolle (1971) and Sacherer (1971) that the K-V equation (4.35) applies when the external focusing is linear even if the beam is not of uniform transverse density. The co-ordinates x and y, however, no longer correspond to the co-ordinates of the beam edge but to twice the r.m.s. values of the co-ordinates of the particle, and the emittance is replaced by the r.m.s. value $\bar{\varepsilon}$. For systems with axial symmetry equation (4.30) applies, where a is now interpreted as twice the r.m.s. radius of the density distribution projected on a plane through the axis. This result is of limited value, however, unless we know how $\bar{\varepsilon}$, which is not an invariant, varies with s.

The most interesting non-uniform density distributions are those which correspond to a Maxwellian velocity distribution. A number of problems involving such distributions are discussed in the following sections.

4.R.1-3 Notes and references The material in section 4.2 is merely a summary of standard classical theory to be found in many texts. Much of section 4.3 was evolved in the 1950's in connection with particle accelerators. The concept of emittance was introduced during this period, and its relation to the older optical concepts of brightness soon became apparent. The precise way in which these quantities are defined and related in general situations has given rise to a certain amount of discussion, referred to in the papers quoted in section 4.3.1.

The phase-space concept and its use in accelerators and other devices is discussed extensively in the book by Lichtenberg (1969); matching problems in both transverse and longitudinal (synchrotron) phase-space are discussed in detail, and the dynamical foundations of the theory are discussed in a general way. Rather more practically oriented are the books on beam-transport systems by Steffen (1965) and Banford (1966) which contain extensions of much of the material in this section and in section 2.5. Important contributions in this field have been made by Kapchinskij (1966) in a book unfortunately not yet available in English translation.

The theory of sections 4.3.2 and 4.3.4 is essentially contained in the review by Courant and Snyder (1958) referred to in section 2.R.6-7, though the presentation in section 4.3.2 follows that of Garren (1969).

4.4. Thermal distributions

<u>4.4.1. Introduction; beams and plasmas</u> Charged-particle beams are clearly not in thermodynamic equilibrium. Nevertheless, it is often convenient to consider gaussian distribution functions, and it is helpful to enquire into the meaning of terms such as temperature and pressure, frequently used in plasma physics. A direct link between beams and thermal distributions arises in the formation of electron beams from thermionic cathodes and ion beams from plasmas. In thermodynamics the concepts of temperature and pressure are defined without reference to the microstructure of the media which exhibit them. In non-equilibrium situations, on the other hand, they can only be consistently defined in terms of

the distribution function in momentum or velocity space. This is done in the next section.

It is worthwhile at this point to enquire in what sense a beam may be considered as a special type of plasma. Certainly it is excluded by Langmuir's original definition of the term, as 'that portion of an arc type discharge in which the densities of ions and electrons are substantially equal'. The idea of a plasma has, however, been gradually extended, and is sometimes used to describe any assembly of charges in which the effect of the self-fields is not negligible in comparison with the external fields. An intense beam, especially when viewed in a frame of reference such that the mean particle velocity is zero, often fulfils this criterion. Already the significance of the plasma frequency has become apparent in section 3.2.7, and many examples of characteristic plasma behaviour will be studied in Chapter 6.

An important plasma parameter, already mentioned in connection with Liouville's theorem, (section 4.1) is the Debye length λ_D. This will be encountered again, and it is convenient here to discuss its physical significance. In an ordinary gas a particle in general only interacts with its nearest neighbour; in a plasma, however, because of the long-range nature of the Coulomb force, a particular particle interacts simultaneously with many others. We consider now a single test charge in a stationary plasma, in which electrons and ions each have a Maxwellian velocity distribution corresponding to a temperature kT,

$$f(\underline{v}) = n\left(\frac{m_0}{2\pi kT}\right)^{3/2} \exp\left[\frac{-m_0 v^2}{2kT}\right] . \qquad (4.44)$$

As we shall show, the other particles readjust their positions slightly in the presence of the test charge so as to 'screen off' the field $q/4\pi\varepsilon_0 r^2$ which would otherwise exist. This characteristic screening distance is the 'Debye length' λ_D.

In order to calculate λ_D we make use of an expression due to Boltzmann which describes the effect of an external field on a Maxwellian distribution. In terms of the potential ϕ the Boltzmann relation is

$$n = n_0 \exp\left(\frac{-q\phi}{kT}\right) . \tag{4.45}$$

This shows that the form of the distribution is unchanged, but the density decreases rapidly as the potential increases. (A familiar application of this expression is to an isothermal neutral atmosphere in a gravitational field.) The proof of equation (4.45), which is true also in the presence of a magnetic field, is mostly conveniently demonstrated with the aid of the Vlasov equation. It is accordingly left until section 4.6.2.

To find λ_D we solve equation (4.45) simultaneously with Poisson's equation. The test charge is assumed positive, the ions are taken as singly charged, and the ion and electron densities n_i and n_e are both equal to n_0 in the absence of the test charge. Applying equation (4.45) to the ions and electrons separately and substituting in Poisson's equation yields

$$\nabla^2 \phi = -\frac{q}{\varepsilon_0}(n_i - n_e) = \frac{2qn_0}{\varepsilon_0} \sinh \frac{q\phi}{kT} \tag{4.46}$$

Equation (4.46) can be solved in closed form when $q\phi \ll kT$, which is true except close to the charge. It may be written

$$\nabla^2 \left(\frac{q\phi}{kT}\right) = \frac{2(q\phi/kT)}{\lambda_D^2} \tag{4.47}$$

where

$$\lambda_D^2 = \frac{\varepsilon_0 kT}{n_0 q^2} . \tag{4.48}$$

The potential varies with r as

$$\begin{aligned} \phi &= \frac{q}{4\pi\varepsilon_0 r} \exp\left(\frac{\sqrt{2}r}{\lambda_D}\right) \qquad (r \gg \lambda_D) \\ &= \frac{q}{4\pi\varepsilon_0 r} \qquad (r \ll \lambda_D) . \end{aligned} \tag{4.49}$$

From this the role of λ_D as a shielding distance can be seen. The number of particles within a 'Debye sphere'

$$N_{\mathrm{p}} = 2n_0 \frac{4}{3} \pi \lambda_{\mathrm{D}}^3 \tag{4.50}$$

is sometimes called the 'plasma parameter'. When $N_{\mathrm{p}} \gg 1$, the effect of the smooth part of the self-field on the motion greatly exceeds that of individual particle scattering, so that, as indicated in section 4.1, Liouville's theorem is a good approximation.

Particle beams frequently have a length which is large compared with λ_{D}, but transverse dimensions which are small. Some consequences of this fact will become apparent when wave motion on beams is studied (section 6.4.7).

4.4.2. Pressure, temperature, and emittance The full distribution function of the particles in a beam or plasma often contains more information than is needed, so that average values are of interest. For a non-relativistic system, the average velocity at a point is known as the fluid or drift velocity $\bar{\mathrm{v}}$, and the second velocity moment defines the pressure tensor p_{ij}. In terms of the local velocity-distribution function $f(v_i)$ normalized such that $\int f(v_i)\mathrm{d}^3v = n$ and $\int v_i\, f(v_i)\mathrm{d}^3v = n\bar{v}_i$ the pressure tensor is defined as

$$\begin{aligned} p_{ij}(x) &= m_0 \int (v_i - \bar{v}_i)(v_j - \bar{v}_j) f \mathrm{d}^3 v \\ &= m_0 \left(\int v_i v_j f \mathrm{d}^3 v - n\bar{v}_i \bar{v}_j \right). \end{aligned} \tag{4.51}$$

In the theory of beams only systems in which this tensor is diagonal are of interest. In general, however, the three diagonal terms are not equal, and these are often referred to as the pressures in a particular direction.

In axially symmetrical systems p_x and p_y are denoted by $p_\perp$, and p_z by $p_\parallel$. The temperature associated with one of these pressure components is defined as

$$kT_i = \frac{1}{n} m_0 \int (v_i - \bar{v}_i)^2 f(v_i) \mathrm{d}v_i. \tag{4.52}$$

It may readily be shown by inserting the one-dimensional Maxwellian distribution that

$$f(v_i) = n\left(\frac{m_0}{2\pi kT}\right)^{1/2} \exp\left\{\frac{m_0(v_i-\bar{v}_i)^2}{2kT}\right\} \tag{4.53}$$

that the corresponding pressure is given by

$$p_i = nkT_i \ . \tag{4.54}$$

When the drift velocity v_i is zero, the kinetic energy density of the particles may be found by averaging over the square of the velocity distribution. It is equal to $\frac{1}{2}\sum_i nkT_i$. The quantity kT is therefore an order-of-magnitude measure of the energy of a typical particle; in a non-relativistic beam observed in a reference frame with velocity $\bar{v}$, the energy densities associated with longitudinal and transverse temperatures are $\frac{1}{2}nkT$ and nkT respectively,

The treatment so far has been non-relativistic; a relativistic generalization of the pressure tensor is in general complicated. An essential feature is that velocity is no longer proportional to momentum, so that momentum flux cannot be identified with energy.

In most relativistic beams of practical importance, however, transverse velocities are non-relativistic and γ is almost the same for all particles, so that it is possible to write $\frac{1}{2}\gamma m_0 v_\perp^2 = kT_\perp$. Seen in the frame of reference moving with the particles

$$kT_{\perp m} = \tfrac{1}{2}m_0(\gamma v_\perp)^2 = \gamma kT_\perp . \tag{4.55}$$

This result follows from the fact that the transverse momentum $\gamma m_0 v_\perp$ is Lorentz invariant. In the longitudinal direction it is readily shown from the relativistic law of addition of velocities that, in the limit of $\gamma \gg 1$, two particles with energies γ and $\gamma + \Delta\gamma$ in the laboratory frame have relative velocities β and $\beta + \Delta\beta$ in the moving frame, where $\Delta\beta = \Delta\gamma/\gamma$. Beams with a small fractional energy spread, even if highly relativistic, appear therefore to have a non-relativistic temperature when observed in the moving frame. This point is examined in more detail in section 5.5.

It is interesting to calculate the transverse temperature $T_\perp$ of the K-V distribution specified by equation (4.26). For simplicity we consider a cylindrical beam with emittance ε and radius a. From equation (4.26) the distribution function may be written

$$\delta\left(\frac{r^2}{a^2} + \frac{a^2(\dot{x}^2 + \dot{y}^2)}{\beta^2 c^2 \varepsilon^2} - 1\right) = 0 \qquad (4.56)$$

where $r^2 = x^2 + y^2$, $\dot{x} = \beta c x'$, and β refers to the velocity of the particles in the z direction assumed constant and uniform. The transverse energy $\frac{1}{2}\gamma m_0(\dot{x}^2 + \dot{y}^2)$ of the particles is a function of r only; substituting $\frac{1}{2}\gamma m_0(\dot{x}^2 + \dot{y}^2) = kT_\perp$ yields

$$kT_\perp = \frac{\beta^2 \gamma m_0 c^2 \varepsilon^2}{2a^2}\left(1 - \frac{r^2}{a^2}\right) \qquad (4.57)$$

on the beam axis $kT_\perp \propto \varepsilon^2/a^2$, falling to zero at the beam edge.

This is obviously rather an academic distribution, as noted earlier; nevertheless, it illustrates the fact that for any finite distribution $T_\perp$ is necessarily zero at the beam edge, since all the particles there have no radial velocity. Because of the relation between temperature and pressure, this analysis suggests that the finite emittance can be alternatively interpreted as indicating a pressure gradient tending to disperse the beam. Thus the uniform pinch represented by equation (4.33) represents a balance between an outward pressure and an inward $\underline{j} \times \underline{B}$ force. The equivalence of the hydrodynamic and optical descriptions of a beam are further explored by Lawson (1975).

In contrast to the above result, the temperature of the rotating laminar flow with emittance ε described in section 4.3.5 is zero. We can conveniently, however, define a 'temperature' $k\overline{T}$ which, instead of applying to the distribution in velocity space at a point, includes all the particles in the cross section of the beam. This quantity which must be distinguished from the temperature averaged across the beam, turns out to be the same ıor both the K-V and laminar distributions. The integration is straightforward and leads to a value

$$k\overline{T}_i = \frac{\beta^2\gamma m_0 c^2\varepsilon^2}{4a^2} . \tag{4.58}$$

Combining this result with the condition for a uniform pinch given in equation (4.34) and substituting $K = -2\nu/\gamma$, $\nu = \mu_0 Nq^2/4\pi m_0$, and $I = Nq\beta c$ yields

$$\frac{\mu_0 I^2}{4\pi} = 2Nk\overline{T} . \tag{4.59}$$

This is the Bennett pinch relation (section 3.2.7), which will be encountered again in section 4.4.5.

Alternative models may be constructed in which the true transverse temperature is independent of radius; this naturally implies a distribution with a 'tail' which never reaches zero. Two models of this type are discussed in the next sections.

4.4.3. Focused beams with finite temperature In section 4.3 the properties of beams with finite perveance and emittance in a paraxial focusing system were analysed. By virtue of the very special distribution function chosen, the equations governing the transverse motion of the particles are of linear form. Some of the complications associated with non-linearities, arising either from the external focusing system or non-uniformities in the charge and current distribution across the beam, were discussed in sections 4.3.8 and 4.3.9. In the remainder of the present section beams with a transverse velocity distribution which is Maxwellian will be discussed.

The problem of finding self-consistent distributions appropriate to a matched beam in a uniform focusing channel or pinch is not difficult. For a mismatched beam, on the other hand, or a more general focusing system which varies with s, it is not easy to find such distributions when self-fields are taken into account; in many practical situations, however, approximations based on the linear model are adequate. The simpler problem of finding how the transverse density distribution of a beam with negligible space charge varies along its length, though tractable, is of limited value, since in practice the effect of aberrations cannot be neglected. It is instructive, however, and will be studied in section 4.4.7.

4.4.4. Matched beam in a uniform external linear focusing field We now study the configuration of a beam with a transverse velocity distribution which is Maxwellian, travelling in a uniform focusing channel in which the external focusing is linear. Self-field effects produce an overall focusing force on the particles which is non-linear. The beam is assumed to be matched, so that there is no variation in the direction of propagation. The problem reduces therefore to that of finding a self-consistent two-dimensional distribution in a potential well. This requires the Boltzmann relation (equation (4.45)) and Poisson's equation.

The potential well could ideally be produced by a uniform stationary background of heavy ions; alternatively it might represent a uniform axial magnetic field viewed in the Larmor frame. Because of the form of the Boltzmann relation the distribution of transverse velocities is Maxwellian at all points in space. At one extreme, where the density is low and space-charge forces weak, the particles form an 'atmosphere' in a potential well. At the opposite extreme, where the temperature is zero, the charge just 'fills up' the potential well to form a uniform sharp-edged cloud with zero electric field inside. If the focusing field is provided by stationary ions, then the beam is just a cold plasma with drifting electrons. Alternatively, the theory applies to beams in an axial magnetic field which in the limit of zero temperature exhibit Brillouin flow (Pierce and Walker (1953)).

In the above discussion we have not considered the effect of the self-magnetic field. In paraxial approximation B_θ is proportional to E_r, and, as calculated in section 3.2.6, the effect of the magnetic field is merely to reduce the force from the electric field alone by $(1-\beta^2)$; this is the same as neglecting the magnetic field and assuming a transverse mass of $\gamma^3 m_0$. In this section we shall treat the problem non-relativistically, but recall that for relativistic z velocities it is only necessary to replace m_0 by $\gamma^3 m_0$.

If the gradient $\partial E_r/\partial r$ of the radial focusing field associated with the external potential well is constant, then the potential arising from the external field is

$$\phi_{\text{ext}} = -\int \frac{\partial E_r}{\partial r}\,\mathrm{d}r = -\tfrac{1}{2}\frac{\partial E_r}{\partial r}\,r^2 . \tag{4.60}$$

The potential associated with the space charge is

$$\phi_{\text{s.c}} = -\frac{q}{\varepsilon_0}\int_0^r \frac{1}{r}\int_0^r n(r_1)r_1\,dr_1\,dr . \tag{4.61}$$

Inserting the total value of ϕ into the Boltzmann relation (section 4.45) yields the integral equation for n:

$$n(r) = n_0 \exp\left[\left(\frac{q}{kT}\right)\left\{\tfrac{1}{2}\frac{\partial E}{\partial r}r^2 + \frac{q}{\varepsilon_0}\int_0^r \frac{1}{r}\int_0^r n(r_1)r_1\,dr_1\,dr\right\}\right] . \tag{4.62}$$

This can be simplified by a change of notation. We introduce dimensionless quantities

$$\begin{aligned} R &= \frac{r}{\lambda_D} = r\left(\frac{n_0 q^2}{\varepsilon_0 kT}\right)^{1/2} \\ D &= \frac{n}{n_0} \\ D_1 &= \frac{n_1}{n_0} = -\frac{2\varepsilon_0}{n_0 q}\frac{\partial E}{\partial r} . \end{aligned} \tag{4.63}$$

This implies that distances are measured in Debye lengths. It is readily verified that n_1 is the density which the beam would have if the temperature were zero. With these new variables equation (4.62) becomes

$$D(R) = \exp\left\{-(\tfrac{1}{4}D_1 R^2 - \int_0^R \frac{1}{R}\int_0^R D(R_1)R_1\,\mathrm{d}R_1\,\mathrm{d}R\right\} . \tag{4.64}$$

This determines the density profile (D as a function of R)

for various values of the parameter $D_1 = n_1/n_0$.

It is more instructive to present the results as a function of radius for a given total number of charges per unit length, $N = \int 2\pi nr\,dr$, with kT as variable. This is done in Fig.4.11, with the temperature expressed in units of

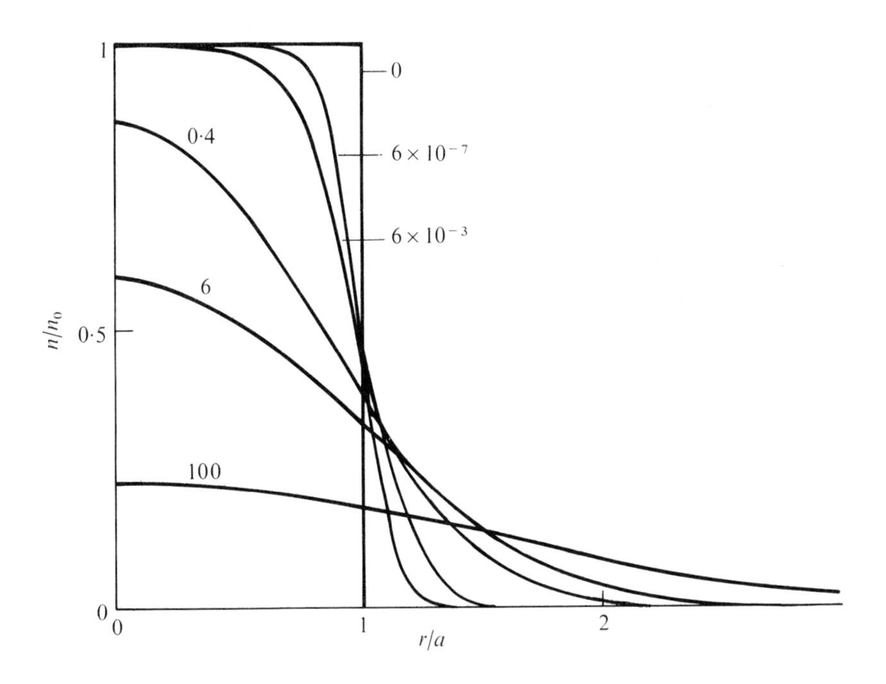

Fig.4.11. Density distribution across a matched beam in a uniform focusing channel in the presence of thermal transverse velocities. The numbers on the curves are $kT/\nu m_0c^2$; this is the ratio of the thermal energy to the potential energy at the radius which corresponds to the beam edge when $kT = 0$.

$kT/\nu m_0c^2$, where ν is Budker's parameter. The parameter $kT/\nu m_0c^2$ can also be expressed as $(2\lambda_D/a)$; it will be seen that $\lambda_D \approx a$ represents the transition from a 'square' $(\lambda_D = 0)$ to 'gaussian' $(\lambda_D = \infty)$ type solution. These units express the ratio of the kinetic energy to the potential energy of the external field, measured at the radius $r = a$ which the beam assumes when $kT = 0$.

4.4.5. Self-focused beams; the Bennett pinch Beams in which the effect of finite emittance is balanced by the inward force arising from the self-magnetic field were discussed in section 4.3.5 for velocity distributions which give rise to a uniform transverse charge and current distribution in the beam. We now find a self-consistent solution for a beam in which the transverse velocity distribution is Maxwellian, following in spirit but not in detail the original calculation by Bennett (1934).

Two components are assumed. Ions, with charge q_i, have no component of velocity along the beam and electrons move with z-velocity $\beta_z c$, which is assumed large compared with the transverse velocity. The neutralization fraction f, equal to $-n_i q_i / n_e q_e$, is assumed independent of beam radius. This is reasonable for a fully neutralized beam, but a somewhat artificial assumption when f is less than unity. The density n is given by equation (4.45) where ϕ is the transverse potential associated with the radial forces on the ions and electrons. In a non-relativistic beam ϕ is the electrostatic potential; for a relativistic beam, however, it may be generalized to take into account the radial magnetic forces also, and is thus different for electrons and ions. The electron and ion temperatures are assumed to be different, and equal to kT_e and kT_i respectively.

The potential at radius r is

$$\begin{aligned} \phi_e &= -\int_0^r (E_r - \beta_z c B_\theta)\,\mathrm{d}r \\ \phi_i &= -\int_0^r E_r\,\mathrm{d}r\,. \end{aligned} \tag{4.65}$$

From equation (4.45), dropping subscripts on E, B, and β,

$$\begin{aligned} \frac{1}{n_e}\frac{\mathrm{d}n_e}{\mathrm{d}r} &= \frac{q_e}{kT_e}(E - \beta c B) \\ \frac{1}{n_i}\frac{\mathrm{d}n_i}{\mathrm{d}r} &= \frac{q_i E}{kT_i}\,. \end{aligned} \tag{4.66}$$

From Poisson's equation and Ampère's law

$$\frac{1}{r}\frac{\mathrm{d}}{\mathrm{d}r}(rE) = \frac{(q_e n_e + q_i n_i)}{\varepsilon_0}$$
$$\frac{1}{r}\frac{\mathrm{d}}{\mathrm{d}r}(rB) = \mu_0 q_e \beta c n_e. \qquad (4.67)$$

Eliminating B and E yields

$$\frac{1}{r}\frac{\mathrm{d}}{\mathrm{d}r}\left(\frac{r}{n_e}\frac{\mathrm{d}n_e}{\mathrm{d}r}\right) = \frac{n_e q_e^2}{\varepsilon_0 kT_e(1-\beta^2-f)}$$
$$\frac{1}{r}\frac{\mathrm{d}}{\mathrm{d}r}\left(\frac{r}{n_i}\frac{\mathrm{d}n_i}{\mathrm{d}r}\right) = \frac{n_i q_i^2}{e_0 kT_i(1/f-1)} \qquad (4.68)$$

Since n_e is assumed proportional to n_i at all radii, these equations may be readily solved to yield the radial density profiles

$$n_e = \frac{n_{0e}}{(1+r^2/a^2)^2}, \quad n_i = \frac{n_{0i}}{(1+r^2/a^2)^2} \qquad (4.69)$$

and the temperatures

$$4\pi\varepsilon_0 kT_e = -\tfrac{1}{2} q_e(q_i N_i + q_e N_e/\gamma_e^2)$$
$$4\pi\varepsilon_0 kT_i = -\tfrac{1}{2} q_i(q_e N_e + q_i N_i) \qquad (4.70)$$

where a is an arbitrary length, and $N = \int_0^\infty 2\pi n r \mathrm{d}r = \pi a^2 n_0$.

Addition of the two equations (4.70) yields Bennett's pinch relation for singly charged ions, in which $q_e = -q_i$:

$$N_e q_e^2 \beta^2 = 8\pi\varepsilon_0 k(T_e + T_i) \qquad (4.71)$$

or, in more familiar form,

$$\frac{\mu_0 I^2}{4\pi} = 2N_e k(T_e + T_i). \qquad (4.72)$$

With T_i set equal to zero this is identical to equation (4.59) for a laminar pinch, with kT_e in place of $k\overline{T}$.

The theory of this section describes a self-consistent system, with no indication of how to produce it in practice. When collisions are included it may be shown that this represents an equilibrium distribution, to which other initial distributions tend. The theory and experimental confirmation are referred to in section 5.3.

Equations (4.68) and (4.69) were also found by Thonemann and Cowhig (1951) for a neutral gas discharge, using the concept of pressure. This has not been used here, though equation (4.66) could have been written

$$\frac{\mathrm{d}p_{\perp}}{\mathrm{d}r} = \frac{\mathrm{d}}{\mathrm{d}r}(n_{\mathrm{e}}kT) = n_{\mathrm{e}}q_{\mathrm{e}}(E-\beta cB). \tag{4.73}$$

Thonemann and Cowhig did not make the assumption in their calculation that $\beta_z \gg \beta_{\perp}$; indeed, in their experiment this condition was certainly not satisfied. Probe measurements gave good agreement with the profiles specified by equation (4.69).

The pinch profile, equation (4.69), can be generalized if a filamentary current flows along the axis (Benford, Book, and Sudan 1970). Under these conditions the density profile of electrons or ions is of the form

$$\frac{n}{n_0} \propto \left[\frac{r}{a}\left\{\left(\frac{r}{a}\right)^{\eta} + \left(\frac{a}{r}\right)^{\eta}\right\}\right]^{-2} \tag{4.74}$$

where η is a positive quantity. The ratio of the current on the axis, I_{a}, to the beam current, I_{b}, is

$$\frac{I_{\mathrm{a}}}{I_{\mathrm{b}}} = \frac{1-\eta}{2\eta} \;. \tag{4.75}$$

When $\eta < 1$ this is positive, and the density is more peaked than in equation (4.69). When $\eta = 1$, $I_{\mathrm{a}} = 0$, and equation (4.69) is obtained. When $\eta > 1$, on the other hand, the current on the axis is oppositely directed to the main current and the density on the axis is zero, giving a hollow beam. Profile curves are plotted in the original paper, and further properties of this type of distribution, in particular the limiting currents obtainable, are explored in the review by Benford and Book (1971). Other pinch models are discussed in this review, and in the book by Davidson (1974).

4.4.6. The planar diode with finite emission velocities In this section we study an example of a beam where a velocity distribution in the direction of propagation plays an important part in determining its detailed behaviour. Emission from a planar diode in the absence of thermal velocities was studied in section 3.2.2. We now review very briefly the more realistic situation when thermal velocities are present. If the emitter is a thermionic cathode, the situation is fairly well defined; if, on the other hand, it is a plasma or operates by the 'field-emission' mechanism the behaviour is somewhat more complicated. We do not attempt a comprehensive or detailed treatment of this extensive field, but merely indicate the main features for future reference.

Although the physical situation in a space-charge limited planar diode may be simply defined, a calculation of the potential distribution between the electrodes and the velocity distribution at any point is by no means straightforward. The type of behaviour to be expected, however, can be anticipated by considering the conditions in the neighbourhood of an isolated plane cathode, and then deducing the effect of applying an external field. Such a cathode is a conductor which contains free electrons, with density and velocity distribution dependent on its structure. These electrons are prevented from escaping by an effective potential barrier at the surface characterized by the 'work-function' ϕ_w which is typically a few electron volts; it is a complicated function of the structure of the conductor and possibly also of temperature. For crystalline materials it depends also on the orientation of the surface. When materials with sufficiently low work function are heated, electrons can penetrate the potential barrier to provide a useful flow of current. Typical emitters in common use are thoriated tungsten, barium and strontium oxides, and lanthanum hexaboride.

We consider first a single, hot, isolated, planar metal surface *in vacuo*. Under steady-state conditions electrons which are emitted must necessarily remain in the neighbourhood of the plate. Initially some may escape, though if this occurs the plate becomes progressively more positively charged until all electrons are turned back. A 'sheath' with thickness of order of a Debye length λ_D is formed near the plate. The variations of potential and number density in this sheath can be found from Poisson's equation and the Boltzmann relation (equation (4.45)). These give

$$\frac{d^2\phi}{dz^2} = \frac{2qn_0}{\varepsilon_0} \exp\left(-\frac{q\phi}{kT}\right) \tag{4.76}$$

which has the solution

$$\frac{q\phi}{kT} = -\ln\frac{z^2}{2\lambda^2} + 1 \tag{4.77}$$

where $\phi = 0$ at $z = 0$, the conductor surface. The potential decreases monotonically with z, becoming logarithmically infinite at infinity where the 'infinite-energy' tail of the Maxwellian is reflected.

If now a second non-emitting conducting plate is placed at $z = d$, maintained at a constant potential with respect to the hot cathode, then electrons will flow from cathode to anode. If the potential of the anode ϕ_a is above that of the cathode, there will be a potential minimum ϕ_m between anode and cathode. The current and potential configuration depend on n_0, kT, and d. For $z < z_m$, where z_m is the co-ordinate of the potential minimum, there are electrons with both positive and negative values of $\dot{z}$; for $z > z_m$, $\dot{z}$ is always positive. If $\phi_a \gg \phi_m$, then the potential minimum is near the cathode and Child's law (equation (3.23)) is a good approximation. This is true for most beams of practical interest, since $\phi_m \approx kT/q < 1$ V.

The accurate analysis of this problem has given rise to many papers. Reference to these, together with a very detailed discussion, may be found in the book by Birdsall and Bridges (1966) and the review by Lindsay (1960). A typical solution is sketched in Fig.4.12.

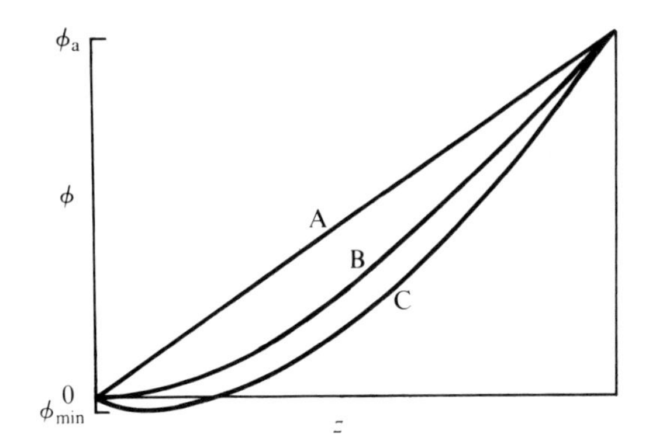

Fig.4.12. Potential distribution across a planar diode in the absence of current (curve A), under space-charge-limited conditions with $kT = 0$ (curve B), and the presence of thermal velocities (curve C).

Sources of ions are essentially more complicated than electron guns; the emitter is a gaseous plasma rather than a solid and there are limitations to the surface shape which can be obtained. The analogue of the planar diode represents a rather unrealistic idealization, and ion-source design in general is somewhat more empirical and less obvious than electron-gun design.

4.4.7. Beam in a general linear external focusing system In section 4.4.4. the properties of a uniform or 'matched' beam with thermal velocity distribution in a linear focusing field was studied. If a beam with a different density profile is launched, for example from a cathode, then the distributions both in x and x' vary along the system. For the general situation, in which space charge is important, it is difficult to calculate how the beam develops. Analytical methods are not applicable, but step-by-step computer methods have been used. A typical calculation of this class, though not with a thermal velocity distribution, has been made by Lapostolle

(1971). In this example an axially symmetrical distribution corresponding to a uniformly filled hyper-ellipsoidal shell in four-dimensional phase space is fed into a uniformly focusing channel. The spatial projection of such a distribution may be shown to be parabolic:

$$\begin{aligned} n &= n_0\left(1-\frac{r^2}{a^2}\right)^{1/2} && r<a \\ &= 0 && r>a \end{aligned} \qquad (4.78)$$

The subsequent behaviour for a given emittance depends on the focusing strength of the channel. For a ratio of 2·3 between the beam size and that of a matched beam of the same emittance a gradual growth of r.m.s. emittance to about 2·3 times the original value was found. This might be expected from the phenomenon illustrated in Fig.4.9.

In the absence of space charge, it is possible to find an analytical form for the variation along the beam of the transverse distribution of density across it, in terms of the known distribution at some point and the principal solutions of the paraxial ray equation. We examine first the special problem of a beam in a uniform focusing channel which originates from a circular uniformly emitting cathode. This is followed by a more general treatment applicable to a focusing system which varies with z.

The system to be analysed first is shown in Fig.4.13. Electrons are emitted uniformly and isotropically from a cathode of radius r_c. At the cathode the paraxial equation is not valid, since the transverse velocities are of the same order as the longitudinal velocities and furthermore there is a continuous spectrum of z velocities ranging from zero to several times $(kT/m_0)^{1/2}$. In the presence of an accelerating field E_z, however, the motion becomes paraxial at a distance z_1 such that $z_1 \gg kT/qE_z$. In practical systems, where kT is of the order of 1 eV or less, this distance is small. If it is small compared with the cathode radius, then it is a fair approximation to use the paraxial ray equation throughout. Under these circumstances the change in radius in moving from $z = 0$ to $z = z_1$ is small, and the distribution function for x' is Maxwellian corresponding to a cathode temperature kT. The normalized distribution in x and x' near such a cathode may be written

$$f_1(x) = (r_c^2 - x^2)^{1/2} / \pi r_c^2 \tag{4.79}$$

$$f_2(x') = \left(\frac{m_0 \beta^2 c^2}{2\pi kT}\right)^{1/2} \exp\left(-\frac{x'^2 m_0 \beta^2 c^2}{2kT}\right) . \tag{4.80}$$

The mean-square values of x and x' are $r_c^2/4$ and $kT/m_0\beta^2c^2$ respectively, and $\langle xx' \rangle = 0$, so that from equation (4.38) the r.m.s. emittance is

$$\bar{\varepsilon} = \beta\bar{\varepsilon}_n = 4(\langle x^2 \rangle \langle x'^2 \rangle)^{1/2} = 2r_c\left(\frac{kT}{m_0\beta^2c^2}\right)^{1/2} . \tag{4.81}$$

The phase-space distribution distribution corresponding to equations (4.79) and (4.80) is shown in Fig.4.13.

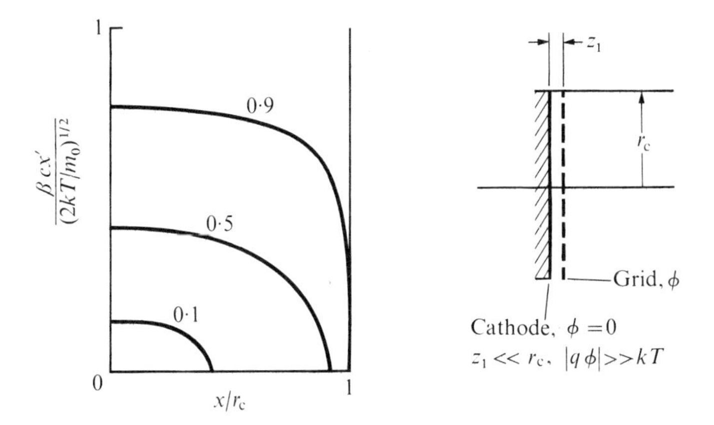

Fig.4.13. Emittance diagram and orbits for a beam accelerated from a circular cathode of radius r_c and temperature kT by a grid placed close to it. The curves are symmetrical about the x and x' axes; the numbers denote the fraction of the beam within the contours.

The beam profile can be calculated by the use of equation (4.30), where r is interpreted as twice the r.m.s.

radius of the current-density distribution projected on a plane through the axis. In the absence of self-fields there is no net increase of emittance along the beam; in the presence of the non-linear forces arising from self-fields, however, $\bar{\varepsilon}_n$ will in general increase with z. For relatively short systems (such as cathode-ray tubes) it is probably nevertheless a good approximation to keep $\bar{\varepsilon}_n$ constant. This is indicated by calculations of Weber (1964) who derived the non-relativistic version of equation (4.30) with $\bar{\varepsilon} = r_c kT/\frac{1}{2}m\beta^2c^2$, where r_c is the effective cathode radius defined as the total current divided by π times the current density at the centre of the cathode.

We study now the behaviour of a beam launched from a cathode through a grid into a focusing channel, as illustrated in Fig.4.13. Self-fields are neglected, and the projections of the particle orbits on a plane through the axis are therefore sinusoidal with wavelength $\lambda = 2\pi\kappa^{-1/2}$. At a distance $\lambda/4$ from the cathode, the emittance diagram has rotated through $\pi/2$, the spatial distribution is gaussian, and the distribution in x' varies as $(r_c^2 - x^2)^{1/2}$. After a further $\lambda/4$ the conditions at the cathode are repeated. Such values of z may be termed 'cross-over planes' and 'image planes'. This behaviour is illustrated in Fig.4.14. The

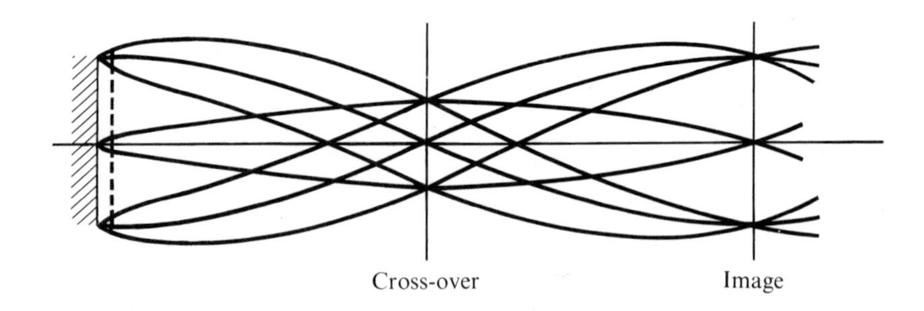

Fig.4.14. Orbits in a uniform focusing channel from the source illustrated in Fig.4.13 showing cross-over and image planes.

current density in the cross-over plane can be found by

replacing x' by $\kappa^{1/2}r$ in equation (4.80) and renormalizing f_2 in such a way that

$$\int f_2(\kappa^{1/2}r)\,2\pi r\,\mathrm{d}r \quad = \quad \pi r_c^2 i_c$$

where i_c is the (uniform) current density at the cathode. Denoting the current at the crossover by i_x, this yields

$$\frac{i_x}{i_c} = \frac{\kappa r_c m_0 \beta^2 c^2}{2kT} \exp\left(-\frac{\kappa r^2 m_0 \beta^2 c^2}{2kT}\right). \tag{4.82}$$

The ratio of the peak current density at the cross-over to that at the cathode is

$$\left(\frac{i_x}{i_c}\right)_{r=0} = \kappa r_c^2 \frac{m_0 \beta^2 c^2}{2kT} = \frac{2\kappa r_c^4}{\bar{\varepsilon}^2}. \tag{4.83}$$

The channel therefore concentrates the beam if $\kappa^{1/2} r_c$ exceeds the ratio of the square of thermal to longitudinal velocities, $2kT/m_0\beta^2c^2$; if this condition does not hold, the beam is more diffuse at the cross-overs than at the images.

This example has been chosen to indicate the type of behaviour to be expected from thermal beams without space charge. Since $\bar{\varepsilon}_n$ remains constant it is clear from equation (4.83) that the maximum current compression possible is of order $kT/m_0\beta^2c^2$. A more accurate calculation of this limit cannot be made within the confines of the paraxial approximation. We return to this problem in the next section.

The beam just discussed is essentially a description in the Larmor frame of a beam launched into a magnetic field from a shielded gun, illustrated (in idealized form) in the central diagram in Fig.3.4. A large number of experimental studies of beams of this class have been made. The behaviour is very much more complicated than might be expected from the analysis presented above; a very pronounced radial structure is often found which varies continuously in a quasi-periodic manner along the beam. This is associated with the essential aberrations of the gun system, which produce groups of crossing electrons and caustics in the beam.

A number of papers in which these effects are studied and analysed, which contain some very elegant diagrams and pin-hole photographs, were published between 1957 and 1959. These are by Harker (1957), Webster (1957), Herrmann (1958), Ashkin (1958), Brewer (1959), and Johnston (1959). Since that date steady progress has been made; a near perfect Brillouin beam is described by Mihran and Andal (1965).

So far we have considered a beam in a uniform focusing channel. The extension to a general system in which the focusing is a function of z has been made by Kirstein, and is described in the book by Kirstein *et al.* (1967). The focusing is defined by the principal solutions of the paraxial equations. Kirstein uses time as the independent variable. The principal solutions measured in the Larmor frame are $M(t)$ and $S(t)$, with initial conditions $(M,\dot{M}) = (1,0)$ and $(S,\dot{S}) = (0,1)$. The cathode has radius r_c and temperature T. To make the expression more compact the following quantities are introduced:

$$
\begin{aligned}
R &= \rho(M^2 + S^2\Omega_{Lc}^2)^{1/2} \\
\sigma &= \left(\frac{kT}{m}\right)^{1/2} S
\end{aligned}
\qquad (4.84)
$$

where Ω_{Lc} is the Larmor frequency corresponding to the magnetic field at the cathode. Both R and σ have the dimensions of length; they represent respectively the radial positions of a particle which left the cathode normally at a radius ρ and a particle which left the centre of the cathode with transverse velocity $(kT/m)^{1/2}$. The derivation of the required expression is straightforward but rather lengthy; it may be found, in slightly different form, in Kirstein *et al* (1967)[†] or Amboss (1969). The radial current distribution at a point z, specified by the values of M and S there, is

---

[†]There is a typographical error in this reference. Delete r^2 in the final bracket of equation (5.22).

$$\frac{i_z(r)}{i_c} = \left(\frac{r_c}{R_c}\right)^2 \exp\left(-\frac{r^2}{2\sigma^2}\right) \int_0^{R_c/\sigma} \frac{R}{\sigma} \exp\left(-\frac{R^2}{2\sigma^2}\right) I_0\left(\frac{rR}{\sigma^2}\right) \mathrm{d}\left(\frac{R}{\sigma}\right) \tag{4.85}$$

where R_c is that value of R for which $\rho = r_c$, and I_0 is the modified Bessel function of zero order. The integral was first derived by Cutler and Hines (1955) in the study of thermal effects in electron guns; Kirstein's work, however, has given it wider interpretation. Fig.4.15, based on the

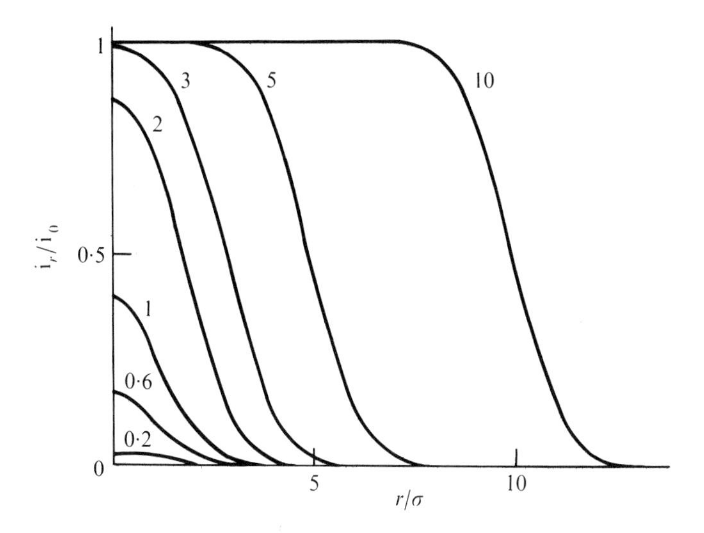

Fig.4.15. Profile of an initially uniform beam dispersed by thermal velocities. The figures on the curves denote R/σ which is defined in equation (4.84). The current density is i_r, and $i_0 = i_c^2/M^2$, where i_c is the cathode current density and M the principal solution of the paraxial equation with values (1,0) at the cathode.

curves of Cutler and Hines, shows the variation of i/i_0 with r/σ for given R/σ, where $i_0 = i_c/M^2$ is the current density for a beam with zero temperature.

It is evident from Fig.4.15 that for $R/\sigma \gtrsim 3$, $i_z(0)$ is almost independent of R/σ. For a given value of M this implies that it is independent of the beam temperature T.

This may be seen also by evaluating equation (4.85) at $r = 0$; this gives

$$\frac{i_z(0)}{i_c} = \frac{1 - \exp(-R^2/2\sigma^2)}{R_c^2} \qquad (4.86)$$

Space charge may be included approximately in equation (4.85) by taking M and S as the solutions appropriate to a sharp-edged beam with uniform current density and zero temperature. As indicated earlier, however, the value of this analysis in all except very simple practical situations is somewhat limited because of the great sensitivity of the current distribution to aberrations.

4.4.8. Limitations to current density in a beam spot arising from thermal velocities at the source In a beam of finite emittance the concentration in real space can be increased at the expense of the concentration in $x'y'$ space. The size of a beam can be decreased by steepening the trajectories. Limitations to this process may be imposed by space-charge forces, but even if these are absent there must be a limitation as the angle of approach of some of the trajectories approaches $\pi/2$. To study this limit it is evidently necessary to move outside the confines of the paraxial equation and take into account also the phase space in the direction along the beam.

We present now a quite general approach to the problem for a beam of non-relativistic particles emitted from a thermionic cathode (or plasma) of temperature kT. The treatment follows that of Pierce (1954a) and leads to a result established by Langmuir (1937). No assumptions are made about the nature of the optical system which is used. The treatment is non-relativistic, so that velocity is proportional to momentum. This restriction makes the calculation considerably simpler; however, it is possible to make a restricted (though rather obvious) generalization after the results have been obtained.

We start by considering the phase-space distribution $f(\underline{x},\underline{v})$ of particles originating from a plane emitter. For a space-charge limited system we take the emitter to be the potential minimum in front of the cathode. The velocity distribution at $\phi = 0$ is 'half-Maxwellian', since only positive values of v_z are allowed. If kT is the temperature of the emitter

$$f(\underline{x},\underline{v}) = 2n_0\left(\frac{m_0}{2\pi kT}\right)^{3/2} \int \exp\left[-\left\{\tfrac{1}{2}m_0 v^2 + \frac{q\phi(x)}{kT}\right\}\right] \mathrm{d}^3x\,\mathrm{d}^3v \qquad (4.87)$$

where ϕ is the potential with respect to the emitting surface. The argument of the exponential is constant for a particular electron, so that $\mathrm{D}f/\mathrm{D}t = 0$ and Liouville's theorem is satisfied. The charge density in real space is $n_0 q$; the factor 2 arises in equation (4.87) because negative values of v_z are excluded in the integration. The half-Maxwellian distribution implies that the current density leaving the cathode in a direction θ to the normal is proportional to $\cos\theta$, a property which has been verified experimentally.

The current density leaving the emitting surface is given by

$$\begin{aligned} i_0 &= q\int v_z f\,\mathrm{d}^3v \\ &= n_0 q(2kT/\pi m_0)^{1/2} \quad . \end{aligned} \qquad (4.88)$$

It remains to calculate the density arriving at a point where the potential is ϕ. We assume axial symmetry and a defining aperture which limits the maximum angle of incidence to α. Spherical co-ordinates centred on the spot are chosen, with $v_2 = v\cos\theta$ and $\mathrm{d}^3v = 2\pi v^2 \sin\theta\,\mathrm{d}\theta\,\mathrm{d}v$. Since all particles have $v^2 \geqslant v_{\min}^2 = -q\phi/2m_0$, the range of integration is from $v_{\min}$ to ∞.

The current density at the centre of the spot is

$$i_s = 2\pi q\int_{v_{\min}}^{\infty} \int_0^{\alpha} v^3 f \sin\theta \cos\theta\,\mathrm{d}\theta\,\mathrm{d}v \qquad (4.89)$$

where f is given by equation (4.87). Inserting f and using the normalizing equation (4.88) leads to the simple result

$$\frac{i_s}{i_0} = \left(1 - \frac{q\phi}{kT}\right) \sin^2 \alpha . \tag{4.90}$$

When $q\phi \gg kT$, and α is a small angle, the r.h.s. can replaced by

$$\alpha^2 \frac{m_0 v_z^2}{2kT} \approx \frac{\alpha^2 \beta_z^2}{\beta_\perp^2} .$$

This implies

$$\frac{i_s}{i_0} = \frac{m_0 v_z^2 \alpha^2}{2kT} = \frac{\alpha^2 \beta_z^2}{\beta_\perp^2} \tag{4.91}$$

which states that the current density cannot be increased by more than the square of the ratio of the transverse velocity to a characteristic thermal velocity $\beta_\perp c$. In this simplified form this result could have been derived from the invariance of the normalized emittance (section 4.3.2). The relativistic generalization is then obvious; γv_z is written in place of v_z.

It can be proved quite generally that it is not possible to concentrate all the current from a uniform cathode in a spot spot with the current density given by equation (4.91). The design of an aperture system to produce the limiting density necessarily reduces the total current. On the other hand, a system which has a wide enough aperture to let through all the current produces a lower current density. This problem has been analysed by Pierce, and he presents a graph showing the fraction of total cathode current obtainable in the spot as a function of the fraction of current density as given by equation (4.91). A reasonable compromise is represented by a value of 0·63 for both quantities.

Further details may be found in Pierce's (1954a) book, and discussions from slightly different viewpoints are given by Lichtenberg (1969) and Ash (1964). Ash considers the effect of a magnetic field, where the angle θ is not clearly defined.

4.4.9. A general survey of factors which limit spot size

In the last section limitations to the current density

obtainable in a beam arising from thermal velocities at the source were analysed. If a given fraction of the current from the source is to be concentrated into a finite area, or spot, this implies a lower limit to the area of the spot. We now survey other limits, imposed by spherical and chromatic aberration of the final lens, the finite de Broglie wavelength of the particles, and space charge. Different limits predominate in different situations. Spherical and chromatic aberration have been discussed in section 2.2.8 and 2.2.9; the diffraction limit may be found from simple optical arguments, and the space-charge limit was analysed in section 3.2.6.

These limits will now be evaluated in terms of the parameters of a typical beam with axial symmetry, and of the final lens and defining apertures. The analysis is relativistic, and the following parameters determine the properties of the beam:

$(\gamma - 1)m_0c^2 = -q\phi$ = kinetic energy of particles

$N = \nu/r_0$ = number of particles per unit length of beam

ε = emittance of beam

$\lambda = h/p = h/\beta\gamma m_0c$ = de Broglie wavelength of particles

kT = cathode temperature

$I = Nq\beta c$ = beam current

$K = 2\nu\,(1-\beta^2-f)/\beta^2\gamma$ = generalized perveance of beam.

The characteristics of the apparatus can be represented by

C_s = spherical aberration coefficient of the final lens

C_c = chromatic aberration coefficient of the final lens

α = angle of convergence of beam, defined by apertures.

To find the required limits, we first consider the imperfections individually. For a beam with emittance ε, the minimum radius for a convergence angle α is simply

$r_{min} = \varepsilon/a$. For a current I, emitted from a cathode with temperature kT, the minimum beam radius can be found from equation (4.91). Since $I = \pi r_{min}^2 i_s$, it follows that

$$r_{min}^2 = \frac{I}{\pi i_0 \alpha^2} \frac{2kT}{\gamma m_0 \beta^2 c^2} \tag{4.92}$$

where a factor γ has been inserted to make the expression relativistic in the manner explained after equation (4.91). The limits imposed by spherical aberration were calculated in section 2.2.8. The radius r_s of the smallest circle through which all the rays pass is equal to $\frac{1}{4}C_s\alpha^3$. Chromatic aberration was discussed in section 2.2.9 and the radius of the corresponding circle found to be

$$r_c = 2C_c\alpha\frac{\Delta p}{p} = 2C_c\alpha\frac{\Delta\gamma}{\beta^2\gamma}. \tag{2.91}$$

In the non-relativistic limit this becomes $C_c\alpha\Delta\phi/\phi$.

Diffraction associated with the finite wavelength of the electron limits the spot size in an otherwise perfect system. The diffraction pattern associated with a lens aperture of finite size is a decreasing oscillatory function of radius. The distance from the maximum to the first zero, the radius of the 'Airy disc', can conveniently be used as an order-of-magnitude measure of the spot size. In terms of the convergence angle α and the de Broglie wavelength λ of the particle

$$r_d = 0{\cdot}61\frac{\lambda}{\alpha} = 0{\cdot}61\frac{h}{\alpha\beta\gamma m_0 c} = 1{\cdot}93\frac{r_0}{\alpha_f\alpha\beta\gamma} \tag{4.93}$$

where h is Planck's constant, r_0 is the classical radius of the particle and α_f is the fine-structure constant equal to 1/137.

The final limiting factor to be considered is that imposed by space charge. An expression for the minimum spot size associated with convergence α and initial aperture of radius R was given in equation (3.37); with change of notation the space-charge-limited minimum radius is given by

$$r = R \exp\left(\frac{-\alpha^2}{2K}\right) \qquad (4.94)$$

Five factors which limit spot size have now been discussed. Which of these is the most important depends of course on the application. In Table 4.2 the dependence of the spot size on various parameters is listed for all five effects. (There are six entries in the table, but the first two are alternative expressions of essentially the same limitation). When the spot size depends on particle energy

TABLE 4.2.

Spot-size limitations in focused beams

Limitation	Formula for spot radius	Eqn	Dependence on α
Finite emittance ε	ε/α	—	$1/\alpha$
Cathode current density i_0	$(2kTI/\pi i_0\alpha^2\gamma m_0\beta^2 c^2)^{1/2}$	4·91	$1/\alpha$
Spherical aberration C_s	$\frac{1}{4}C_s\alpha^3$	Fig. 2.10	α^3
Chromatic aberration C_c	$2C_c\alpha\Delta\gamma/\beta^2\gamma$	2·91	α
Diffraction λ	$0{\cdot}61\lambda/\alpha, (\lambda = 2\pi r_0/\alpha_f\beta\gamma)$	4·93	$1/\alpha$
Space charge K	$R\exp(-\alpha^2/2K)$	4·94	$\exp(-\alpha^2)$

or current it is always decreased by increasing the energy or by reducing the current. The only common parameter which acts in an opposite sense for different limitations is α, the convergence angle. Aberrations demand small α, whereas in the presence of other limiting features it is advantageous to make α large. Under these circumstances an optimum angle, and a corresponding relation between the determining parameters, may be found.

As an example we calculate the optimum value of α in the presence of spherical aberration and diffraction in the limit of small current and zero emittance. Denoting the spot radius by $r \approx (r_d^2 + r_s^2)^{1/2}$, from the table

$$r^2 \approx \left(r_1 \alpha^3\right)^2 + \left(\frac{r_2}{\alpha}\right)^2 \qquad (4.95)$$

where $r_1 = \frac{1}{4}C_s$ and $r_2 = 0{\cdot}61\lambda$. Setting $dr/d\alpha = 0$ yields

$$\alpha = (r_2/3r_1)^{1/4}$$

$$r = (3^{-3/4}+3^{1/4})^{1/2}\, r_1^{1/4}\, r_2^{3/4} \approx 0{\cdot}64\; C_s^{1/4}\, \lambda^{3/4} \quad . \qquad (4.96)$$

This calculation is very rough; other simple treatments use slightly different conventions (for example, adding radii rather than their squares, using $r_1 = C_s$ instead of $\frac{1}{4}C_s$). A more careful argument (Grivet and Septier 1972, chap. 16) suggests that the constant 0·64 should be near unity.

The above calculation is presented in most texts on electron microscopy. In intense beams, however, the other effects in the table must also be considered. The space charge is a local property of the beam; the emittance and chromatic aberration are determined by the source or by scattering of beam particles one by another at a waist. The latter phenomenon, the 'Boersch effect', is outside the scope of effects considered so far, since it is not governed by Liouville's theorem in μ-space. It is discussed in Chapter 5. An extension of the result of equation (4.96) to include these effects has been given by Pfeiffer (1972).

When only spherical aberration and particle wavelength are significant, spot sizes of the order of a few nanometres can be obtained with currents of order 10^{-12} A and voltages of tens of kilovolts. For situations in which finite emittance and spherical aberration are the limiting factors, equation (4.96) still holds, but with $r_2 = \varepsilon/\alpha$:

$$r = 0{\cdot}93\; C_s^{1/4}\, \varepsilon^{3/4} \quad . \qquad (4.97)$$

If the emittance arises from thermal velocities at the cathode only, then it may be expressed in terms of the cathode temperature and radius, equation (4.81). Alternatively, ε can be written in terms of the brightness B and the current I. From equation (4.12), $\varepsilon^2 = \eta I/\pi^2 B$; combining this with equation (4.97) and setting $\eta = 2$ gives the result

$$I \approx \frac{5Br^{8/3}}{C_s^{2/3}} \qquad (4.98)$$

The numerical factor must not be taken too seriously in this very much simplified treatment. It is the dependence on r and C_s which is of interest. A more detailed analysis of this problem with particular reference to the design of probe-forming systems has been given by Mulvey (1967).

4.R.4. Notes and references The material in this section has been developed in a number of contexts. The properties of the Maxwellian distribution, the theory of sheaths, and discussions of the concepts of pressure and temperature may be found in many texts on plasma physics.

Focused beams, and an understanding of the limit to the size of the spot which can be obtained, are of importance in a large number of electron devices, such as cathode-ray tubes, electron microscopes, apparatus for microprobe analysis, and for electron machining. More detailed theory appropriate to particular applications may be found in the specialized texts. Examples of these are the book on beam design with special emphasis on cathode-ray tubes by Moss (1968), and books on welding by Meleka (1971) and on various technical applications by Bakish (1962).

The subject of ion beams and ion sources represent a very extensive field of study, mainly experimental in nature. The literature is vast and scattered; a good starting point for further information is one of the various groups of conference proceedings related to particular applications, such as particle accelerators, mass spectrometers, fusion research, and ion implantation.

Studies of the pinch effect in plasma columns have for a long time been made in connection with proposed fusion devices. More recently they have become relevant to beams, as discussed in section 3.2.6. The configurations studied analytically often represent considerable idealizations of practical situations, and much detailed design is carried out either empirically or by use of numerical methods.

4.5. Ring beams and cylindrical current sheets

4.5.1. Introduction High-current ring beams and cylindrical sheets (or 'layers') have been studied in connection with collective accelerators and proposed fusion devices. Consideration of such rings is also of importance in assessing the maximum current which can be held in particle accelerators. Although in practice limitations to the performance of such rings frequently arise from dynamic instabilities, it is nevertheless important to study limits associated with static self-fields. The effect of space charge has been taken into account in the rather artificial example of laminar flow in a betatron (section 3.4.4). In this section we consider beams with finite emittance, and include the effect of image currents in the walls.

For sufficiently low currents in paraxial beams self-field and image effects can be treated in a straightforward manner; at higher currents, however, the problem of making the solution self-consistent can lead to considerable analytical complexity. A standard procedure exists for tackling such problems, but we leave it until after studying some of the simpler situations in a more direct way.

It is evident that in a betatron-type field a number of very different self-consistent current and charge distributions can be constructed. Such a field, which consists essentially of an axially symmetrical arrangement in which the radial field decreases away from the axis but increases in the axial direction on both sides of a central median plane, is shown in Fig.4.16. Even in the absence of wall and space-charge effects, a wide variety of types of orbit are possible. The standard 'betatron' orbit (section 2.5.3) is almost in the symmetry plane and makes only small excursions in r and y from the equilibrium orbit. Other trapped orbits do not encircle the axis and make large ex-

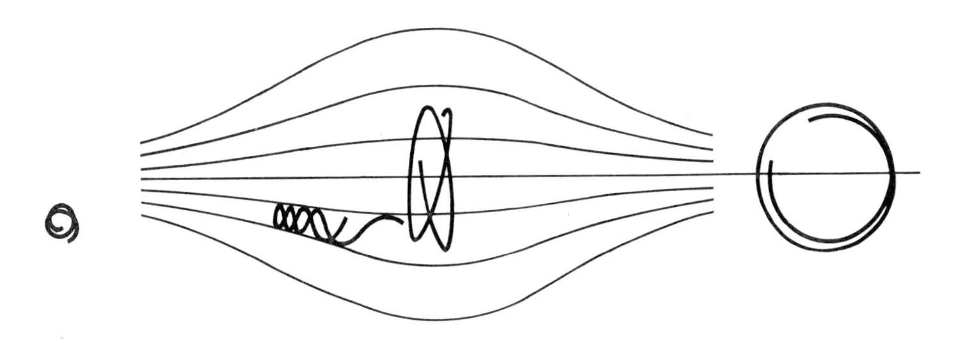

Fig.4.16. Different types of orbit in a 'magnetic bottle'. The central orbit illustrates vertical and radial betatron oscillations; the off-centre orbit illustrates reflection in a magnetic mirror. (The diagram is schematic only.)

cursions in the y-direction. These are often described as being reflected by the magnetic-mirror effect where the field is high and the whole system is known as a 'magnetic bottle'.

The proper description of orbits in such fields and the identification of suitable invariants may be found in many texts on plasma physics, such as those of Lehnert (1964) and Northrup (1963). We do not pursue this topic here, but concentrate on distributions which can be said to form a beam or current sheet. These systems, however, merge into a class of distribution most appropriately studied as a branch of plasma physics. A subclass consisting of particles of one sign, which includes both beam-like and non-beam-like configurations, forms the subject of a book by Davidson (1974). This gives a good view of the wide variety of possible configurations and an extensive list of references.

We now study some particular examples of actual or potential practical interest. The examples are chosen and presented in such a way as to emphasize the essential physical features, rather than analytical technique.

4.5.2. Ring beam in a betatron field in the absence of walls

In the limit of a small number of particles motion in a

betatron field is straightforward; individual particles merely follow orbits of the type described in section 2.5.3. In general the flow is non-laminar and the cross-section of the beam is determined by the distribution in momentum and in betatron-oscillation amplitude. Momentum spread without betatron oscillations results in a disc-shaped beam with laminar flow, having spread of orbit radii $\Delta r/r = \alpha\Delta p/p$ where α is the momentum compaction equal to $(1-n)^{-1}$. Betatron oscillations without energy spread give rise to a non-laminar beam with finite emittance. If the current is increased until self-fields become important, two additional effects occur. First, the repulsion between any two parts of the ring arising from its electric and magnetic field increases the radius of the ring; secondly, the self-fields affect the Q-values of the individual particle orbits.

We calculate first the change in mean radius of a ring with elliptical cross-section assuming that the minor radii a and b are small compared with the major radius R. With this assumption it is not necessary to enquire into the detailed structure of the beam, which is regarded essentially as a current thread. Nevertheless, it must be appreciated that an arbitrarily thin beam is not necessarily physically realizable; clearly for a given current, the values of a and b cannot be less than that allowed by equation (3.83). The ring is illustrated in Fig.4.17; as in the beam considered in section

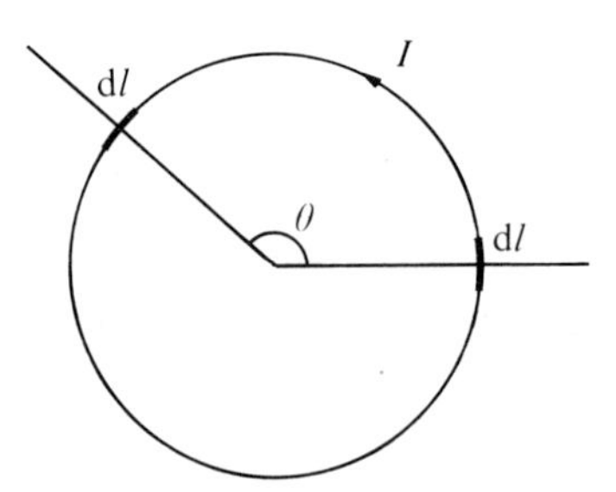

Fig.4.17. Interaction between two elements of a ring current. By symmetry the forces on an element dl are radial. Both electric and magnetic forces between moving charges are repulsive. Electric forces between moving charges and stationary neutralizing charges are attractive.

3.2.6 we allow partial neutralization by a fraction f of charges of opposite sign trapped in the beam. The following forces then act per unit length of the beam:

F_L = Lorentz force
F_c = centrifugal force
F_e = electric self-force between beam particles
F_{es} = electric self-force arising from stationary particles
F_m = magnetic self-force between beam particles.

For equilibrium, these forces must balance. They are first evaluated, and then the electric and magnetic forces are equated to the centrifugal force. To accord with our earlier convention we use curvilinear co-ordinates xys, with $r = R_0 + x$ where R_0 is the radius of the equilibrium orbit. The angle θ is measured as $-s/R_0$ (section 1.4). Expressions for F_L and F_c are straightforward; for a beam with N particles per unit length

$$F_L = -Nq\beta cB_y \tag{4.99}$$

where B_y is the external field, and

$$F_c = \frac{\gamma Nm_0\beta^2c^2}{R} . \tag{4.100}$$

In the absence of the other forces the equation $F_c+F_L = 0$ yields the familiar result

$$R_0 = \frac{qB}{\beta\gamma m_0 c} \tag{4.101}$$

where R_0 denotes the equilibrium orbit radius for a single particle.

The other forces are found by calculating the electric or magnetic repulsion or attraction between two small elements of the ring at an angular separation θ and integrating over θ, as indicated in Fig.4.17. By symmetry all forces are radial. The radial component of electric field at $\theta = 0$ arising from the charge $NRq\,d\theta$ at angle θ is, from Coulomb's law,

$$\Delta E_r = \frac{NRq\sin\frac{1}{2}\theta\,d\theta}{4\pi\varepsilon_0(2R\sin\frac{1}{2}\theta)^2} \quad (4.102)$$

whence the force per unit length may be found, as

$$F_e = \frac{(1-f)N^2q^2}{8\pi\varepsilon_0}\int_0^\pi \frac{d\theta}{R\sin\frac{1}{2}\theta} = \frac{N^2q^2}{4\pi R\varepsilon_0}\ln(\tan\tfrac{1}{4}\theta)\Big|_0^\pi \quad (4.103)$$

This integral is divergent; this arises because the minor radius has been taken as zero. For small values of θ the correct expression is complicated; nevertheless, because of the logarithmic form it is a reasonable approximation to take $\theta_{min} = \alpha a/R$, where α is a factor of order unity. With this substitution

$$F_e = (1-f)\frac{N^2q^2}{4\pi\varepsilon_0 R}\ln\frac{4R}{\alpha a} \quad (4.104)$$

A more accurate calculation (Laslett 1969) shows that when $R \gg a$ then $\alpha \approx \frac{1}{2}$; this value will be used hereafter.

The force arising from the stationary charges is equal to $-fF_e$. In equation (4.104) there are two components to F_{es}, since the stationary particles repel the other stationary particles but attract the moving ones. A consideration of the magnetic interaction, using Ampère's law, shows that F_m is essentially of the same form, with

$$\frac{F_e}{(1-f)} = -\frac{F_m}{\beta^2} \quad (4.105)$$

The equation expressing the balance of forces may now be written

$$\frac{\gamma N m_0\beta^2c^2}{R} - Nq\beta cB + \frac{Nq^2}{4\pi\varepsilon_0 R}\ln\frac{8R}{a}\{1-f-f(1-f)+\beta^2\} = 0. \quad (4.106)$$

Writing B in terms of R_0, the equilibrium radius for a single particle in the field B (equation 4.101), gives the required result:

$$\frac{R}{R_0} = 1 + \frac{\nu}{\beta^2\gamma}\left\{\beta^2 + (1-f)^2\right\} \ln\frac{8R}{a} \quad . \tag{4.107}$$

The conditions that the change in orbit radius arising from self-forces should be small are evidently

$$\begin{aligned} \nu L &\ll \beta^2, \; (\beta \ll 1, \; f = 0) \\ \nu L &\ll \gamma \;, \; (\gamma \gg 1, \; \text{all } f) \end{aligned} \tag{4.108}$$

where L has been written for the logarithmic term. It is of interest to compare these with criteria found earlier which relate to properties of straight beams, for example equation (3.12). When $\nu \gg 1$, a beam is no longer possible. Currents in plasma rings can, however, exist; indeed such rings have been extensively studied in connection with proposed fusion devices. The equilibrium of such current rings in an external field has been studied by Mukhovatov and Shafranov (1971), who give extensive references to earlier work.

An alternative method of obtaining equation (4.107) is from considerations of energy. A small change in R changes the electric and magnetic stored energy associated with the ring. By assigning an appropriate capacity and inductance per unit length of the ring, the principle of virtual work can be used to equate the change in stored energy to the work done by the centrifugal force. The logarithmic term represents essentially half the inductance or capacity per unit length of the beam.

So far we have not considered the internal structure of the current thread. If the external field is a betatron field with field index n, we might expect that a possible distribution is the uniform elliptical distribution of Kapchinskij and Vladimirskij (section 4.3.3), with individual particles making betatron oscillations at frequencies determined by a combination of the external field and the self-fields. For $f = 0$ this is a fair description. If, on the other hand, there is a neutralizing background present, it is not possible to produce a transverse density distribution which matches that of the circulating particles. The reason is that the radial electric field $(F_e+F_s)/q$ displaces the point of zero radial field from the centre of the ellipse, and this moves the centre of mass of the neutralizing particles to a greater radius than that of the circulating beam; the ring is polarized. The restoring force for betatron oscillations is not now strictly proportional to amplitude for all particles, and the system is essentially non-linear. This situation is illustrated in Fig.4.18.

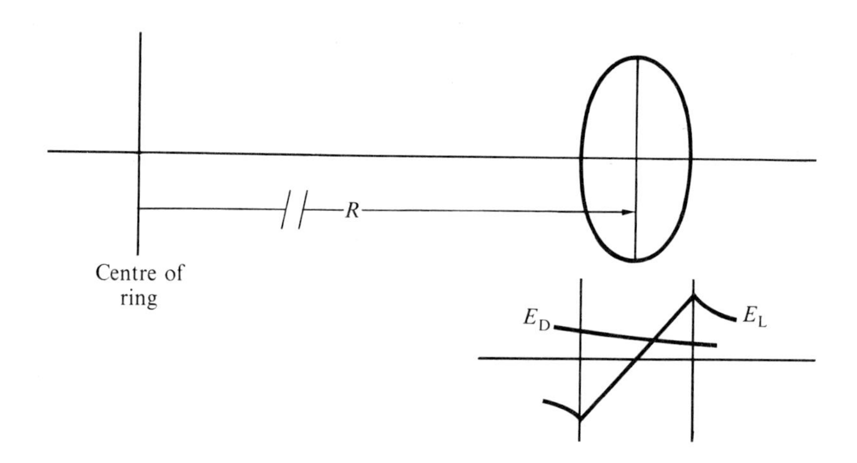

Fig.4.18. Components of electric field in the symmetry plane of a ring current with elliptical minor cross-section. The local component E_L is symmetrical about the centre of the section, but the component E_D arising from the distant part of the ring is not. Partial neutralization of the ring results in a configuration with radial polarization.

In practical beams of this type, studied intensively in connection with the electron ring accelerator, the transverse structure depends very much on how the beam has been generated. In general the transverse density distribution is more nearly gaussian in shape than uniform. A further point which must be appreciated is that the type of configuration discussed is by no means always stable. Indeed, in situations where the self-forces become significant it is likely to exhibit both negative-mass instability and transverse two-stream instability in which the ions and electrons move in opposite directions. These effects are analysed in Chapter 6.

Despite these reservations, it is of interest to evaluate the betatron oscillation frequencies in the structure studied so far, neglecting the non-linearities produced by the polarization. The betatron frequencies are determined essentially by three components of the fields, these are as follows: (1) the externally applied magnetic field; (2) the local self-fields, identical to those in a straight beam; (3) self-fields from distant parts of the ring, arising from its toroidal shape. The effect of the first two alone has been considered in section 4.3.6. The effect of the third on beam radius, but not on focusing, is described by equation (4.107). In order to calculate the Q-values it is necessary to find also the gradients of the distant self-fields at the equilibrium orbit. The derivation is straightforward but rather lengthy; we quote here from a paper by Reiser (1973) in which the problem is reviewed. He gives (in our notation)

$$Q_x^2 = (1-n) + \frac{2KR^2}{a(a+b)} + \frac{\nu}{\gamma}\left\{\ln\left(\frac{16R}{a+b}\right)\right\}\left\{1 + \frac{(1-f)^2\nu}{\beta^4\gamma^3}\ln\frac{16R}{a+b} - \frac{n(1+\beta^2-f)}{\beta^2}\right\}$$

$$Q_y^2 = n + \frac{2KR^2}{b(a+b)} + \frac{\nu}{\gamma}\left\{\ln\left(\frac{16R}{a+b}\right)\right\}\left\{-1 + \frac{n(1+\beta^2-f)}{\beta^2}\right\}. \tag{4.109}$$

The three terms correspond to the three focusing components listed above. The perveance K can be expressed in terms of ν, γ, and f through equation (3.30). These formulae are rather complicated, and do not simplify in any obvious way in the extreme relativistic or non-relativistic limits.

4.5.3. Ring beam in a betatron field in the presence of walls; Q-shifts In the previous section the properties of paraxial ring currents in the absence of walls were evaluated. In the presence of walls the problem is in general more complicated; currents and charges are induced which modify the forces on the beam, and hence its configuration. If the fields associated with the beam terminate on surfaces which are relatively close compared with R, then the problem is simplified somewhat in that the self-fields of the distant parts of the ring are not important. The terms arising from the near self-field, which correspond to cylindrical rather than toroidal geometry are retained, and additional image terms appear.

It is important to estimate these image terms, since they determine how much current can be held in a strong focusing accelerator before the Q-value moves to an unacceptable resonant value (section 2.7.3). This problem has been analysed by Laslett (1963) in connection with injection into synchrotrons, and such changes in Q are sometimes known as 'Laslett Q-shifts'. In an 'ideal' system, consisting of a beam of elliptical cross-section with boundary conditions such that the configuration of fields described in section 3.4.2 for a straight elliptical beam is preserved, the Q-shifts would be given simply by the second term in equation (4.109):

$$\Delta Q_x^2 = \frac{2KR^2}{a(a+b)}$$
$$\Delta Q_y^2 = \frac{2KR^2}{b(a+b)} \qquad (4.110)$$

(This applies to an azimuthally uniform field; in an actual strong-focusing system the expression is of course more complicated. Assuming that the beam profile is known, the additional focusing terms can be calculated and the change in Q found using the methods outlined in section 2.7.2. Alternatively the 'smooth approximation', in which the focusing field is replaced by a hypothetical uniform field with the same Q values, can be used.)

Most accelerators have a metallic vacuum chamber located between the poles of a magnet. The electric field lines terminate on the walls of the chamber, and under truly static conditions the magnetic field lines pass through the chamber and into the magnet. Under transient conditions, however, or when the beam is bunched the magnetic field can also terminate on the vacuum chamber.

We consider first a uniform, unbunched beam. The geometry is specified by h and g, the half-heights of the vacuum chamber and magnet gaps, and by ε_1 and ε_2, form factors which depend on the specific geometry. The boundary may for example be rectangular, elliptical, or trapezoidal. The effect of boundaries symmetrically placed with respect to the median plane is to produce a field at the orbit. If the beam height is small compared with h and g, then the field depends only on the beam charge and current, not on its size. At the orbit the field may be considered as an expansion into multipole components (Fig.4.19). The dipole component produces a

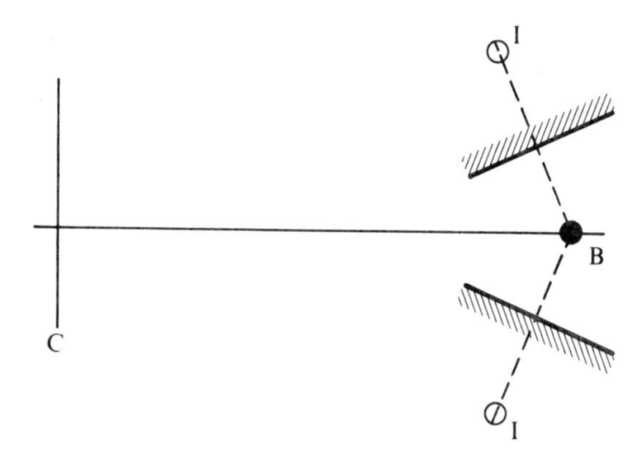

Fig.4.19. Image fields I of the beam B produce a field at the orbit which can be resolved into a spectrum of multipoles. Of these, the dipole field produces a shift in orbit radius, and the quadrupole a shift in Q-value.

small radial displacement of the equilibrium orbit, the quadrupole component adds equal focusing and defocusing terms respectively in the equations of radial and vertical motion, and the higher multipoles introduce non-linear terms. If the boundaries are remote and nearly parallel, only the quadrupole term is significant. Since the boundary conditions for electric and magnetic fields differ, the near cancellation between the electric effect and magnetic self-fields which occurs for $f = 0$ when β is large is destroyed. Instead of the self-field term of an un-neutralized beam varying as γ^{-3} (since $K = 2\nu/\beta^2\gamma^3$), we find a γ^{-1} dependence.

Details of the calculation and the variation of the form factors ε with geometry may be found in the original paper by

Laslett. Here we quote his result for the Q-values, in terms of the parameters defined earlier (and in Fig.4.19). His studies refer to a beam at injection, since this is where the effect is of most importance. During the short injection period, neutralizing charges do not have time to accumulate; f is accordingly taken equal to zero. The expressions for Q, using a smooth approximation so that n is constant, are, in our notation,

$$Q_x^2 = 1-n-\frac{4\nu}{\beta^2\gamma^3}\cdot\frac{R^2}{a(a+b)}\left\{1-\gamma^2 a(a+b)\left(\frac{\varepsilon_1}{h^2}+\frac{\varepsilon_2}{g^2}\right)\right\}$$

$$Q_y^2 = n-\frac{4\nu}{\beta^2\gamma^3}\cdot\frac{R^2}{b(a+b)}\left\{1+\gamma^2 b(a+b)\left(\frac{\varepsilon_1}{h^2}+\frac{\varepsilon_2}{g^2}\right)\right\}. \tag{4.111}$$

For an order-of-magnitude calculation, we set $a = b$, $g = h = 5a$, $\varepsilon_1 = \pi^2/48$, and $\varepsilon_2 = \pi^2/24$. These values of ε_1 and ε_2 are those appropriate to a plane gap, as calculated by Laslett. With these numbers the second term in the braces becomes $(0\cdot22\gamma)^2$. Equation (4.111) is not directly comparable with the formulae quoted by Laslett, since it refers to a continuous rather than a bunched current. As explained above, the boundary condition for a harmonic current is the same for magnetic as for electric fields. Defining a 'bunching factor' B representing the ratio of the average to the maximum linear particle density, the Q values for those particles at the point where the current density is a maximum is obtained by replacing the braces in the first equation (4.111) by

$$\frac{1}{B}\left[1-\gamma^2 a(a+b)\left\{\left(\frac{1}{B\gamma^2}+\beta^2\right)\frac{\varepsilon_1}{h^2}+\frac{\varepsilon_2}{g^2}\right\}\right] \tag{4.112}$$

(and in the second by the same quantity but with $\gamma^2 b$ in place of $-\gamma^2 a$) where ν now represents the *average* value round the orbit.

The foregoing analysis applies to the situation in which the betatron oscillation of the individual particles are 'incoherent'; at any azimuth all phases of oscillation are present, so that the beam is everywhere matched in the sense discussed in section 4.3.2. The system has a time structure when observed at a given azimuth, but not if one travels with the particle velocity along the beam. If all phases of betatron oscillation are not uniformly present, however; an observer moving with the beam also sees a time structure. Such oscillations complicate the situation as described so far; they may be classed as dynamic effects, and will be discussed later in Chapter 6. The appropriate image coefficients have been discussed by Laslett and Resegotti (1967). It is found in some practical situations that these 'coherent Q-shifts' impose a more severe limitation than the 'incoherent' effects analysed in this section.

4.5.4. Cylindrical current sheets Closely related to the ring beams just described is the cylindrical current sheet. A particular case of such a system, with laminar flow, was described in section 3.4.3. Many models of non-laminar sheets can be devised, especially if the axial length is taken as finite and some axial velocity is allowed. Many essential physical features are, however, exhibited by an infinite system with uniform charge density, and this will now be analysed.

For a system infinite in the axial direction, the external magnetic field must be uniform. A low-intensity ring of finite thickness in such a field consists of an assembly of circular orbits, with centres confined to a small circle, as shown in Fig.4.20. If the mean radius of the ring is R, its thickness is $2a$, and the fractional momentum spread of the particles, assumed small, is $\pm\Delta p/p$, then the centres of the circles lie within a radius $r = a + (\Delta p/p)R$. The charge density within the radial range $R \pm a$ is clearly arbitrary. We study a special case where $\alpha = 0$ and the charge density is uniform. Taking $a \ll R$ implies the paraxial approximation, so that the current density is uniform also. The orbits may all be considered as oscillating about the equilibrium orbit, of radius R, with $Q = 1$. The emittance diagram is bounded by an ellipse with semi-axes of lengths a and a/R, giving an emittance $\varepsilon = a^2/R$. (The density distribution within this boundary is not, however, uniform.)

If now the number of particles is increased, electric

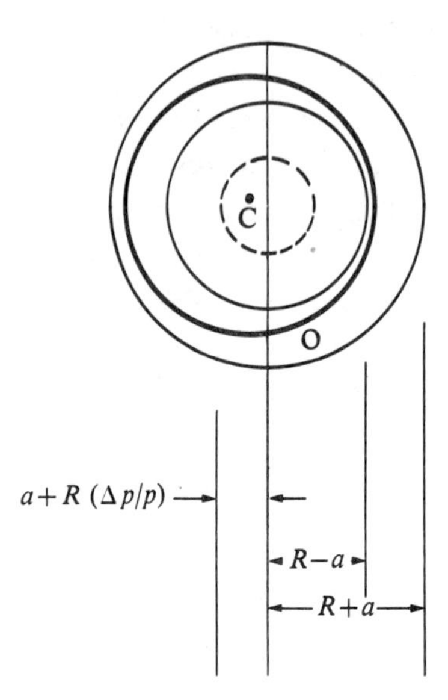

Fig.4.20. Geometry of a cylindrical current sheet. Orbits lie within $r = R \pm a$, and centres where $r \leq a + (\Delta p/p)$. A typical orbit O and its centre C are shown. In the analysis it is assumed that $a/r + \Delta p/p \ll 1$.

and magnetic self-fields appear. For a cylinder in free space the radial electric field is zero inside the ring, increases linearly from inside to outside, and falls off as $1/r$ outside. The axial self-magnetic field, on the other hand, is constant inside the cylinder, falling linearly to zero at the outer radius. If N_a is the number of particles per unit area of the cylinder and B_e the external magnetic field, the fields at $r = R$ are

$$E_1 = \frac{N_a q}{2\varepsilon_0}$$
$$B_1 = B_e - \tfrac{1}{2}\mu_0 N_a q\beta c \ . \tag{4.113}$$

Writing $r = R + x$, the fields between $x = -a$ and $+a$ are

$$E = E_1(1+\frac{x}{a})$$
$$B = B_1+(B_e-B_1)\frac{x}{a} \;. \qquad (4.114)$$

The radius R of the equilibrium orbit can be found by setting the sum of the outward electric and magnetic forces equal to the centrifugal force:

$$\frac{\beta^2\gamma m_0 c^2}{R} = q(\beta c B_1 - E_1). \qquad (4.115)$$

Writing $\nu_y = RN_a q^2/2\varepsilon_0 m_0 c^2$, and substituting for E_0 and B_0 yields

$$\frac{R}{R_0} = 1+\frac{\nu_y}{\gamma}(1+\frac{1}{\beta^2}) \qquad (4.116)$$

where R_0 is the orbit radius in the limit of $\nu_y = 0$. This is analogous to equation (4.107) for ring beams. Note here, however, that ν_y is measured in terms of the number of particles per unit length in the *axial* direction, rather than azimuthally. As with rings, for conditions of partial neutralization the layer becomes polarized; as in equation (4.107) $1/\beta^2$ is replaced by $(1 - f)^2/\beta^2$.

The diamagnetic factor, defined as the ratio of the self-field B_s to B_e, is, from equations (4.113) and (4.116),

$$-\frac{B_s}{B_e} = \frac{2(B_e - B_1)}{B_e} = \frac{\mu_0 N_a q\beta\gamma}{B_e} = \frac{2\nu_y/\gamma}{1+(\nu_y/\gamma)(1+1/\beta^2)} \;. \qquad (4.117)$$

For an un-neutralized layer this is always less than unity, so that the field reversal cannot be obtained. For a neutral layer, on the other hand, $B_s = -B_e$ when $\nu/\gamma = 1$. Such field reversal has been obtained in practice. (Andrews *et al.* 1971).

In the presence of self-fields the orbits are no longer circles, and, furthermore, in a charged system the kinetic energy of the particles varies across the layer. An interesting parameter is the ratio of potential difference across the layer to the kinetic energy of a particle on the equilibrium orbit:

$$\int_{-a}^{a} \frac{qE\,\mathrm{d}x}{(\gamma-1)m_0c^2} = \frac{Nq^2a}{(\gamma-1)\varepsilon_0 m_0 c^2} = \frac{2a}{R}\frac{\nu}{\gamma-1} \tag{4.118}$$

The betatron oscillation frequencies can be found in terms of the field gradients (section 2.5.3):

$$n_{\mathrm{M}} = -\frac{R}{B_1}\frac{\mathrm{d}B}{\mathrm{d}r} = -\frac{R}{a}\frac{\mu_0 N_{\mathrm{a}} q\beta c}{2B_{\mathrm{e}} - \mu_0 N_{\mathrm{a}} q\beta c} \tag{4.119}$$

Substituting $B_{\mathrm{e}}R$ from equation (4.116) yields

$$n_{\mathrm{M}} = -\frac{R}{a}\left(\frac{\nu_y}{\gamma + \nu_y/\beta^2}\right). \tag{4.120}$$

For a neutralized beam the second term in the denominator is absent. The electric index n_{E} is found directly from equation (4.114):

$$n_{\mathrm{E}} = -\frac{R}{a}. \tag{4.121}$$

From equation (4.113) we define

$$\beta_0 = \frac{-\mu_0 E_1}{B_1} = \frac{-\nu_y}{\beta(\gamma-\nu_y)}. \tag{4.122}$$

An explicit expression for Q^2 may now be found by substituting n_{M}, n_{E}, and β_0 into equation (2.121). This yields a somewhat unwieldy expression for a charged layer; on the other hand, for a neutral layer

$$Q_{\mathrm{n}}^2 = 1 + \frac{R}{a}\frac{\nu_y}{\gamma}. \tag{4.123}$$

For a charged layer the expression simplifies considerably when $\nu \ll \beta^2\gamma$ to

$$Q^2 \approx 1 + \frac{R}{a}\frac{\nu_y}{\gamma}\left(1 + \frac{1}{\beta^2}\right). \tag{4.124}$$

Since $\beta < 1$, it follows that $Q < 1$ and hence the orbit precession is in the opposite direction from that in a neutralized beam. Expressions have now been found for the mean radius of the cylinder, the diamagnetic effect, and the orbital frequencies in terms of the parameters of the system. The dimensionless parameters ν/γ and a/R play a significant role.

As pointed out earlier, this model is highly idealized. Practical layers have ends, and furthermore the particles always have a component of axial velocity. Image forces, as with ring beams, can be important. If there is an inner conductor centred on the axis, then E is not necessarily zero for $r < R-a$. If, for example, such a conductor is maintained at the same potential as an external cylinder, the equilibrium conditions are different. Simple models may readily be constructed, however, on the same lines as those just studied.

Extensive experimental studies of cylindrical electron-current sheets of finite length have been carried out in connection with the 'Astron' fusion reactor concept (Christofilos 1958), and many theoretical analyses of equilibrium configurations with varying assumptions have been made. Most of these use the Vlasov technique, described below in section 4.6. This is a specialized field, treated in detail by Davidson (1974) where an exhaustive list of references may be found. Work on the Astron project has now ceased, but layers of even greater intensity have been produced (see earlier reference to Andrews *et al.* 1971). Studies of proton layers have also been made (Sudan 1975, Fleischmann 1976).

At the time of writing this is a developing field; intense transient electron layers and rings are being produced by trapping of intense relativistic beams of the type described in section 4.7.

4.5.5. Adiabatic variation of ring beams and cylindrical current sheets; betatron 2:1 condition If the magnitude of the external magnetic guide field is changed slowly compared with the orbital rotation frequency, then the radius of the ring, the energy of the particles, and the orbital oscillation frequencies change adiabatically. We analyse first the behaviour of a thin low-current ring of radius R in a betatron field. The variation in radius is found from the condition that the canonical angular momentum is conserved (equation (2.6)):

$$\frac{\mathrm{d}}{\mathrm{d}t}\,(\gamma m_0 R^2\dot{\theta} + qA_\theta R) = 0. \tag{4.125}$$

Setting $R\dot{\theta} = -\beta c$, $qBR = \beta\gamma m_0 c$, $A_\theta = (\int_0^R Br\mathrm{d}r)/R$ we find

$$\frac{\mathrm{d}}{\mathrm{d}t}\left(\int_0^R Br\,\mathrm{d}r - B_1 R^2\right) = 0 \tag{4.126}$$

where B_1 and R are the values at the orbit. In order to determine how R varies with t, we need to know how B varies with both r and t. As a particular example we choose

$$B = B_0(t)\left(\frac{r}{r_0}\right)^{-n} \tag{4.127}$$

where $B = B_0$, $R = R_0$ at $t = 0$ and $0 < n < 1$. Substituting equation (4.127) in equation (4.126) yields the adiabatic invariant B_1R^2 = const. It is interesting to note that this is independent of the value of n. The field B_1 is of course the value at the orbit; B_1/B_{10} and R/R_0 do depend on n.

It is interesting to note in passing that if instead of an orbiting particle we consider a stationary particle in the field, it also moves radially inwards such that B_1R^2 = constant. This follows immediately from the conservation of canonical angular momentum, $qA_\theta R$, since RA_θ is equal to $1/2\pi$ times the flux inside a circle passing through the particle. This illustrates the idea that the particle is 'tied' to the magnetic field line.

Another important situation is when the two terms in equation (4.126) are equal. The radius then remains constant when the field varies. The condition for this may be written

$$\int_0^R 2\pi rB\,\mathrm{d}r = \Psi = 2\pi BR^2. \tag{4.128}$$

This states physically that the flux through the orbit is twice what it would be if the field were uniform, having everywhere the value that it has at the orbit. This is known as the 'betatron 2:1 condition' or the 'Wideröe flux condition' (Livingston 1966, p.92). Betatron accelerators have an iron core linking the orbit to enable this condition to be satisfied.

When the current in the ring is sufficient, the self-fields need to be considered. The calculation is essentially the same except that equation (4.107) must be used to relate radius to momentum, and the component of A_θ arising from the self-magnetic field included in equation (4.126). This is directly related to l, the inductance per unit length in the beam through the relation

$$\Psi_s = 2\pi R l \mu_0 N q \beta c = 2\pi R A_s \, . \tag{4.129}$$

For a ring current $l \sim 2\ln(8R/a)$, where a is the mean cross-sectional radius of the beam. Adding A_s to the value A_θ associated with the external field in equation (4.126), and substituting as before for $\dot{\theta}$ yields

$$\frac{\mathrm{d}}{\mathrm{d}t}\{\beta(\gamma+\nu l) m_0 c R - q A_\theta R\} = 0. \tag{4.130}$$

The form of this equation is interesting; the particles behave as though their mass is enhanced. The increased energy at a given velocity is accounted for by the energy stored in the self-magnetic field. To find how R varies with B, equation (4.130) must be solved simultaneously with equation (4.107), where the logarithmic term can be identified with $\frac{1}{2}l$. (As explained in section 4.5.2, equation (4.107) can be derived from the principle of virtual work in such a way that the logarithmic term appears directly as the inductance per unit length of beam.)

For an un-neutralized beam ($f = 0$) and $\beta \approx 1$ the situation is as for a single particle, with the enhanced mass $(\gamma+\nu l) m_0 c^2$. It may be readily verified that under these circumstances the 2:1 condition for constant radius applies. If $\beta < 1$ or $f > 0$, on the other hand, it does not apply. (Indeed, if f is not equal to zero the situation becomes more complicated; the dynamics of the ions, which is different from that of the electrons, needs to be taken into account.)

Similar arguments apply to cylindrical layers. Since B is uniform the vector potential of the current layer $A_s = \frac{1}{2}B_s R$ has the simple form $\frac{1}{2}\mu_0 q \beta c N_a R$. This may be rewritten as $\nu_y m_0 \beta c/q$. Furthermore, the component of vector potential associated with the external field is $\frac{1}{2}B_e R$. Equation (4.130) may then be written

$$\frac{d}{dt}\{(1+\nu/\gamma)\beta\gamma m_0 cR - \tfrac{1}{2}qB_e R^2\} = 0 \qquad (4.131)$$

Expressing B_e as $\beta\gamma m_0 c/qR_0$ and eliminating R_0 between this equation and equation (4.116) gives the variation of R as a function of γ. It is evident from the form of these equations that in the presence of self-fields $B_e r^2$ is not conserved.

In the presence of walls and magnet poles the behaviour described in this section is, of course, influenced by image currents. Furthermore, in practical situations instabilities often occur which limit the value of ν/γ which can be achieved. This is the situation for example in the electron ring accelerator, where ν/γ is only of order 10^{-3}. It is not difficult to exceed this, however, for rings which have a larger ratio of minor to major radius and a larger energy spread. As explained in the introductory section 4.5.1, it is not always clear when a particular configuration ceases to become an 'electron ring' and is better described as a 'non-neutral plasma'. The formation of beams of 'runaway' electrons by applying an inductive electric field to a toroidal plasma is briefly described in section 5.8.

4.5.6. Crossed-field flows The rings studied in the previous few sections represent examples of the very general class of 'crossed-field flows', where the essential ingredients are an electric field, a magnetic field, and a drift velocity which are mutually perpendicular, or nearly so. The examples studied are restricted by the conditions that the particle velocity is essentially the same as the drift velocity and tha the electric field arises entirely from space-charge forces rather than from an axial charged conductor. Dropping the first of these restrictions leads to systems which are not beams in the sense defined in Chapter 1; dropping the second (but not the first) permits the inclusion of the types of flow in an important range of microwave devices related to the magnetron and also the 'Penning gauge' used for pressure measurement. Somewhat arbitrarily perhaps we exclude consideration of these rather special devices. Details of the properties of crossed-field flows applied to microwave devices may be found in the book by Okress (1961); the theory of the Penning gauge, together with earlier references, is given by Jepsen (1961).

Equilibrium in a somewhat different geometry has been analysed by Daugherty and Levy (1967) and studied experimentally by Daugherty *et al.* (1969b) in their 'Hipac' device. Instead of a uniform magnetic field, the field lines are closed to form a torus. Because of the essential non-uniformity of such a field with radius, discussed for example in section 2.7.7 in connection with the betatron with azimuthal magnetic field, a toroidal system will not confine a single charged particle. A particle with a component of velocity in a plane through the symmetry axis experiences a drift parallel to the axis arising from the radial gradient of the toroidal field. If it has a velocity component along the lines of force, the curvature of the lines in the plane perpendicular to the axis again causes a drift in the same direction. The presence of other charges, however, and a conducting toroidal chamber give rise to additional forces which modify these drifts in such a way that a charge can be contained. Experimental verification of the existence of such equilibria is given in the reference quoted.

4.R.5. Notes and references Little of the material in this section is in standard texts, though many papers on the topics covered have appeared in the literature and in conference proceedings. A few of these have been quoted to illustrate specific points. Some of the motivation for studies of ring beams and layers has been in connection with cyclic accelerators and storage rings, where image and self-field effects are small compared with external focusing (sections 4.5.2 and 4.5.3). Attention was drawn to the properties of rings with ν/γ of order unity by Budker (1956) in a remarkable paper proposing an intense electron beam as a guide field for a betatron-type accelerator. (This work is referred to again in Chapter 5). The use of a layer for plasma containment was suggested by Christofilos (1958), and gave rise to an extensive experimental programme (Christofilos *et al.* 1971). Interest in electron rings of moderate (though not extreme) intensity was revived by the suggestion by Veksler (1967) of the electron ring accelerator, which makes use of the principle of collective acceleration. Much of the early work on the properties of rings of this type was done in the Soviet Union.

The validity of formulae relating to effects calculated in this section are often limited by the onset of various types of instability, which in many circumstances can set in for quite small values of K or ν/γ. For example, rings in betatron fields with no spread in particle energy are liable to negative-mass instability, and to instabilities arising from wall interactions. These effects are discussed in Chapter 6. An assessment of the way that the various limits arise in the electron ring accelerator, for example, is given by Möhl, Laslett, and Sessler (1973). An appreciation of the status of this proposal at the time of writing may be found in the group of papers in the 1974 Stanford Accelerator Conference.

Applications of cross-field flows are varied and somewhat different; references have already been given in section 4.5.6.

4.6. A more general approach; the Vlasov equation

4.6.1. Introduction In Chapters 3 and 4 a large number of different beam configurations, most of them paraxial, have been described. For laminar flows the beam characteristics have normally been obtained by considering the balance of forces on a particular particle. In the case of non-laminar flows in which self-fields are important, self-consistency of the particle orbits and the fields must be ensured; in the examples studied different approaches have been used for different problems, and the special properties of some particular systems, such as the K-V distribution, have been exploited.

We now consider a more powerful and general method of

constructing self-consistent equilibria, not confined to paraxial systems, which is applicable to all situations where Liouville's theorem is obeyed. Although powerful, the method often results in mathematical complexities even if approximations and simplifications are made in the initial model which is being analysed. For this reason, simplified approximate treatments are often to be preferred to an exact but complicated treatment of an oversimplified model. With this reservation, we now introduce the method.

The solution we seek consists of a configuration of charges and currents in real space which takes into account not only the external fields, but the fields which arise from the charges and currents themselves. The form of this configuration is determined by constraints placed on the density distribution in phase space of the particles, $f(P_i, x_i)$. (If more than one type of particle is present, summation over species is implied.) The fields arising from a given distribution of charges and currents can be found from Maxwell's equations. In terms of the phase-space density f, the charge and current densities in real space are $q\int f d^3v$ and $q\int fv d^3v$ respectively. We now study the complete problem of finding the distribution function which is consistent with the fields, starting from Liouville's theorem, and confining attention to systems in which collisions between particles are unimportant. If P and x are canonically conjugate co-ordinates, then Liouville's theorem may be written

$$\frac{\partial f}{\partial t} + \dot{x}_i \frac{\partial f}{\partial x} + \dot{P}_i \frac{\partial f}{\partial P_i} = 0. \tag{4.9}$$

By expressing $\dot{P}$ in terms of the external fields a relation is obtained which is known as the 'Vlasov equation' or 'kinetic equation' (Vlasov 1945).

In cartesian co-ordinates P may be taken *either* as the linear momentum $\underline{p} = \gamma m_0 \dot{\underline{x}}$ of a particle acted upon by electric and magnetic fields *or* as the canonical momentum $\underline{P} = \gamma m_0 \dot{x} + q\underline{A}$ of a particle moving in an electrostatic potential. For the moment we take the first of these alternatives. Writing

$$\dot{\underline{p}} = \frac{d}{dt}(\gamma m_0 \underline{v}) = q\{\underline{E}(\underline{x},t)+\underline{v}\times\underline{B}(\underline{x},t)\}$$
$$\dot{\underline{x}} = \underline{v} = \frac{\underline{p}}{\gamma m_0} \tag{4.132}$$

Equation (4.9) can be written in vector form

$$\frac{\partial f}{\partial t} + \frac{\underline{p}}{\gamma m_0}\frac{\partial f}{\partial \underline{x}} + q\left(\underline{E}+\frac{\underline{p}\times\underline{B}}{\gamma m_0}\right)\frac{\partial f}{\partial \underline{p}} = 0. \tag{4.133}$$

This is the Vlasov equation. In the non-relativistic limit it simplifies to

$$\frac{\partial f}{\partial t} + \dot{\underline{x}}\frac{\partial f}{\partial \underline{x}} + \frac{q}{m_0}(\underline{E}+\dot{\underline{x}}\times\underline{B})\frac{\partial f}{\partial \dot{\underline{x}}} = 0 \tag{4.134}$$

Alternatively, expressed in terms of the canonical momentum $\underline{P}$ and the scalar and vector potentials ϕ and $\underline{A}$,

$$\frac{\partial f}{\partial t} + \frac{(\underline{P}-q\underline{A})c}{\{(\underline{P}-q\underline{A})^2+m_0^2c^2\}^{1/2}}\frac{\partial f}{\partial \underline{x}} - \left(\underline{\nabla}\phi-\frac{\underline{P}-q\underline{A}}{\gamma m_0}\times\text{curl }\underline{A}\right)\frac{\partial f}{\partial \underline{P}} = 0 \tag{4.135}$$

where f is now $f(\underline{P})$ rather than $f(\underline{p})$. This equation (or equation 4.133) together with Maxwell's equations and the relations

$$\rho = q\int f d^3v$$
$$i = q\int v f d^3v \tag{4.136}$$

(summed over species if more than one type of charge is present) form a set of closed integro-differential equations which determine the self-consistent dynamics of an assembly of charges obeying Liouville's theorem. We recall that this is an approximation insofar as it represents the 'smoothed-out' behaviour of an ensemble containing many particles. The force on a given particle arising from the other particles represents a smoothed-out average and is not affected by the particular position of any of the other particles in the ensemble. Collisions between individual particles are thus excluded.

Finding solutions of the Maxwell-Vlasov equations is in general difficult, though there are two classes of problem which we shall tackle using this technique. The first is the establishment of self-consistent equilibria with certain specified properties; the second is the study of the stability of such equilibria using a perturbation technique. The method of finding self-consistent equilibria is described in the next section.

4.6.2. The determination of self-consistent equilibria Equilibrium configurations can often be found if some of their properties are specified and others left as undetermined parameters. If a proper choice is made, then substitution into the Vlasov and Maxwell equations (hopefully) yields a tractable relation between the undetermined parameters which ensure self-consistency. Some skill is required in selecting equilibria which are both analytically tractable and physically realistic. The method of solution makes use of the fact that any function of the constants of motion is a solution of the Vlasov equation. That this must be so can be seen by considering the trajectory of a single particle; Liouville's theorem (equation 4.10) states that $Df/Dt = 0$, and by definition this must be satisfied when f is expressed in terms of constants of the motion.

A very simple application of this property leads to the Boltzmann relation between particle density and potential for a gas with a Maxwellian velocity distribution quoted in section 4.4.2. For a non-relativistic gas in an electrostatic potential ϕ, a constant of the motion is the energy

$$H = \tfrac{1}{2}m_0 v^2 + q\phi(x). \tag{4.137}$$

The distribution function

$$f = f_0 \exp\left(\frac{-H}{kT}\right) = f_0 \exp\left\{\frac{-(\tfrac{1}{2}m_0 v^2 + q\phi)}{kT}\right\} \tag{4.138}$$

is a function of the constants of the motion, and is therefore a solution of equation (4.134). This may be verified by substitution, setting $\underline{E} = -\underline{\nabla}\phi$. If $\int f_0 \exp(-\frac{1}{2}mv^2)\mathrm{d}^3v$ is written as n_0, then

$$\int f \mathrm{d}^3 v = n = n_0 \exp\left(\frac{-q\phi}{kT}\right) \tag{4.45}$$

which is the required result.

4.6.3. Relativistic beam of Hammer and Rostoker The method of finding self-consistent solutions to the Maxwell-Vlasov equations will now be illustrated by a more complicated example, a cylindrical self-constricted partly neutralized beam. This example is not restricted to the paraxial approximation, though by imposing restrictions later the paraxial pinch described in section 4.3.5 is recovered. The equilibrium which we study, and the method of calculation, are essentially those of Hammer and Rostoker (1970), though a simple non-relativistic treatment of the same problem is given as an example of the method in the book by Longmire (1963).

The beam is assumed to have the following properties

(1) uniformity in the z-direction
(2) axial symmetry
(3) partial neutralization by a stationary ion background, such that the ratio f of beam to background charge is everywhere constant
(4) all moving charges with the same energy
(5) all moving charges with the same axial canonical momentum.

Properties (4) and (5) can be expressed by the relations

$$H_0 = \gamma_0 m_0 c^2 = \gamma m_0 c^2 + q\phi(r) \tag{4.139}$$

where γ_0 is the value on the axis ($r = 0$), $\phi = 0$ at $r = 0$, and

$$P_{z0} = \beta_z \gamma m_0 c + qA_\theta = \beta_{z0}\gamma_0 m_0 c \tag{4.140}$$

where $\beta_z c$ is the z-component of velocity and β_{z0} is the value of β_z on the axis.

We use the fact that any constant of the motion satisfies the Vlasov equation and write for the distribution function

$$
\begin{aligned}
f(\underline{x},\underline{P}) &= f(r,P_x,P_y,P_z) \\
&= C\,\delta(H-H_0)\,\delta(P_z-P_{z0})
\end{aligned}
\qquad (4.141)
$$

where C is a normalizing constant as yet undetermined. The six phase-space variables are reduced to four since the system is independent of z and θ. To find the density $n(r)$ we integrate f over P_x, P_y, P_z, and θ. First, we note that f is zero unless P_z is equal in both δ-functions, both of which have zero argument. Expressing H in terms of p and ϕ, with $P_z = p_z + qA_z$, $P_x = p_x$ and $P_y = p_y$

$$
f(r,p_x,p_y,P_z) = C\,\delta[c\{m_0^2c^2+p_x^2+p_y^2+(P_{z0}-qA)^2\}^{1/2}+q\phi-H_0]\,\delta(P_z-P_{z0}). \qquad (4.142)
$$

To integrate f, it is convenient to write $p_x^2+p_y^2 = p_\perp^2$, so that $\mathrm{d}p_x\mathrm{d}p_y = \frac{1}{2}\mathrm{d}p_\perp^2$; this follows because of axial symmetry. Since the distribution in P_z is a δ-function and integration over θ contributes 2π, the required integral is

$$
n(r) = \int f\,\mathrm{d}p_x\mathrm{d}p_y\mathrm{d}P_z\,\mathrm{d}\theta = \pi C\int \mathrm{d}p_\perp^2\,\delta(F) \qquad (4.143)
$$

where F is the argument of the δ-function in equation (4.142). To evaluate the integral we need the δ-function in the form $\delta(p_\perp^2-p_{\perp 0}^2)$. Setting $F = 0$ in equation (4.142), we have

$$
p_{\perp 0}^2 = \left\{\frac{(H_0-q\phi)^2}{c^2} - (P_{z0}-qA)^2 - m_0^2c^2\right\}. \qquad (4.144)
$$

Making use of the relation

$$
\delta\{F(p_\perp^2)\} = \frac{\delta(p_\perp^2-p_{\perp 0}^2)}{\mathrm{d}F/\mathrm{d}p_\perp^2} \qquad (4.145)
$$

from equation (4.43) we obtain

$$n(r) = \pi C \int \frac{\delta(p_\perp^2 - p_{\perp 0}^2)\,\mathrm{d}p_\perp^2}{\mathrm{d}f/\mathrm{d}p_\perp^2}$$
$$= \frac{2\pi C}{c^2} \int \delta(p_\perp^2 - p_{\perp 0}^2)\{m_0^2c^4 + c^2p_\perp^2 + c^2(P_{z0} - qA)^2\}^{1/2}\,\mathrm{d}p_\perp^2 \,. \tag{4.146}$$

The last term may be written as $(H_0 - q\phi)^{1/2}$ (equation 4.144). Writing $H_0 = \gamma_0 m_0 c^2$ and integrating

$$n(r) = 2\pi C\gamma_0 m_0 \left(1 - \frac{q\phi}{\gamma_0 m_0 c^2}\,\mathrm{U}(p_\perp^2)\right) \tag{4.147}$$

where U is the Heaviside step function, which is zero for negative argument and unity for positive argument. This takes into account whether or not the argument of the δ-function is within the range of integration. Physically, the condition $p_\perp^2 = 0$ determines the maximum extent of the particle motion allowable from energy considerations. Particles are in a well of depth $\int q\phi\,\mathrm{d}r$, and when $p_\perp^2 = 0$ the motion is parallel to the axis. The constant C may be expressed in terms of n_0, so that, writing $p_\perp^2$ explicitly, we have finally

$$n(r) = n(0)\left\{1 - \frac{q\phi\ r}{\gamma_0 m_0 c^2}\right\}\mathrm{U}\left\{\frac{(H_0 - q\phi)^2}{c^2} - (P_{z0} - qA)^2 - m_0c^2\right\}. \tag{4.148}$$

This equation gives the density as a function of the potential and the constants of the motion H_0 and P_{z0}. The charge density is found by solving equation (4.149) simultaneously with Poisson's equation. Before doing this, however, it is of interest to find also the current distribution.

This is found by calculating $n\langle\beta_z c\rangle(r)$ rather than $n(r)$. Instead of equation (4.143), we have

$$n\langle\beta_z\rangle = \int \frac{f\ \mathrm{d}p_x\ \mathrm{d}p_y\ \mathrm{d}P_z \mathrm{d}\theta(P_{z0} - qA)}{\gamma m_0 c}$$
$$= \pi C \int \frac{\mathrm{d}p_\perp(P_{z0} - qA)\,\delta(F)}{\gamma m_0 c} \tag{4.149}$$

Following through the algebra in an analogous way, we find

$$n\langle\beta_z\rangle = \left[n(0)\ \frac{\{P_{z0} - qA_z(r)\}}{\gamma_0 m_0}\right]\mathrm{U}(p_\perp^2). \tag{4.150}$$

To complete the problem it now remains to solve equations (4.148) and (4.149) simultaneously with Maxwell's equations to determine the spatial distribution of $\underline{E}$ and $\underline{B}$. Once these fields are known, orbits may be calculated if required. Further properties such as density and current distributions can also be established.

Before evaluating the fields we suppose that there is a neutralizing background that everywhere reduces the charge density in the ratio $1-f$. In principle we can postulate stationary massive particles as a background, though these could alternatively be replaced by a further self-consistent dynamic distribution. The fields are best found by evaluating the potentials

$$\operatorname{div}\underline{E} = -\nabla^2\phi = -\frac{1}{r}\frac{\mathrm{d}}{\mathrm{d}r}\left(r\frac{\mathrm{d}\phi}{\mathrm{d}r}\right) = \begin{cases} \dfrac{q(1-f)}{E_0}\, n(0)\left(1 - \dfrac{q\phi}{\gamma_0 m_0 c^2}\right) & r \le a \\ 0 & r > a \end{cases} \tag{4.151}$$

$$\operatorname{curl}\underline{B} = -\nabla^2 A_z = -\frac{1}{r}\frac{\mathrm{d}}{\mathrm{d}r}\left(r\frac{\mathrm{d}A_z}{\mathrm{d}r}\right) = \begin{cases} \dfrac{qn(0)}{\gamma_0 m_0}\,(P_{z0} - qA_z) & r \le a \\ 0 & r > a \end{cases} \tag{4.152}$$

where a is the radius at which the function U in equations (4.147) and (4.150) becomes zero. The potentials ϕ and A_z are taken as zero at $r = 0$, and the collisionless skin depth D is introduced:

$$D^2 = \frac{\varepsilon_0 \gamma m_0 c^2}{n_e(0) q^2} = \frac{\gamma}{4\pi n_e(0) r_0}\,. \tag{4.153}$$

In terms of these quantities the solution of equations (4.149) and (4.151) can be written

$$\phi(r) = \frac{\gamma m_0 c^2}{q}\left[1 - \mathrm{I}_0\left\{\frac{r}{D}(1-f)^{1/2}\right\} + \mathrm{U}(r-a)\frac{a}{D}(1-f)^{1/2}\,\mathrm{I}_1\left\{\frac{a}{D}(1-f)^{1/2}\right\}\ln\frac{r}{a}\right] \tag{4.154}$$

$$A(r) = \frac{P_{z0}}{q}\left[1 - \mathrm{I}_0\left(\frac{r}{D}\right) + \mathrm{U}(r-a)\,\frac{a}{D}\,\mathrm{I}_1\left(\frac{a}{D}\right)\ln\frac{r}{a}\right] \tag{4.155}$$

where I_0 and I_1 are the modified Bessel functions and U is again the Heavside step function. From these potentials the fields may be found by integration of equations (4.151) and (4.152):

$$E_r(r) = \begin{cases} \dfrac{\gamma_0 m_0 c^2}{q} \dfrac{(1-f)^{1/2}}{D} \; I_1\left\{\dfrac{r}{D}(1-f)^{1/2}\right\} & r<a \\ \dfrac{\gamma m_0 c^2}{q} \dfrac{(1-f)^{1/2}}{D} \dfrac{a}{r} \; I_1\left\{\dfrac{a}{D}(1-f)^{1/2}\right\} & r>a \end{cases} \tag{4.156}$$

$$B_\theta(r) = \begin{cases} \dfrac{P_{z0}}{qD} I_1\left(\dfrac{r}{D}\right) & r<a \\ \dfrac{P_{z0}a}{qDr} I_1\left(\dfrac{a}{D}\right) \; . & r>a \end{cases} \tag{4.157}$$

The variation of density and axial velocity $\beta_z c$ with radius may be found by combining equations (4.147) and (4.150) with equations (4.154) and (4.155), making use also of equation (4.142). This yields

$$\frac{n(r)}{n(0)} = I_0\left\{\frac{r}{D}(1-f)^{1/2}\right\} = \frac{\gamma(r)}{\gamma_0} \tag{4.158}$$

$$\frac{\beta_z(r)}{\beta_z(0)} = \frac{I_0\ (r/D)}{I_0\{(r/D)(1-f)^{1/2}\}} \; . \tag{4.159}$$

Equations (4.156) - (4.159) show how the fields, particle density, kinetic energy, and velocity vary with radius, in terms of the values on the axis. What is not immediately evident, however, is the beam radius a; this is found by substituting for A_θ and ϕ from equations (4.154) and (4.155) into equation (4.156), with $U = 0$,

$$H_0^2 I_0^2\left\{\frac{r}{D}(1-f)^{1/2}\right\} = P_{z0}^2 c^2 I_0^2\left(\frac{r}{D}\right) + m_0^2 c^4 + p_\perp^2 c^2 . \tag{4.160}$$

Setting $p_\perp^2 = 0$ gives the required value for a.

Having obtained these equations, we now look at the nature of the solutions. For a neutralized beam, in which $f = 1$, n and γ are independent of radius, but β_z and hence

i_z increase with radius. This increase of β_z is to be expected, since, as we observed earlier, particles at the edge of the beam are moving in the z-direction. For beams in which a/D is large $\beta_z(r)/\beta_z(0)$ is large. Since γ is independent of r this implies that particles crossing the axis make a large angle with it, as illustrated in Fig.4.21. If f

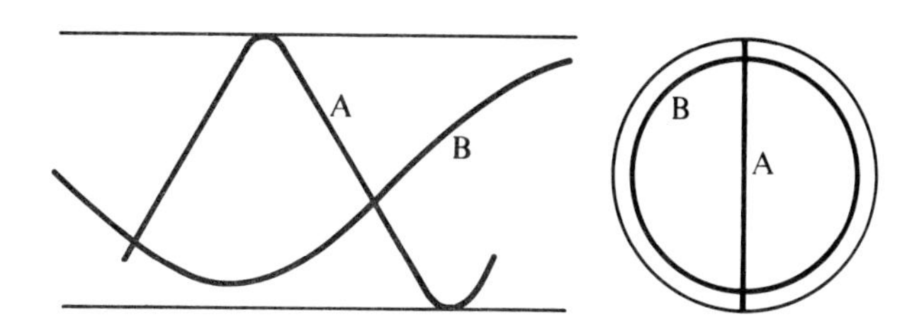

Fig.4.21. Orbits in the beam of Hammer and Rostoker. Orbit A is planar and reaches the edge of the beam; orbit B is helical.

is decreased, then both n and γ increase with radius; the beam tends to become hollow with fewer, slower electrons near the axis. No value of a exists for f = 0. This is physically obvious, since the electrostatic repulsive force always exceeds the magnetic attractive force. For small f the solutions in which a/D is large have most of the charge at large radius. Solutions exist, however, for arbitrarily large current for all values of f other than zero.

It may be argued that the configuration represented by this beam is highly artificial and unlikely to be produced in practice. Certainly if one considers launching it from some form of gun and cathode it becomes difficult to provide appropriate matching conditions at the boundary. It does, however, show some features which have been observed experimentally, in particular the absence of any 'limit' at $\nu/\gamma = 1$, and a hollow shape. Simpler (but even more artificial) models with these properties can be devised; some of these have been reviewed by Lawson (1973).

For $a/D \gg 1$ asymptotic forms can be used for the Bessel functions $I_0(x) \approx 1 + \frac{1}{4}x^2$, $I_1(x) = \frac{1}{2}x$. It is readily verified that in this limit the transverse distributions of axial velocity and density are independent of radius, that $\nu/\gamma \ll 1$ and $\beta_z \ll \beta_\perp$, and the distribution of transverse velocities is that of Kapchinskij and Vladimirskij. The beam is identical with that described in section 4.3.3. Toroidal versions of essentially the same equilibrium have been studied, both with and without the paraxial approximation. In one of these treatments (Davidson and Lawson 1972) the description of a paraxial ring obtained by this technique is compared with that given in section 4.5.2. In practical ring beams the assumption that P_θ and H are constant for all particles is not a good one. There is a spread in both these quantities and the profile is typically not far from being gaussian.

Many theoretical studies of beam configurations, linear, toroidal, and in the form of cylindrical layers, have been made using this technique. A much studied distribution function is the 'rigid rotor'

$$f(H, P_\theta, p_z) = f(H - \omega_R, P_\theta, p_z) \tag{4.161}$$

which has a mean angular velocity of ω_R about the axis. The functional form is commonly taken to be a δ-function, or exponential, and this class of distribution represents a generalization of that studied in section 3.2.8. More details are given by Davidson (1974).

4.R.6. Notes and references A great deal of experimental activity directed towards the study of intense relativistic beams of energy several Mev with ν/γ of order unity is continuing at the present time. Linear and toroidal configurations with many different environments, such as conducting walls, cavities, and axial magnetic fields in the presence or absence of plasma or residual gas, are being explored. It is difficult at this stage to give a succint summary of the state of knowledge; this will not be attempted, though some references to the discovery of previously unobserved phenomena have been made in earlier sections. A short popular review with emphasis on applications has been given by Fleischmann (1975), but for more details reference should be made to recent conference proceedings.

The detailed behaviour of these beams is often very complicated, and the rather precise correlation between theory and experiment to be expected in the theory of accelerators or microwave tubes is not often achieved. Theoretical treatments, especially of time-dependent phenomena, are often sketchy or somewhat empirical, and, while they may indicate the type of behaviour to be expected, they seldom allow close qualitative predictions to be made. Progress towards better understanding will doubtless continue to be made.

5

BEAMS WITH SCATTERING OR DISSIPATION

5.1. Introduction

In all situations studied so far, energy loss and scattering of the beam particles on one another or in background gas or plasma have been neglected. Interaction between an individual particle and the smoothed-out effect of the fields of the other particles manifested as self-electric and self-magnetic fields has been treated in Chapter 4. The motion of particles in these fields satisfies Liouville's theorem; in the present chapter we consider some processes of practical importance which are not governed by Liouville's theorem in six-dimensional phase space (μ-space).

Four classes of process will be considered. These are scattering and radiation in the fields of external agents, and the same two processes arising from fields within the beam. In the first category is scattering on residual gas in particle accelerators and storage rings; this gives rise to a continuous rise in normalized emittance. Closely related to this is the cooling of one beam by another, mingled with it and travelling at the same velocity. The second category is exemplified by the dynamics of beams in electron synchrotrons, where the radiation arising from the curvature of the orbits gives rise both to energy loss and scattering associated with the quantum nature of the emitted radiation. Scattering between particles within a single beam illustrates the third category. This occurs in the cross-over in electron-microscope beams and in electron storage rings. In these two situations it is known as the 'Boersch effect' and 'Touschek effect' respectively. Finally, radiation associated with the self-field of the beam occurs in the somewhat unrealistic but conceptually interesting radiation-limited stream described by Budker (1956).

There is one important area in the application of beams in microwave tubes which will not be treated; this is the very extensive field covered by the word 'noise'. Although of great interest and importance, it is more appropriately studied in connection with the whole system rather than just the beam. It is worth remarking, however, that it is only their superior noise characteristics which gives low-power travelling-wave tubes an advantage over corresponding solid-state devices.

5.2. Multiple scattering of a beam in a background gas or plasma

The scattering of an initially parallel beam of particles in a uniform gas is well understood. When the energy loss of the particles from radiation or ionization of the background gas can be neglected and the scattering angle is small, so that $\sin\theta \approx \theta$, the behaviour is particularly simple. For a very thin gas layer relatively few particles are scattered, and those which are have an angular distribution characteristic of the particular scattering centre. For a thicker layer, assuming that the scattering cross-section is a rapidly decreasing function of θ, particles near $\theta = 0$ will have made several collisions, and the distribution becomes approximately gaussian with a 'single scattering tail'; at large angles the distribution still has the form appropriate to single scattering. As the path through the gas increases in length, the gaussian part of the distribution spreads further out, until the gaussian distribution becomes a good approximation for the whole angular range. This behaviour is described for example in the book by Rossi (1952).

We now analyse this effect for a beam of particles colliding with heavy stationary scattering centres, represented, for example, by the nuclei of atoms in nitrogen gas. The cross-section for scattering into a solid angle $d\Omega$ of particles of unit electronic charge is given by the Rutherford scattering formula

$$\sigma(\theta)\,d\Omega = \frac{Z^2 r_0^2}{4\beta^4\gamma^2}\,\operatorname{cosec}^4(\tfrac{1}{2}\theta)\,d\Omega \qquad (5.1)$$

where r_0 is the classical radius of the incident particle and Z is the charge of the scattering nucleus. For small angles

$$\sigma(\theta) \approx \frac{4Z^2 r_0^2}{\beta^4 \gamma^2 \theta^4} . \tag{5.2}$$

This tends to infinity as $\theta \to 0$, and, furthermore, the expression for the r.m.s. scattering angle

$$\langle \theta^2 \rangle = \frac{\int \sigma(\theta) 2\pi\theta^3 d\theta}{\int \sigma(\theta) 2\pi\theta d\theta} \tag{5.3}$$

diverges logarithmically at large and small angles. To resolve this, upper and lower limits to θ must be found. Different limits are appropriate to un-ionized media, where the nucleus is screened by orbital electrons, and plasma, where Debye screening (section 4.4.1) occurs. Furthermore, it is necessary to enquire whether a classical or quantum treatment of the scattering is appropriate. (Although both theories give the same result for an unscreened point nucleus, they differ when screening is present and the nucleus is of finite size).

Fortunately, since $\theta_{max}/\theta_{min}$ is a large number, which occurs as the argument of a logarithm, a rough estimate of the angles suffices. In a classical description the maximum scattering angle, associated with a head-on collision, is π. The approximation $\theta \approx \sin\theta$ is no longer valid, however, and θ_{max} is conventionally set equal to $\frac{1}{2}\pi$. Quantum mechanically, the maximum angle is the diffraction angle associated with the nuclear radius r_n,

$$\theta_{max} = \frac{\lambda\!\!\!^{-}}{r_n} = \frac{\lambda\!\!\!^{-}}{0\cdot 57 Z^{1/3} r_e} = \frac{1\cdot 75\, r_0}{\alpha_f \beta\gamma r_e Z^{1/3}} \tag{5.4}$$

where λ is the de Broglie wavelength of the incident particle and $\alpha_f = q^2/2\varepsilon_0 hc = 1/137$ is the fine-structure constant. When β is sufficiently large this gives a value of θ less than $\frac{1}{2}\pi$, which should therefore be used.

Similar arguments can be applied to the minimum angle, θ_{min}. The appropriate quantum-mechanical and classical expressions correspond to the diffraction angle associated with the screening radius and the classically calculated scattering angle at the same radius. For a radius a these two expressions are as follows:

$$\theta_{min} \text{ (q.mech.)} = \frac{\lambda\!\!\!^{-}}{a} = \frac{r_0}{\alpha_f \beta\gamma a} \tag{5.5a}$$

$$\theta_{min} \text{ (classical)} = \frac{2Zr_0}{\beta^2 \gamma a} \tag{5.5b}$$

From these expressions it is evident that the diffraction formula applies when $\beta > 2Z\alpha_f$. These expressions refer to the scattering of a singly charged particle of classical radius r_0 on a massive target particle with charge Zq. For scattering in an un-ionized medium the quantum-mechanical formula is normally applicable, and the radius is set equal to the screening radius

$$r_s = \frac{\lambda\!\!\!^{-}\alpha^2 Z^{1/3}}{r_e} = \frac{\alpha Z^{1/3}}{\beta\gamma}\left(\frac{m_0}{m_e}\right) \tag{5.6}$$

When the scattering medium is a plasma, it is plausible that the Debye shielding distance λ_D should be substituted for a, though a rigorous justification for this is not simple. A more careful discussion of the problem of finding suitable values for θ_{max} and θ_{min} is given by Sivukhin (1966). Having established values sufficiently accurate for our present purpose we now leave the subject and return to the problem of multiple scattering in a gas. If now there are $N_A/10A$ scattering centres per kg m^{-2}, where N_A is Avogadro's number, 6×10^{23} mol^{-1}, and A is the atomic weight of the scattering centre, the increase in the mean-squared scattering angle after passing through Δl kg m^{-2} is

$$\Delta\langle\theta^2\rangle = \frac{N_A}{10A}\Delta l \int_{\theta_{min}}^{\theta_{max}} \theta^2 \sigma(\theta) 2\pi\theta \, d\theta \, . \tag{5.7}$$

Substituting for σ from equation (5.2) and integrating, this becomes

$$\Delta\langle\theta^2\rangle = \frac{8\pi Z^2 N_A r^2}{10A\beta^4\gamma^2} \ln\left(\frac{\theta_{max}}{\theta_{min}}\right) \Delta l. \tag{5.8}$$

When the scattering material is un-ionized, θ_{max} and θ_{min} may be substituted from equations (5.4), (5.5) and (5.6) to give

$$\Delta\langle\theta^2\rangle = \frac{16\pi N_A Z^2 r_0^2}{10\beta^4\gamma^2 A} \ln(183Z^{-1/3})\,\Delta l. \tag{5.9}$$

The form of equation (5.9) suggests that we measure length in dimensionless radiation lengths. One radiation length is defined as

$$L_0 = \frac{10A}{4\alpha_f N_A Z^2 r_e^2 \ln(183Z^{-1/3})} \text{ kg m}^{-2} \tag{5.10}$$

where r_e is the classical electron radius and α_f is the fine-structure constant. For hydrogen, carbon, and lead L_0 is respectively (in kg m^{-2}) 1380, 520, and 59. If S is a distance measured in radiation lengths and ρ the density, then

$$\Delta S = \frac{\Delta l}{L_0} = \frac{\rho\Delta s}{L_0} \text{ m}. \tag{5.11}$$

Furthermore, from equations (5.8) - (5.10) it follows that

$$\Delta\langle\theta^2\rangle = \frac{4\pi}{\alpha_f}\left(\frac{m_e}{m_0}\right)^2 \frac{\Delta S}{\beta^4\gamma^2}. \tag{5.12}$$

So far we have calculated the mean-squared scattering angle. It is of interest now to calculate both the angular and lateral distribution functions, assuming sufficient thickness to ensure that they are gaussian. This should be a good approximation when the mean-squared scattering angle greatly exceeds that for a single collision. We use now a method due to Fermi, described by Rossi and Greisen (1941). We retain the small-angle approximation, but work now with angles projected on the XS plane. Capital letters denote dimensionless lengths measured in radiation lengths; lower-case letters denote actual lengths. If θ_p is the projected angle on the XS plane, $\theta_p = dX/dS = dx/ds = x'$, and $\langle\theta^2\rangle = 2\langle\theta_p^2\rangle = 2\langle x'^2\rangle$. If now $F(S,X,x')$ is the distribution function in projected angle and lateral displacement X after traversing a thickness S, it may be shown that

$$\frac{\partial F}{\partial S} = -x'\frac{\partial F}{\partial X} + \frac{1}{w^2}\frac{\partial^2 F}{\partial x'^2} \tag{5.13}$$

where

$$\frac{4}{w^2} = \frac{\Delta\langle\theta^2\rangle}{\Delta S} = \frac{2\Delta\langle x'^2\rangle}{\Delta S} = \frac{4\pi}{\alpha_f\beta^4\gamma^2}\left(\frac{m_e}{m_0}\right)^2 . \tag{5.14}$$

It may be readily verified that

$$F(S,X,x') = \frac{\sqrt{3}}{2\pi}\frac{w^2}{S^2}\exp\left\{-w^2\left(\frac{x'^2}{S}-\frac{3x'X}{S^2}+\frac{3X^2}{S^3}\right)\right\} \tag{5.15}$$

is a solution of equation (5.13). For $X \neq 0$ the angular distribution is gaussian centred on an angle $3X/2S$. Integrating over X and x' respectively yields

$$G(S,x') = \frac{1}{2\sqrt{\pi}}\frac{w}{S^{1/2}}\exp\frac{-w^2x'^2}{4S} \tag{5.16}$$

$$H(S,X) = \frac{\sqrt{3}}{2\sqrt{\pi}}\frac{w}{S^{3/2}}\exp\frac{-3w^2X^2}{4S^3} . \tag{5.17}$$

The mean-squared angle $\langle x'^2\rangle$ increases as S, but $\langle X^2\rangle$ increases as S^3, and the ratio of the r.m.s. values of X and x' is $S/\sqrt{3}$.

The r.m.s. emittance $\bar{\varepsilon}$ can be found as a function of S from equations (4.38), (5.16), and (5.17):

$$\bar{\varepsilon} = \frac{4L_0(\langle X^2\rangle\langle x'^2\rangle)^{1/2}}{\rho} = \frac{8S^2L_0}{\sqrt{3}w^2\rho} = \frac{8s^2\rho}{\sqrt{3}w^2L_0} = \frac{8\pi s^2\rho}{\sqrt{3}\alpha\beta^4\gamma^2L_0}\left(\frac{m_e}{m_0}\right)^2 . \tag{5.18}$$

In the absence of any focusing, the emittance grows as the square of the distance traversed by the beam.

5.3. Multiple scattering in the presence of focusing

In the previous section scattering of a beam of particles in a homogeneous slab of material was calculated. In an accelerator there is focusing about an axis, and the behaviour is somewhat different. First, provided that the increase in scattering angle per betatron wavelength is small, the distribution in x' is directly related to that in x. The ratio x/x', equal to $L_0X/\rho x'$, is simply $\lambda/2\pi$, where λ is the focusing wavelength equal to C/Q in an orbital machine. Secondly the rate of increase of $\langle x'^2\rangle$ with distance is only half as much as it is in the absence of focusing. This is because the occurrence of a scattering event which deflects the particle through an angle $\delta x'$ does not in general result in a change $\delta x'$ in the angle x'_0 at which the trajectory crosses the axis. It is readily verified from geometrical considerations that for a trajectory for which $x' = x'_0 \sin 2\pi x/\lambda$, the change in x_0 is $\delta x' \cos 2\pi x/\lambda$. From this it follows that $\langle (\delta x_0)^2 \rangle = \frac{1}{2}\langle (\delta x)^2 \rangle$, and therefore that the angular spread builds up just half as fast as for an unfocused beam.

In the presence of a focusing field, therefore, equation (5.18) is replaced by

$$\bar{\varepsilon} = 4(\langle x^2 \rangle \langle x'^2 \rangle)^{1/2} = \frac{4\lambda}{2\pi}\langle x'^2 \rangle = \frac{2\lambda S}{\pi w^2} = \frac{2\lambda \rho s}{\alpha_f \beta^4 \gamma^2 L_0}\left(\frac{m_e}{m_0}\right)^2. \tag{5.19}$$

The emittance is now proportional to the distance travelled by the beam and the beam diameter varies as the square root of s. This expression does not depend on λ, and for a beam with different focusing strengths in the x and y directions is true for either. The rate of change of emittance with distance is given by

$$\left(\frac{d\bar{\varepsilon}}{ds}\right)_{\text{scatt.}} = \frac{2\lambda\rho}{\alpha_f \beta^4 \gamma^2 L_0}\left(\frac{m_e}{m_0}\right)^2. \tag{5.20}$$

If the beam is accelerated, there is a negative contribution to $d\varepsilon/ds$ which arises from the fact that it is $\bar{\varepsilon}_n$ rather than $\bar{\varepsilon}$ which is invariant. Since $\bar{\varepsilon}_n = \beta\gamma\bar{\varepsilon}$

$$\left(\frac{d\bar{\varepsilon}}{ds}\right)_{\text{acc.}} = -\frac{\bar{\varepsilon}_n}{\beta^2\gamma^2}\frac{d}{ds}(\beta\gamma) = -\frac{\varepsilon_n \gamma'}{\beta^3\gamma^2}. \tag{5.21}$$

Adding these two contributions gives

$$\left(\frac{d\bar{\varepsilon}}{ds}\right)_{\text{tot.}} = \frac{1}{\beta^4\gamma^4}\left\{\frac{2\lambda\rho}{\alpha_f L_0}\left(\frac{m_e}{m_0}\right)^2 - \beta\bar{\varepsilon}_n\gamma'\right\}. \tag{5.22}$$

At low energies the second term is small and the net effect is an increase of emittance. At high energies damping predominates provided that γ' is large enough; this is the situation encountered in cyclic accelerators.

Equation (5.22) specifies what happens in an accelerator in the absence of walls. In practice one wishes to design for a pressure which will not result in excessive loss on the walls. Even in the steady-state condition, where damping just balances scattering, it is not simple to estimate the rate of loss of beam. The distribution function is no longer gaussian, and is necessarily zero at the walls. If the beam expands and then contracts during the accelerating cycle, it is only in this balanced condition for a short while, and if the walls are never less than a few standard deviations from the beam axis the loss should be small. To determine precisely what 'a few' means requires detailed calculations. The problem of loss shortly after injection was important in the early days of synchrotrons, since it was the determining factor in specifying the gas pressure. With the advent of better vacuum techniques and higher injection energies, however, the loss is quite negligible, though it is of interest to estimate the growth in emittance.

The loss problem is treated rather generally in the book by Kolomenskij and Lebedev (1966), and references are given to early papers in this field. It is also analysed in detail by Bruck (1966). Emittance growth and loss are also calculated by Hardt (1968) who gives curves in a practically convenient form. A more general calculation leading to the envelope equation (4.37) but including the effect of scattering, has been made by Lee and Cooper (1976). The transverse density distribution is not uniform in the presence of scattering, so that some approximations involving averaging are required. The self-consistent Bennett distribution for a pinched beam with Maxwellian velocity distribution was analysed in section 4.4.5. It has been shown by Lee (1976) that a pinched beam with arbitrary initial transverse velocity distribution and density profile passing through a scattering background tends to the distribution given by equation (4.69). Experiments are reported by Briggs *et al.* (1976) in which this profile is attained in a 5 MeV electron beam of 220 A, passing through nitrogen and argon in the pressure range 5-100 torr.

5.4. Scattering between beam particles in storage rings

In very many situations the scattering between individual particles in beams is completely negligible. It is of importance, however, in particle storage rings, where beams are stored for many hours, and the object is to observe collisions between particles in beams moving in opposing directions.

Several classes of collision may be distinguished in storage rings; collisions can be between particles in the same beam or opposing beams, and they can be such as to remove the particles from the beam or merely to change their oscillation amplitude about the equilibrium orbit or stable phase. Further, there is a considerable difference of behaviour between proton machines, where synchrotron radiation is negligible, and electron machines, where it dominates the dynamics.

We consider first proton machines, in which there is no damping. Head-on collisions between the particles in the two beams result in processes in which the incident particles and others which may be created leave the beam. It is this type of process that the apparatus is designed to study, and the effect is to produce a decay rate in stored current proportional to the beam density. Small-angle elastic scattering does not give rise to loss, but produces a small continuous increase of emittance. Both these effects, however, are small compared with that of collisions between particles in the same beam, where particles remain relatively close for longer periods, allowing for larger exchange of momentum. In proton machines this type of scattering produces a steady growth of emittance, but a negligible number of particles are scattered out of the beam.

In electron storage rings conditions are somewhat different. In collisions between two beams the production of bremsstrahlung can give rise to loss of energy; this can be the dominant mechanism by which particles are lost from the machine. If, however, the particles are not lost from the machine, the outward diffusion from this process is counteracted by the strong radiation damping described below (section (5.10)).

For scattering between particles in the same beam, the diffusion process of importance in proton machines is again overcome by damping. Collisions involving loss of particles are, however, sometimes important. This phenomenon, known as the 'Touschek effect', depends on the momentum distribution in the beam, the emittance which can be contained by the vacuum chamber, and the Rutherford scattering cross-section. This decreases rapidly with energy (equation (5.1)), and it is found that the phenomenon is only of importance at low energies. It should be noted that it is not necessary that to be lost a particle should be scattered through a large angle; if it suffers simply an increase or decrease in momentum, its equilibrium orbit radius can change sufficiently for it to strike the walls of the vacuum chamber. Further discussion and a detailed analysis, together with references, may be found in the book by Bruck (1966).

We conclude by noting that in proton storage rings diffusion processes are of most interest, whereas in electron machines only single scattering events are of importance. This is because the rate of growth of the beam cross-section arising from diffusion processes is small compared with damping associated with the radiation.

Before studying any further processes in detail, we look at some properties of a beam with finite temperature.

5.5. Some properties of a beam with finite temperature

In chapter 4 several beam configurations with transverse Maxwellian velocity distribution were studied. In the paraxial approximation it is a good assumption in most situations to neglect the longitudinal velocity distribution. If the beam is produced from a cathode and then accelerated, the longitudinal velocity spread, and hence the longitudinal energy measured by an observer moving with the beam, is reduced also. Thus, if two electrons are emitted with velocities 0 and v, and they are accelerated by a field which brings the first up to a (non-relativistic) energy $\frac{1}{2}m_0V^2$, the second has energy $\frac{1}{2}m_0(V^2+v^2)$, which is equal to $\frac{1}{2}m_0(V+\Delta V)^2$, where ΔV is the relative velocity between the electrons. The second electron then has energy with respect to the first in the frame in which it is at rest:

$$\begin{aligned} \Delta E &= \tfrac{1}{2} m_0 (\Delta V)^2 \\ &\approx \frac{m_0 v^4}{8V^2} \quad \text{when } v \ll V. \end{aligned} \tag{5.23}$$

The transverse temperature is unaltered by the acceleration in the longitudinal direction, so that in a beam accelerated from a cathode the transverse temperature greatly exceeds that in the longitudinal direction. The effect of collisions in a continuously focused beam is to redistribute the energy uniformly between the transverse and longitudinal directions in the frame of reference of the beam. This results in an increased energy spread in the laboratory frame. This phenomenon, responsible for the Touschek effect, was first discovered experimentally by Boersch (1954) where it occurs in the cross-over of high-brightness electron beams. It is discussed below in section 5.6.

Equation (5.23) was introduced in connection with emission from a cathode. In accelerators and storage rings, where the beam is injected, the ratio of transverse to longitudinal energies observed when moving with the beam depends on the particular situation. Indeed, in circular machines the concept of 'moving with the beam' needs careful examination. It is found that in a sense the (non-relativistic) temperature in the longitudinal direction is m^*v^2, where m^* is the effective mass. When this is negative, this implies a 'negative temperature' which is hotter than all positive temperatures so that energy feeds continuously from the longitudinal to the transverse motion without equilibrium ever being established. This has been shown by Piwinski (1974).

In orbital machines, particles with different energies have different equilibrium orbits, and a general analysis of diffusion effects is rather complicated.

To conclude this section we return to a rectilinear system, and find the relativistic relation between energy differences in the stationary and moving frames. We consider two particles with normalized energies γ and $\gamma+\Delta\gamma$ in the stationary frame, where $\Delta\gamma \ll \gamma$. These two particles have normalized velocities

$$\begin{aligned}\beta_1^2 &= 1 - \frac{1}{\gamma^2} \\ \beta_2^2 &\approx 1 - \frac{1}{\gamma^2} + \frac{2\Delta\gamma}{\gamma^3} \approx \beta_1^2\left(1+\frac{2\Delta\gamma}{\beta_1^2\gamma^3}\right) .\end{aligned} \tag{5.24}$$

In the moving frame, denoted by subscript m, the relative velocities can be found from the relation for the addition of velocities

$$\begin{aligned}\beta_{2m} &= \frac{\beta_2 - \beta_1}{1-\beta_1\beta_2} \\ &= \beta_1\gamma^2\left\{\left(1+\frac{2\Delta\gamma}{\beta_1^2\gamma^3}\right)^{1/2} - 1\right\}.\end{aligned} \tag{5.25}$$

When γ is sufficiently large that $\gamma^2 - 1 \approx 2\Delta\gamma/\gamma$ this may be simplified to

$$\beta_{2m} \approx \frac{\Delta\gamma}{\beta_1\gamma} \approx \frac{\Delta\gamma}{\beta_2\gamma} . \tag{5.26}$$

For large γ this depends only on $\Delta\gamma/\gamma$ and is always non-relativistic. When $\gamma \approx 1$ the expression (5.25) reduces to $\beta_{2m} = \beta_2 - \beta_1$.

5.6. The Boersch effect

In the previous section beams with different longitudinal and transverse temperatures were described. It was shown that a beam accelerated from a thermionic cathode has different transverse and longitudinal temperatures when measured in a frame of reference moving with the beam. If collisions occur this causes the longitudinal temperature to increase until it comes into equilibrium with the transverse temperature. Whether the effect is important thus depends on the relaxation time for the distribution to become isotropic in the moving frame.

In many applications the effect is negligible. In electron microscopy, however, where the electrons pass through a cross-over of very high density, it can produce observable energy broadening of the beam and reduction in brightness. Observed first experimentally by Boersch (1954), the effect has been the subject of a great deal of experimental and theoretical study, and also a certain amount of controversy. A simple physical description of the effect and estimate of its magnitude in a parallel beam has been given by Zimmermann (1969). The more complicated situation at a cross-over has been analysed by Loeffler (1969). These papers contain references to earlier contributions. The relaxation time in a parallel beam can be estimated by considering the beam as a drifting gas; this method is applied below. For the cross-over, however, the calculation involves the detailed statistics of the scattering with the appropriate distribution functions and boundary conditions. We do not reproduce this rather special calculation, but later quote the results of Loeffler.

To illustrate the nature of the effect and the relevance of ideas developed in the previous section, we now calculate the energy broadening per unit length in a parallel beam. Seen in a frame moving with the beam, the process is essentially that of relaxation to an isotropic Maxwellian distribution of a gas with transverse temperature $kT_{\perp}$ and longitudinal temperature $kT_{\parallel m}$, where $kT_{\perp} \gg kT_{\parallel m}$. If τ is the relaxation time, then

$$\frac{\mathrm{d}T_{\parallel m}}{\mathrm{d}t} = \frac{T_{\perp}}{\tau} \tag{5.27}$$

where the temperatures and time t_{m} are measured in the moving frame. From the relation $z = \beta ct = \beta\gamma ct_{m}$, where z is the distance measured in the stationary frame, it follows that

$$\frac{\mathrm{d}T_{\parallel m}}{\mathrm{d}z} = \frac{T_{\perp}}{\beta\gamma c\tau} \tag{5.28}$$

and hence that

$$T_{\parallel m} = T_{\parallel m0} + \frac{T_{\perp}z}{\beta\gamma c\tau}\,. \tag{5.29}$$

where the subscript 0 denotes conditions at $z = 0$. It is now required to express $T_{\|}$ in terms of the energy spread in the stationary frame. This may be done by squaring equation (5.26) and writing $\beta_m^2 \approx kT_{\| m}/m_0c^2$ to give

$$(\Delta\gamma)^2 m_0 c^2 \approx \beta^2\gamma^2 \, kT_{\| m} . \tag{5.30}$$

At $z = 0$, the energy spread $\Delta\gamma m_0 c^2$ is equal to kT_c, the cathode temperature, and $kT_{\| m0}$ can be found from equation (5.30). The change in energy spread which occurs when the beam drifts a distance z_0 is therefore found by substituting into equation (5.29) to give

$$\Delta\gamma m_0 c^2 \Big|_0^{z_0} = \left(k^2 T_c^2 + \frac{\beta\gamma m_0 c^2 kT_{\perp} z_0}{c\tau} \right)^{1/2} . \tag{5.31}$$

It now remains to evaluate τ. This has been calculated by Spitzer (1962) as

$$\tau = \frac{0 \cdot 29}{n r_0^2 c \ln \Lambda} \left(\frac{kT}{m_0 c^2} \right)^{3/2} \tag{5.32}$$

where n is the density; the numerical constant $0 \cdot 29$ is equal to $3^{3/2}/(8\pi \times 0 \cdot 71)$ and $\ln \Lambda$ is the logarithm corresponding to that which appears in equation (5.7). Since, however, there is a distribution of particles with different velocities, rather than particles of fixed velocity deflected by a fixed scattering centre, it is somewhat more difficult to evaluate. Details of how this may be done are given by Spitzer (1956) and may be found in standard texts. If the Debye length is less than the beam radius, as is usually the case, the factor λ_D in Λ should be reduced to a value between λ_D and the beam radius. Because Λ is large, however, the uncertainty causes little error in $\ln \Lambda$, which is typically between 6 and 20. Substituting the value of τ from equation (5.32) into equation (5.31) gives the final result, in dimensionless form,

$$\Delta\gamma = \left\{ \frac{3 \cdot 4 \nu\beta\gamma r_0 z_0}{\pi a^2} \left(\frac{m_0 c^2}{kT_{\perp}} \right)^{1/2} \ln \Lambda + \left(\frac{kT_c}{m_0 c^2} \right)^2 \right\}^{1/2} . \tag{5.33}$$

It is interesting to note that this depends on the current density ($i = \nu q \beta c/\pi a^2 r_0$), but for a given current density the only dependence on beam energy at non-relativistic energies is through the logarithmic term. As a typical example, for a beam with current density 10^4 A m^{-2}, $kT_c/q = 0{\cdot}1$ V, $\gamma m_0 c^2/q = 1$kV, and $z_0 = 0{\cdot}1$ m we find $\Delta\gamma m_0^2 c^2/q \approx 2$ V.

In order to deal with a beam in which the radius, and hence the transverse temperature, varies with z it might at first sight appear that all that is required is to insert the value of τ in equation (5.28) and integrate over z, using the fact that $T_\perp$ is inversely proportional to the cross-sectional area of the beam. This would be permissible if the beam diameter were large compared with the Debye length; if, however, this is not so, and furthermore the angle of convergence or divergence of the beam envelope is large compared with the angles associated with the thermal velocities, then the model for thermal relaxation is not valid, and a more detailed model of the interaction region needs to be considered.

Such a calculation has been made by Loeffler (1969); unfortunately, as in other problems of this type, the detailed physics of the scattering, which can only be treated approximately, appears in logarithmic terms. It is interesting to note that even if one considers a converging beam with zero temperature, both longitudinal and transverse, then the effect still exists, and so the relaxation model is not appropriate for a converging system. Loeffler finds for the energy spread an expression

$$\Delta\gamma m_0 c^2 = \frac{1}{\varepsilon_0}\left(\frac{q^3}{2\beta c}\right)^{1/2} (Ba_0)^{1/2}\{3+\ln 2 + 2\ln(8Na_0) + \tfrac{1}{4}\ln^2(8Na_0)\}^{1/2} \tag{5.34}$$

where B is the beam brightness, a_0 the radius at the waist, and N the number of particles per unit length. The correlation between energy spread and the quantity $(B_0 a)^{1/2}$ has been experimentally confirmed by Pfeiffer (1970). He has also calculated the chromatic aberration introduced by the effect, and evaluated its role in determining the minimum obtainable spot size (Pfeiffer 1972).

Further references, especially to earlier work, may be found in the review by Zimmermann (1970).

5.7. Electron cooling

A further phenomenon involving collisions in beams is 'electron cooling', suggested by Budker (1967) as a method of decreasing the emittance of a stored proton beam. An electron beam of lower temperature is made to travel with the proton beam, and energy is exchanged between them. The effect has been demonstrated experimentally (Budker *et al.* 1975), but for useful application to storage rings the relaxation times are rather long. The method of estimating these is similar to that used in the previous section; the system is examined in a frame of reference moving with the beams and the relaxation theory for a two-component Maxwellian system applied. This has been done by Budker; his result is expressed as the rate of change of θ_p^2, where θ_p is the angle which a proton orbit makes with the axis. Rearranging somewhat, this can be expressed in terms of the emittance ε_p of the proton beam with $\nu = \nu_p$:

$$\frac{d}{dt}\left(\varepsilon_p^2\right) \approx \frac{2\nu_p r_e c \ln \Lambda}{\beta^2\gamma^5\beta_{\perp e}} \tag{5.35}$$

where r_e is the classical radius of the electron and $\beta_{\perp e}$ is the characteristic perpendicular velocity of the electrons. Evidently the effect decreases very rapidly with energy.

5.8. Beams formed from runaway electrons

The beams discussed so far have been produced either from cathodes or from ion sources. To produce a ring beam some sort of injection system into a suitable vacuum vessel is required. An alternative possibility is to apply an inductive electric field to a plasma in a toroidal vacuum chamber, in the presence of a suitable guiding field. In such a situation it is possible for the faster electrons in the initial Maxwellian distribution in the plasma to be continuously accelerated. This arises because the scattering cross-section decreases rapidly with velocity, so that the effective frictional force which represents momentum transfer in the collisions decreases with velocity. At some critical velocity, therefore, the energy and momentum which an electron gains from the accelerating field exceeds that lost by collisions and it 'runs away'.

Even in a system with idealized infinite geometry the process is complicated and difficult to analyse satisfactorily. This may be seen in the paper by Connor and Hastie (1975), in which earlier work is briefly reviewed. In practical systems it depends very much on the total geometry of the system and the nature of the confining fields. Runaway electron currents of over 100 kA with energies in the MeV range have been observed in toroidal plasma devices for fusion research which have a strong azimuthal magnetic field. In these circumstances they are an embarrassment rather than an asset. Experimental observations and theory are given, for example, by Spong *et al.* (1974) in a paper which gives references to earlier work.

Attempts to make constructive use of this phenomenon in a 'plasma betatron' have several times been made. In such a device the space-charge limitations to the current discussed in section 3.4.4 should be absent. The many efforts in this direction, however, have not succeeded in producing a useful device; the behaviour is very complicated, and even if an intense beam is produced it rapidly disappears because of the negative mass or other types of instability. Conditions for confinement in a betatron-type field are clearly more stringent than in the intense azimuthal fields associated with toroidal fusion devices. Furthermore, in toroidal fusion devices there is a continuous velocity distribution of electrons, which decreases monotonically with energy, whereas in a betatron only a finite range of energies is represented. As will be seen from arguments developed in section 6.3, the continuous distribution is more stable in the presence of a background plasma.

Typical papers describing experimental work with plasma betatrons and also giving references to earlier work in this field are those of Ferrari and Rogers (1967), Ferrari, Rogers, and Landau (1968), and Bermel (1970).

It is found in practice that the value of ν for the plasma is a significant parameter for determining whether runaway is likely to occur. In typical Tokamak experiments the ratio of thermal velocity to drift velocity; equal to about $\nu^{\frac{1}{2}}$, must not be less than a factor of order 50. Many practical systems can be considered therefore as 'beams' in which $\nu \ll 1$, or 'plasmas' in which $\nu \gg 1$. This distinction is developed in detail in a paper by Lawson (1959).

5.9. Budker's relativistic self-constricted beam

A remarkable quasi-steady-state electron beam in a longitudinal electric field has been described by Budker (1956). Although such a beam is liable to instabilities, and even in the absence of instabilities there would be formidable obstacles to its practical realization, it is of considerable pedagogical interest. Here we present a simplified description which retains the essential features.

In earlier discussions of neutralized electron beams both scattering of electrons on the ions and radiation arising from the curvature of electron orbits in the self-field have been neglected. Under extreme conditions both must be taken into account. The effect of the ions is to cause build-up of betatron oscillations in a manner similar to that described in section 5.2; the radiation, on the other hand, imposes a damping force. A steady state is achieved when these effects balance.

In his original analysis of the problem Budker considered a partially neutralized electron beam in a betatron field. Changing flux through an iron core provided a steady electric field to provide energy to the electrons, but the acceleration of ions was neglected. A Bennett distribution (equation 4.69) was assumed for the beam profile, and the steady state calculated with the aid of a Lorentz transformation to a frame of reference with zero mean electron velocity.

In the present simplified treatment a beam with uniform density profile is assumed, and the calculation performed in the stationary reference frame. The beam is specified by four parameters, E, ν, γ, and the beam radius a, and relations between them are determined by writing down equations for energy and momentum balance. Extreme relativistic conditions are assumed, so that $\beta \approx 1$; ν/γ is somewhat less than unity, but not necessarily very small. The situation to be analysed is illustrated in Fig.5.1.

The power loss of a relativistic electron in a magnetic field B is given by the expression (Jackson 1963)

$$W_r = qEc = \frac{8\pi c r_e^2 \beta^2 \gamma^2 B^2}{3\mu_0} . \tag{5.36}$$

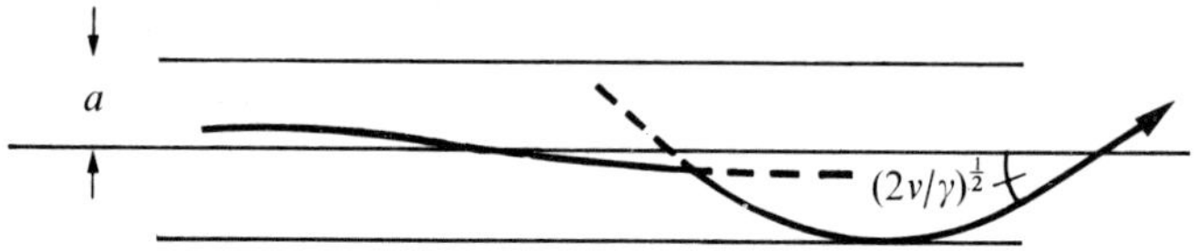

Fig.5.1. Typical electron trajectory in Budker's relativistic self-constricted beam. One scattering event is shown. The angle at which a trajectory reaching the beam edge passes through the axis is $(2\nu/\gamma)^{1/2}$. (Not all trajectories reach the beam edge, nor pass through the axis.)

Setting $\beta \approx 1$ and replacing B by a rough average value across the beam cross-section of $\mu_0 I/4\pi a = \mu_0 Nqc/4\pi a$ yields

$$Eq = \frac{q^2\gamma^2\nu^2}{6\pi\varepsilon_0 a^2} \tag{5.37}$$

The momentum-balance equation is formed by equating qE, the momentum gained in unit time, to $\dot{p}_s + \dot{p}_r$, where $\dot{p}_s$ is the rate of momentum loss by scattering, and $\dot{p}_r$ the rate at which momentum is carried away by radiation. We now calculate these two quantities. The cross-section for momentum transfer by Rutherford scattering may be written

$$\sigma = \left(\frac{r_e}{2\gamma}\right)^2 \int \frac{2\pi \sin\theta(1-\cos\theta)}{(\sin\frac{1}{2}\theta)^4} d\theta . \tag{5.38}$$

The angle θ is small over most of the range which contributes to the integral, so that

$$\begin{aligned}\sigma &\approx \frac{4\pi r_e^2}{\gamma^2}\int\frac{d\theta}{\theta}, \quad \theta \ll 1 \\ &\approx \frac{4\pi r_e^2}{\gamma^2}\ln\frac{\theta_{max}}{\theta_{min}} .\end{aligned} \tag{5.39}$$

In terms of σ,

$$\begin{aligned}\dot{p}_s &= (\gamma m_0 c^2)\frac{N\sigma}{\pi a^2} \\ &= \frac{\nu q^2 L}{\pi\varepsilon^2 a\gamma}\end{aligned} \tag{5.40}$$

where L has been written for $\ln(\theta_{max}/\theta_{min})$.

The momentum carried by the radiation field can be calculated from the radiation reaction force. This may be found from the expression for the radiated power (equation (5.36)), assuming that the force acts against the direction of electron motion. We require the component in the s direction; a crude average is found by noting that the electrons make an angle of order $r' = (\nu/\gamma)^{1/2}$ to the axis. (This may be seen by balancing the emittance and perveance terms in the paraxial ray equation (4.33)). Now the radiation force along the orbit is $8\gamma^2\nu^2 q^2/3\pi\varepsilon_0 a^2$ (equation 5.36), so that the s component is

$$\begin{aligned}\dot{p}_r &= \frac{\gamma^2\nu^2 q^2\cos\theta}{3\pi\varepsilon_0 a^2} \\ &= \frac{\gamma^2\nu^2 q^2(1-\nu/2\gamma)}{3\pi\varepsilon_0 a^2}\end{aligned} \tag{5.41}$$

where $\cos\theta$ has been approximated by $1 - \frac{1}{2}r'^2$. Equating the sum of $\dot{p}_r$ and $\dot{p}_s$ to qE yields, from equations (5.40) and (5.41),

$$4\pi\varepsilon_0 Eq = \frac{2\gamma^2\nu^2 q^2}{3a^2}\left(1-\frac{\nu}{2\gamma}\right)+\frac{4\nu q^2 L}{a^2\gamma} . \tag{5.42}$$

From equations (5.37) and (5.42) it follows that

$$\gamma\nu = (12L)^{1/2}$$
$$a = \left(\frac{2qL}{\pi\varepsilon_0 E}\right)^{1/2} . \qquad (5.43)$$

It is remarkable that each equation contains only two of the four variables. In the equilibrium state the energy of the electrons depends only on the line density N, and the beam radius depends only on the applied field.

Taking $L = 30$, a typical set of parameters for an equilibrium beam is, in practical units, energy = 20 MeV, current = 8200 A, E = 100 V m^{-1}, and $a = 3{\cdot}7 \times 10^{-5}$ m. The self-magnetic field of the beam is 44 T, a very high value. It is this property which led Budker to propose the use of such beams to provide the guide field for particle accelerators.

The numerical factors in equation (5.43) are somewhat different from those given by Budker. Some confusion exists in the literature owing to failure to distinguish between the values of γ associated with individual electrons, and the value associated with the frame of reference with respect to which the mean longitudinal velocity of the electrons is zero. Although for ν/γ small the values of β in both cases are near to unity, the values of $1-\beta^2$ and hence γ can be very different.

The sketchy calculation above illustrates the essential physics of the steady-state relativistic beam; it is greatly oversimplified, however, and, even if such beams did not suffer from instabilities, they would be very difficult to set up.

It is implicit in the above analysis that $\nu/\gamma < 1$. Toroidal self-constricted ring currents in non-relativistic plasma in which $\nu/\gamma \gg 1$ have been studied in connection with fusion research. In such plasmas collisions are important, and the electron velocity distribution is essentially isotropic and Maxwellian but with a small drift component. For a typical value $\nu = 10^4$, $\beta_{\parallel}/\beta_{\perp} \approx \nu^{-1/2} = 0{\cdot}01$. Energy- and momentum-balance calculations analogous to those for the Budker beam have been carried out independently by Pease

(1957) and Braginskij (1958). There is a difference, however, in that the energy loss occurs mainly by bremsstrahlung, and the momentum carried by it is negligible since it is no longer peaked in the forward direction but is emitted more or less isotropically. A detailed analysis is complicated, as reference to the above papers shows; nevertheless it is again possible to simplify very considerably to obtain an order of magnitude result. The only additional physical fact required is the non-relativistic cross-section for bremsstrahlung (Heitler 1954).

$$\sigma = \frac{16}{3}\alpha_f r_e^2 \tag{5.44}$$

where α is the fine-structure constant. Proceeding as before and making use of this result the equations corresponding to equation (5.38) become

$$\begin{aligned} \nu\beta &= \left(\frac{3\pi}{4}\,\frac{\ln\Lambda}{\alpha_f}\right)^{1/2} \\ a &= \left(\frac{4\alpha_f q}{3\pi^2\varepsilon_0 E}\right)^{1/2}\nu^{3/4} \end{aligned} \tag{5.45}$$

The variables do not separate out into pairs as before; the second equation contains a, E, and ν. The first of these equations is the more interesting; in practical units it is simply $I \approx 10^6$ A. It must be emphasized again that this is a highly idealized result, which assumes that the plasma is stable and isolated from the walls of the container. In practice such plasmas are limited by many types of instability. Some further comments are given by Lawson (1959).

5.10. Radiation effects in electron synchrotrons and storage rings

In high-energy electron synchrotrons radiation effects arising from transverse acceleration of the electrons in the guiding magnetic field have an important influence on the beam dynamics. In storage rings the behaviour is dominated by such effects and differs in several essential ways from that in proton storage rings. The problem is rather a special one and will not be analysed in detail. Nevertheless, the essential physical effects will be described and references to more detailed work given. The effects to be described, unlike those in the previous section, do not depend on the intensity of the beam. They represent the interaction between a single electron and the *external* magnetic focusing and guiding fields. They are of course modified at high intensities, where self-field effects of the kind described in Chapter 3 occur.

The most obvious effect of the synchrotron radiation is to reduce the energy of the circulating beam. In an accelerator this involves supplying additional energy from the accelerating gap; indeed, in high-energy machines this can greatly exceed the energy required for acceleration. In a storage ring energy must be continually supplied to replace that which is radiated. The power radiated (equation (5.36)) can be expressed in various forms, of which a convenient one is

$$W_{\mathrm{r}} = \frac{2\gamma^2 r_{\mathrm{e}}}{3m_0 c} F_{\perp}^2 \tag{5.46}$$

where

$$F_{\perp} = \frac{\gamma m_0 \beta^2 c^2}{R} = q\beta c B_z. \tag{5.47}$$

In these expressions R is the radius of curvature of the orbit and $F_{\perp}$ is the force on an electron in a direction perpendicular to its motion.

For the moment we assume that the radiation reaction force acts smoothly and continuously. (Later we consider the actual situation, that the radiation occurs in small steps; for our present purposes, however, the assumption of continuous, steady radiation is adequate.) It is evident that the radiation decreases the energy of the electrons, but it is not obvious whether the betatron and phase oscillations are necessarily damped. Indeed, as we shall see it is possible for antidamping of either radial betatron oscillations or phase oscillations to occur. The physical considerations for phase oscillations, and for vertical and radial betatron oscillations, are all different. We outline the factors in each case and quote the results. A detailed physical discussion leading to these results has been given by Sands (1971). An earlier, more formal, derivation was obtained by Robinson (1958) in a paper which refers to earlier work in this field. The material of this section is taken from these two papers, with minor changes of notation. Extreme relativistic conditions ($\beta \approx 1$) are assumed throughout, since the effects discussed are quite negligible in the non-relativistic region.

We consider first phase oscillations. In the absence of radiation these have been analysed in section 2.8. A particle makes excursions both in phase with respect to the accelerating field and in energy; these two quantities are $\pi/2$ out of phase, the energy excursion being a maximum or minimum when the rate of change of phase is greatest. In the presence of radiation the two halves of the cycle, in which the energy is either greater or less than the energy of a particle on the equilibrium orbit, are not the same. (These will be referred to as the 'positive' and 'negative' half-cycles.) In the absence of betatron oscillations there are three factors to be taken into consideration.

(1) Because of the dependence of the radiation intensity on γ, the power radiated on the positive half-cycle in a given magnetic field exceeds that radiated on the negative half-cycle.

(2) The path length in making a revolution of the machine is longer during the positive half-cycle.

(3) If field gradients are present, the fields experienced by the electron in the two half-cycles are different.

If the loss is greater during the positive half-cycle than during the negative half-cycle, the effect will be to damp the oscillations and *vice versa*. The first two factors therefore tend to cause damping; for a field which falls off with radius, as occurs in a constant-gradient machine, the third effect tends to produce growth. Betatron oscillations change the range of fields which are sampled, but do not, to first order, affect the average.

A detailed calculation of all these effects yields a damping constant for the amplitude of the phase oscillations:

$$\alpha_s = \frac{(2+D)W_r}{4\gamma m_0 c^2} \tag{5.48}$$

where W_r is the radiated power and D is a function of the lattice parameters. The parameter D may conveniently be expressed in terms of $R(s)$, the radius of curvature of the orbit, the focusing term $\kappa(s)$ defined in equation (2.103), and the momentum compaction factor α_p defined in equation (2.137) (written here as X_p to avoid confusion with the damping constants)

$$D = \oint \frac{X_p}{R}\left(\frac{1}{R^2} - 2n\right)ds \Big/ \oint \frac{ds}{R^2} \tag{5.49}$$

where the integral is taken round the orbit. For a 'separated-function' machine, in which the bending occurs in regions of uniform field, it is found that D is positive and small compared with unity, so that

$$\alpha_s \approx \frac{W_r}{2\gamma m_0 c^2} \,. \tag{5.50}$$

For a weak-focusing machine with no straight sections, on the other hand, $D = (1-2n)/(1-n)$ where n is the field index. If $n > \frac{3}{4}$, α_s is negative, corresponding to growth.

We consider now the effect of damping on vertical betatron oscillations. The positive and negative half-cycles of such oscillations are symmetrical, and the damping effect can be thought of as arising from the impulse along the orbit direction received when the particles pass an accelerating gap. This can be considered an analogous mechanisms to that which produces adiabatic damping of the motion in a continuously accelerated particle. Detailed calculations yield a damping coefficient

$$\alpha_y = \frac{W_r}{2\gamma m_0 c^2} \tag{5.51}$$

For oscillations in the orbit plane a further effect occurs.

As the particles lose energy the equilibrium orbit shrinks. For a particle at a radius greater or less than the equilibrium orbit this shrinkage respectively increases or decreases the oscillation amplitude. Which of these effects is the more important depends, as with phase oscillations, on the characteristics of the guide field. Detailed arguments yield

$$\alpha_x = \frac{(1-D)W_r}{2\gamma m_0 c^2}. \tag{5.52}$$

When $D>1$, this is negative, indicating growth. For accelerators this is normally counteracted by the adiabatic damping arising from the acceleration. In storage rings, however, it is necessary to design for D to be less than unity. For all modes to be damped, $-2<D<1$.

From equations (5.48);(5.52), the quite general relation

$$\alpha_x+\alpha_y+2\alpha_s = \frac{2W_r}{\gamma m_0 c^2} \tag{5.53}$$

may be deduced. This may be shown to be true in the even more general situation when the orbit is not planar. When coupling occurs the normal modes and damping coefficients are different, but equation (5.48) is still true. Equation (5.53) is sometimes divided through by the r.h.s. and written

$$J_x+J_y+J_s = 4 \tag{5.54}$$

where the J's are known as the damping partition numbers

$$\begin{aligned} J_x &= 1-D \\ J_y &= 1 \\ J_s &= 2+D. \end{aligned} \tag{5.55}$$

If the damping were smooth, then the oscillation amplitudes in storage rings would damp to very small dimensions. Because of the quantum nature of the radiation, however, the damping is not smooth; the radiation occurs as a succession of small pulses which act as a source of noise. This tends to increase the amplitude of oscillation, and an equilibrium is achieved when the excitation balances the damping. A calculation of the amplitude at which the effects balance requires a knowledge of the spectrum of the radiation. It is straightforward but somewhat lengthy to calculate the statistical amplitude of the noise-excited damped oscillators which correspond to phase and betatron oscillations. This calculation has been done in the references quoted; the distribution function for the oscillations is, as might be expected, gaussian.

It is found that the equilibrium oscillation amplitudes increase rapidly with energy. For storage rings of energy a few GeV this means first that the cavity voltage to supply the losses must be sufficient to prevent particles diffusing out of the effective potential well, and second that the aperture of the vacuum chamber must be sufficiently large to prevent excessive loss to the walls through the 'tail' of the gaussian. Because of the large storage times, of several hours, required in electron storage rings, it is found that the horizontal wall must not be nearer to the beam centre than about six times the r.m.s. width of the beam.

The amplitudes of the phase and radial oscillations depend in a complicated way on the machine parameters. For a 'separated-function' machine with uniform field-bending magnets and quadrupole focusing the following order-of-magnitude expressions apply. The amplitude of phase oscillations is expressed in terms of the r.m.s. energy excursion σ_γ rather than phase angle, and the radius of curvature in the magnets ρ_0 is included as a parameter. For machines operating at constant magnetic field, then $\rho_0 \propto \gamma$. For phase oscillations

$$\left(\frac{\sigma_\gamma}{\gamma}\right)^2 = \frac{C_q \gamma^2}{J_s \rho_0} \tag{5.56}$$

where $C_q = (55/32\surd 3)(\hbar/m_0 c) = 3{\cdot}83 \times 10^{-13}\,\text{m}$. (The curious numerical constant comes from integration involving the radiation spectrum, which is of complicated analytical form.)

For radial betatron oscillations

$$\left(\frac{\sigma_x}{\beta_0}\right)^2 \approx \frac{\lambda_c \gamma^2}{J_x Q_x^2 \rho_0} \tag{5.57}$$

where β_0 is the amplitude function defined in section 4.3.2. It is of order $C/2\pi Q_x$, where C is the machine circumference, but is considerably lower in the special 'low-β' sections of a storage ring where the collisions occur. The constant C_q has been replaced by λ_c, the Compton wavelength of the electron, from which it differs only by the numerical factor, which is of order unity. The quantity σ_x represents deviation from the equilibrium orbit. There is also an energy spread, and consequently a range of equilibrium orbits, so that the total width of the beam is somewhat greater than σ_x, typically by a factor between 1 and 1·5. The equilibrium amplitude for vertical oscillations is of order γ^2 less than that for horizontal oscillations; this would make the vertical beam height negligible in the absence of coupling, which inevitably occurs in a practical machine. Indeed, this is a convenient variable parameter; allowing the vertical height to become too small produces an excessive shift in Q_y.

It is not appropriate to discuss here the many factors to be considered in designing a storage ring. Enough has been described to show the essential physical consequences of the radiation. Further details may be found in the references already quoted, and in the books by Kolomenskij and Lebedev (1966) and Bruck (1966). Practical details of particular machines may be found in the proceedings of accelerator conferences around 1970 and later.

5.11. Concluding remarks

The problems studied in this chapter represent a few typical and special situations in which collisions and dissipation in beams are important. Further examples, involving ionizing collisions, have already been discussed in section 3.2.8. Indeed, many other examples of practical interest may be found. It is difficult to make interesting generalizations, however, and many situations are best studied individually as special examples. This applies especially to the many recent experiments with intense relativistic beams.

6

WAVES AND INSTABILITIES IN BEAMS

6.1. Introduction

Dynamic effects which arise from the finite charge and current in beams will now be introduced. Previously we have studied static effects associated with the self-fields and some dynamic effects associated with time varying external fields. Many phenomena which are found in beams are essentially the same as those studied in plasma physics. Longitudinal and transverse waves may be propagated, and these can couple to the environment or background plasma to cause growth or decay either in time or space. There is a wide variety of possibilities. Some effects have been exploited profitably, to build microwave tubes for example, whereas others, such as instabilities in accelerators, are undesirable and need to be suppressed.

We confine our study to systems exhibiting linear behaviour. Instability thresholds and growth rates are calculated, but, apart from occasional qualitative comment, we do not examine the limitation to the build-up of oscillations. This is a subject which is difficult to treat in a general way; particular situations must be considered individually. Also excluded is the theory of parametric coupling between two waves. Once the essential properties of beams are understood, however, it is relatively straightforward to apply the theory of this type of interaction. This may be seen, for example, in the book by Louisell (1960).

Since only linear phenomena are considered, involving harmonic variations of the fields, charge densities and displacements can be considered as harmonic. By solving Maxwell's equations together with the dynamical equations appropriate to the beam particles the dispersion relation for a particular system may be derived. This relates the frequency ω and wavenumber k of the disturbance. Frequently the dispersion relation has complex roots and the appropriate sign of the imaginary part of ω or k has to be determined.

6.2. Waves in unbounded plasma

6.2.1. Introduction Before studying beams it is profitable to summarize some of the properties of waves in infinite plasma. In this section we introduce the plasma frequency and study drifting cold plasma, the concept of positive and negative energy waves, and Landau damping. In later sections these ideas are applied to beams.

It is convenient at this point to comment on notation and sign conventions. Harmonic quantities are represented by the real part of $\exp j(\omega t - \underline{\underline{k}}\underline{x})$. This factor will often be omitted for brevity, but where there is possibility of confusion its virtual presence will be indicated by a tilde, thus

$$\tilde{A} \equiv A \exp j(\omega t - \underline{\underline{k}}\underline{x}). \tag{6.1}$$

The phase velocity of such a wave, moving in the z-direction, is ω/k_z. By analogy with the notation for relativistic particles we denote this by $\beta_w c$, and define

$$\gamma_w = (1-\beta_w^2)^{-1/2}. \tag{6.2}$$

Changing the sign of either ω or k_z reverses the velocity; changing both leaves it unchanged. Allowing both quantities to have either sign leads to redundancy. Restriction of ω to positive values might seem the more natural choice; as we see later, however, it is often convenient to adopt the opposite convention.

In this chapter we start by studying the simplest types of wave in the simplest type of medium, progressing by stages to more complex situations. Concepts useful in understanding and classifying wave motion are introduced and discussed where required. Frequently ideas are first introduced informally, and later a more complete and rigorous treatment is presented. The basic geometry to be considered is that of a cylindrical beam of radius a in a tube of radius b. We shall see how such parameters as the impedance of the tube walls and momentum spread of the beam particles affect the characteristics of of the longitudinal waves. The behaviour of such a system shows the same kind of characteristics as those of a drifting electron plasma in a stationary neutralizing background, but modified on account of the finite geometry. This modification is analogous to that experienced by free-space waves when they are confined in a waveguide.

6.2.2. Waves in cold stationary plasma The very simplest system is a cold infinite plasma of electrons and ions. All are assumed stationary in the absence of wave motion, and the ion inertia is assumed to be so large that, when fields arising from the electron wave motion appear, the ions do not move. We study first the simplest type of wave. To do this we postulate a disturbance in the z-direction which is harmonic in velocity, density perturbation, and electric field. Of these, only the density $n_0 q$ has a non-varying component. Denoting the electric field and electron velocity as $\tilde{E}$ and $\tilde{\beta}c$, and the charge density as $(n_0+\tilde{n})q$, the (non-relativistic) equation of motion is

$$\begin{aligned} q\tilde{E} &= m_0 c\frac{\mathrm{d}\tilde{\beta}}{\mathrm{d}t} \\ &= \mathrm{j}m_0\omega\tilde{\beta}c \end{aligned} \tag{6.3}$$

and the equation of continuity

$$\tilde{\beta}c\frac{\partial}{\partial z}(n_0+\tilde{n}) + \frac{\partial}{\partial t}(n_0+\tilde{n}) = 0 \tag{6.4}$$

which simplifies, when only first-order terms are included, to

$$n_0 k\tilde{\beta}c = \omega\tilde{n}. \tag{6.5}$$

To find the dispersion relation we need to use Maxwell's equations; in this very simple example Poisson's equation is adequate;

$$\frac{\partial\tilde{E}}{\partial z} = \frac{\tilde{n}q}{\varepsilon_0} \tag{6.6}$$

from which

$$-j\varepsilon_0 k\tilde{E} = \tilde{n}q. \tag{6.7}$$

Eliminating $\tilde{n}$ and $\tilde{\beta}$ between equations (6.3), (6.5) and (6.6) it is found that k and E disappear to leave

$$\omega^2 = \frac{n_0 q^2}{m_0 \varepsilon_0} = \omega_p^2 \tag{6.8}$$

This is the plasma frequency encountered earlier in section 3.1. The absence of k in the expression for ω indicates that the frequency associated with the disturbance is independent of the wavelength. The group velocity $\partial\omega/\partial k$ is zero and the phase velocity is inversely proportional to k.

The reason for this behaviour may be seen in physical terms by considering a slice within the plasma. If all the electrons in the slice are displaced across it by an amount Δz, then layers of negative and positive net charge appear at the faces of the slice and these charges produce a restoring force on the displaced electrons. This force is proportional to the displacement but independent of slice thickness, so that the frequency of oscillation is independent of the dimensions. As we see later this observation does not apply to a plasma finite in the transverse direction where fields penetrate through the boundary.

The magnetic field associated with these waves in an infinite medium is identically zero. It may readily be verified that the conduction current density $nq\beta c$ is equal and opposite in sign to the displacement current density $\varepsilon_0\dot{\underline{E}}$, so that curl $\underline{B}$ and hence $\underline{B}$ are zero.

In the calculation of plasma frequency the ions were assumed stationary. The inclusion of their motion in the calculation is straightforward; they move out of phase with the electrons and make smaller excursions. The plasma frequency is modified to

$$\omega_p^2 = \omega_{pe}^2 + \omega_{pi}^2 \tag{6.9}$$

where ω_{pe}^2 and ω_{pi}^2 are given by equation (6.8) with m_0 equal to the electron and ion mass respectively.

The waves studied so far are longitudinal, and have no existence when the plasma is absent. Transverse waves, which become free-space waves in the absence of the plasma, also exist. The dispersion relation is readily found from Maxwell's equations

$$\begin{aligned} \text{curl}\ \underline{B} &= \mu_0\left(\varepsilon_0\frac{\partial \underline{E}}{\partial t} + \underline{i}\right) \\ \text{curl}\ \underline{E} &= -\frac{\partial \underline{B}}{\partial t} \end{aligned} \tag{6.10}$$

and the relation $\underline{i} = nq\underline{v}$ (neglecting again the ion motion). Using equation (6.3), the expression for the current becomes

$$\tilde{\underline{i}} = \frac{nq^2\tilde{\underline{E}}}{j\omega m_0}\ . \tag{6.11}$$

Since in this example the velocity is transverse to the direction of propagation n does not vary with time. Substituting back into equation (6.10) the first equation, with frequency ω, becomes

$$-\underline{k}\times\tilde{\underline{B}} = \omega\tilde{\underline{E}}\ \frac{(1-nq^2/\omega^2 m_0\varepsilon_0)}{c^2}\ . \tag{6.12}$$

This is the same as the free-space equations (in which $\underline{i} = 0$) but with $\mu_0\varepsilon_0$ (or $1/c^2$) multiplied by the factor in parentheses. This factor, the dielectric constant, is equal to $1 - \omega_p^2/\omega^2$. The dispersion relation for such waves is therefore

$$c^2k^2 = \omega^2-\omega_p^2\ . \tag{6.13}$$

This defines a hyperbola, with the 'free-space' lines $\omega = \pm ck$ as asymptotes. When $\omega < \omega_p$, k is imaginary; this represents evanescent waves decaying exponentially away from the source. When $\omega \ll \omega_p$,

$$\frac{1}{|k|} = \frac{c}{\omega_p} = \left(\frac{1}{4\pi n r_0}\right)^{1/2}\ . \tag{6.14}$$

This quantity is known as the 'collisionless skin depth'.

In the presence of a magnetic field the behaviour becomes more complicated, and the structure of the wave depends on the direction of propagation. The relation between current and electric field is no longer so simple since the electrons do not in general move parallel to the applied electric field, and, furthermore, the motion depends on the relative orientation of the electric and magnetic fields. The analysis is still straightforward, however, and is facilitated by the introduction of a tensor permittivity κ_{ij} relating $\underline{D}$ and $\underline{E}$. Details may be found in many texts; a comprehensive treatment is given by Allis, Buchsbaum, and Bers (1963). They explain the method of calculation and treat a wide range of wave types. Here we simply quote some standard results.

The conductivity tensor σ_{ij} is first found by relating the current to the field in the plasma, making use of continuity and the equation of motion of the charges. From this the permittivity tensor κ_{ij}, equal to $\delta_{ij} - j\sigma_{ij}/\varepsilon_0\omega$, is derived. In the presence of a magnetic field parallel to the z-axis this may be shown to be

$$\kappa_{ij} = \begin{vmatrix} \kappa_{\perp} & -j\kappa_{\times} & 0 \\ j\kappa_{\times} & \kappa_{\perp} & 0 \\ 0 & 0 & \kappa_{\parallel} \end{vmatrix} \tag{6.15}$$

where, neglecting the ion motion

$$\begin{aligned} \kappa_{\perp} &= 1 - \frac{\omega_p^2}{\omega^2 - \omega_c^2} \\ \kappa_{\parallel} &= 1 - \frac{\omega_p^2}{\omega^2} \\ \kappa_{\times} &= \frac{\omega_c\omega_p^2}{\omega(\omega^2 - \omega_c^2)} \quad . \end{aligned} \tag{6.16}$$

For a harmonic disturbance, two of Maxwell's equations may be written

$$
\begin{aligned}
-j\underline{k}\times\underline{E} &= -j\omega\underline{B} \\
-j\underline{k}\times\underline{B}c^2 &= j\omega\underline{\underline{\kappa}}.\underline{E}
\end{aligned}
\tag{6.17}
$$

which, on elimination of $\underline{B}$, yields

$$
-\underline{k}\times(\underline{k}\times\underline{E}) = k_0^2\,\underline{\underline{\kappa}}.\underline{E} \tag{6.18}
$$

where $k_0 = \omega/c$, the value of k in free space. This represents three equations for the three components of E as a function of k, ω, and κ. Dividing through by one of them leaves three equations for two variables. Eliminating those variables yields a relation between ω, k, and κ, which is the dispersion relation.

Even the physically simple system of a cold plasma with immobile heavy ions when in the presence of a uniform external magnetic field B_0 possesses a fairly complicated dispersion relation

$$
\frac{c^2k^2}{\omega^2} = 1 - \frac{\omega_p^2(\omega^2-\omega_p^2)}{\omega^2(\omega^2-\omega_p^2)-\frac{1}{2}\omega_c^2\omega^2\sin^2\theta\pm\omega\{(\frac{1}{2}\omega_c^2\omega\sin^2\theta)^2+(\omega^2-\omega_p^2)\omega_c^2\cos^2\theta\}^{1/2}} \tag{6.19}
$$

where θ is the angle between $\underline{k}$ and $\underline{B}_0$. The two values implied by the $\pm$ sign correspond to two polarizations. For $\theta = \pi/2$ these are plane waves with $\underline{E}$ perpendicular and parallel to $\underline{B}_0$; for $\theta = 0$, on the other hand, they represent right- and left-handed circular polarization. If the second term exceeds the first then k^2 is negative and no propagation occurs (cf. equation 6.13). When $\theta = 0$ or $\pi/2$ the expression simplifies considerably.

When $\theta = 0$

$$
\frac{c^2k^2}{\omega^2} = 1 - \frac{\omega_p^2}{\omega(\omega\pm\omega_c)}\,. \tag{6.20}
$$

The two signs indicate that the phase velocity is zero at frequencies $\pm\,\omega_c$. This is true for plane-polarized radiation. If, however, the radiation is circularly polarized, the sign depends on whether the direction of rotation of the polarization is the same as or opposite to the direction of rotation of the particles. If the two rotations are in the same direction, the negative sign is taken. The dispersion relation is no longer symmetrical, and the sign of ω denotes the direction of rotation of the plane of polarization compared with that of electrons in the magnetic field.

When $\theta = \pi/2$, there are two relations, for $\underline{E}$ parallel and perpendicular to $\underline{B}_0$ respectively. Not surprisingly, the former does not contain $\underline{B}_0$, and is the same as the expression for free space:

$$\frac{c^2k^2}{\omega^2} = 1 - \frac{\omega_p^2}{\omega^2}, \quad (\underline{E} \parallel \underline{B}_0)$$
$$\frac{c^2k^2}{\omega^2} = 1 - \frac{\omega_p^2(\omega^2-\omega_p^2)}{\omega^2(\omega^2-\omega_p^2-\omega_c^2)}, \quad (\underline{E} \perp \underline{B}_0). \tag{6.21}$$

The frequency $(\omega_p^2+\omega_c^2)^{1/2}$ which occurs in the second of these equations is known as the 'upper-hybrid' frequency. These dispersion relations are plotted in Fig.6.1 for a range of ω_c/ω_p.

In terms of the notation introduced in equation (6.2), the last two equations may be written

$$1 = \beta_w^2\gamma_w^2\left(\frac{-\omega_p^2}{\omega(\omega\pm\omega_c)}\right) \tag{6.22}$$

and

$$1 = \beta_w^2\gamma_w^2\left(\frac{-\omega_p^2}{\omega^2}\right), \quad (\underline{E} \parallel \underline{B}_0)$$
$$1 = \beta_w^2\gamma_w^2\left(\frac{-\omega_p^2(\omega^2-\omega_p^2)}{\omega^2(\omega^2-\omega_p^2-\omega_c^2)}\right), \quad (\underline{E} \perp \underline{B}_0) \tag{6.23}$$

The quantities β_w and γ_w correspond to the values of β and γ for a particle moving with phase velocity equal to that of the wave; when $\beta_w>1$, γ_w is imaginary, so that this notation is not particularly useful. In later sections, however, when we deal with waves for which β_w is always less than unity, its value will become apparent.

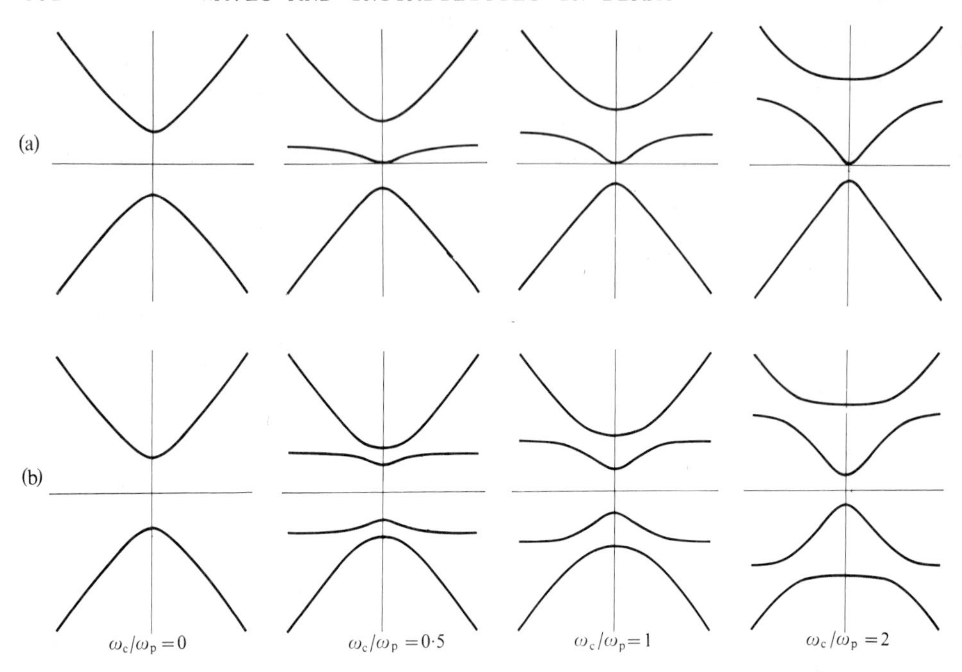

Fig.6.1. (a) Curves of ω versus k illustrating the dispersion relations for transverse waves in cold plasma in a magnetic field. The top row is for propagation along the direction of the field. The frequencies ω_0^+ and ω_0^-, where $k = 0$, are given by

$$\frac{\omega_0}{\omega_p} = \frac{\omega_c}{2\omega_p} \pm \left(\frac{\omega_c}{4\omega_p^2} + 1\right)^{1/2} .$$

The middle branch of the curve is asymptotic to $\omega = \omega_c$
(b) In the second row the direction of propagation and the electric field are perpendicular to the magnetic field. The frequencies where $k = 0$ are defined by the four roots of

$$\left(\frac{\omega_0}{\omega_p}\right)^2 = \tfrac{1}{2}\left\{\left(\frac{\omega_c}{\omega_p}\right)^2 + 2 \pm \left(\frac{\omega_c}{\omega_p}\right)\left(\frac{\omega_c^2}{\omega_p^2} + 4\right)^{1/2}\right\} .$$

The middle branches are asymptotic to $\omega = (\omega_p^2 + \omega_c^2)^{1/2}$. Since the magnetic field is perpendicular to the direction of propagation, The wave is not circularly polarized, and both positive and negative frequencies are present. The curve is therefore symmetrical abou the frequency axis.

6.2.3. Plasma surface waves Because of the existence of a dielectric constant which is negative below the plasma frequency, it is possible to excite surface waves of a kind essentially different from those at a boundary of a dielectric. In a dielectric, surface waves occur in the less 'optically dense' medium beyond the critical angle of total internal reflection. The condition for the existence of such waves is that the phase velocity along the surface in the less-dense medium should be less than the velocity of light in that medium. If the less-dense medium is free space, then the wave amplitude decays away from the surface as $\exp(-y/\gamma_{w}\lambda\!\!\!^{-})$. Such a wave is not localized near the surface; it is not possible to find solutions which decay exponentially from both sides of the surface. At a plasma surface, however, this is possible, and we find a solution by matching impedances and phase velocities across the surface. This is taken as the xz plane, with propagation in the z-direction.

We denote the media for which y is positive and negative respectively by subscripts 1 and 2. If the dielectric constant is κ and the permeability is unity, then for both media

$$k_y^2 + k_z^2 = \frac{\omega^2\kappa}{c^2} \tag{6.24}$$

Denoting the impedance by Z,

$$\begin{aligned} Z_{y1} &= \frac{Z_1 c k_{y1}}{\omega} \\ Z_{-y2} = -Z_{y2} &= \frac{Z_2 c k_{y2}}{\omega} \end{aligned} \tag{6.25}$$

where $Z_1 = \mu_0/\varepsilon_0\kappa_1$. Setting $k_{z1} = k_{z2}$ in equation (6.24) and $Z_{y1} = Z_{y2}$ in equation (6.25), and eliminating k_{y1} and k_{y2} yields the dispersion relation

$$\frac{\omega^2 k_z^2}{c^2} + \frac{\kappa_1\kappa_2}{\kappa_1+\kappa_2} = 0 \tag{6.26}$$

from which it is evident that real k only exists for negative κ_1 or κ_2. Writing $\kappa_1 = 1 - \omega_p^2/\omega^2$, the value appropriate to a plasma in the absence of a magnetic field, and omitting the subscript on k, this becomes

$$k^2 = \frac{c^2}{\omega^2} \frac{(\omega_p^2 - \omega^2)\kappa_2}{\omega^2(1+\kappa_2) - \omega_p^2} . \tag{6.27}$$

There is a solution for real k only when numerator and denominator are both positive:

$$1 > \left(\frac{\omega}{\omega_p}\right)^2 > \frac{1}{1+\kappa_2} . \tag{6.28}$$

The dispersion curve is shown in Fig.6.2. It will be seen that the group velocity $\partial\omega/\partial k$ is negative. The normal component of $\underline{E}$ reverses at the surface, and this means that

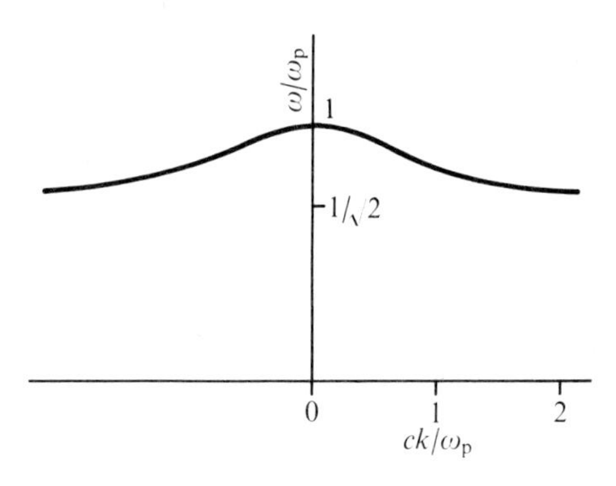

Fig.6.2. Dispersion curve for a surface wave at the interface between a plane uniform plasma and free space.

the power flow on the two sides of the surface is oppositely directed; the negative flow inside the plasma exceeds the positive flow along the surface. Surface waves of the same physical nature, but with somewhat different properties because of the finite geometry, will later be encountered in beams.

<u>6.2.4. Steady-state properties of a drifting plasma in a stationary neutralizing background</u> Intermediate in complexity between a stationary plasma and a beam is a plasma of particles of one sign drifting through a stationary, immobile, neutralizing background. This is an idealization, albeit a useful one in many situations. It is evidently more

appropriate to an electron plasma, and we assume for the moment that it is electrons which are streaming.

Some implications of the assumptions made in setting up this model will now be studied. First, if the plasma is truly infinite in the transverse direction, then there can be no magnetic field, since there is an accumulation of negative charge at infinity and a depletion at minus infinity which gives rise to a uniform rate of change of displacement $\underline{D}$ whose associated magnetic field 'cancels' that arising from the convection current. On the other hand, the presence of $\underline{D}$ implies an electric field in such a direction as to retard the charges so that the continuous free streaming which has been postulated is not possible; in fact, the system is effectively that described in the previous section, which oscillates with frequency ω_p. If, however, there is a return path for the current, however remote, then a magnetic field does exist, and this exerts a transverse force on the electrons. To be specific, we consider a system with axial symmetry and very large transverse dimensions. The self-magnetic field produces a radial inward force proportional to r which causes the electrons to move towards the axis. A criterion for the effect to be unimportant is that the Larmor radius at a distance r from the axis should be large compared with r. This condition has already been calculated in section 3.2.7. It is simply that

$$2\nu/\gamma \ll 1 \tag{6.29}$$

or, writing ν in terms of n and r,

$$\frac{r^2 nq^2}{2\varepsilon_0 \gamma m_0 c^2} \ll 1. \tag{6.30}$$

Rearranging,

$$r^2 \ll \frac{2\varepsilon_0 \gamma m_0 c^2}{nq^2} = \frac{2\gamma c^2}{\omega_p^2}. \tag{6.31}$$

Before commenting on this result, we note that n refers to the number density of electrons as measured in the laboratory frame. In a frame moving with the electrons, lengths appear greater by a factor γ than they do in the stationary frame; the number per unit length, and hence the density, appears less therefore by a factor γ. Using this fact equation (6.28) becomes

$$r \ll \frac{\sqrt{2}c}{\omega_{pm}} \tag{6.32}$$

where ω_{pm} is the plasma frequency measured in the moving frame. It will in future be assumed always that the plasma frequency is measured in the moving (proper) frame of the electrons so that the subscript m will be omitted. The r.h.s. of this inequality will be recognized as $\sqrt{2}$ times the collisionless skin depth.

If the system is such that the extent in the z-direction is very large compared with the radius, the inward force can be balanced by letting n_e exceed n_i, so that there is a radial electric force which can (in principle!) be made equal and opposite to the magnetic force. This corresponds to zero generalized perveance for the beam (equation (3.30)), for which the Budker condition $n_e/n_i = 1/\gamma^2$ holds. This represents a large ratio for highly relativistic particles. This model is very unrealistic; the ions are no longer in equilibrium, being subjected to inward transverse electric forces arising from the space-charge field. Only our previously unrealistic assumption that the ions are 'infinitely heavy' presents them from moving towards the axis. A more convincing way out of the difficulty with an infinite drifting plasma, however, is to postulate a very strong magnetic field in the direction of drift.

In a frame moving with the electrons n_e is decreased by a factor γ and n_i is increased by the same factor, so that for a plasma governed by equation (6.32) we find in the moving frame that $n_{e2}/n_{i2} = 1$ and the plasma is neutral. The inward force on the ions in this frame is to be interpreted as a 'pinch' force arising from the magnetic field of the ion current.

In this section we have considered a drifting plasma of very large cross-section. Later we shall be concerned with beams of finite cross-section, where only some of the factors discussed here are relevant.

6.2.5. Longitudinal waves in a cold drifting plasma We now study the properties of plasma oscillations in a drifting medium, such as that described in the last section. One approach is clearly to make a co-ordinate transformation to a moving frame. Instead of this, however, we solve the problem directly using the method employed at the beginning of section 6.2.2. We assume the drift velocity to be relativistic.

The equation of motion is

$$q\tilde{E} = m_0 c\frac{d}{dt}(\beta\gamma) . \tag{6.33}$$

This differs in two respects from equation (6.3). In the first place the electrons are moving, so that d/dt is replaced by $\partial/\partial t + \beta_0 c\partial/\partial z$, where β_0 is the mean value of β; secondly, relativistic dynamics must be used. Equation (6.33) then becomes

$$q\tilde{E} = m_0 c\left(\frac{\partial}{\partial t}+\beta_0 c\frac{\partial}{\partial z}\right)\left(\beta_0\tilde{\gamma}+\gamma_0\tilde{\beta}\right) . \tag{6.34}$$

Performing the differentiation (with the condition that $\tilde{\beta}/\beta_0$, $\tilde{\gamma}/\gamma_0 \ll 1$) yields

$$q\tilde{E} = j\gamma^3 m_0\tilde{\beta}c\ (\omega-\beta_0 ck) . \tag{6.35}$$

We now express the equation of continuity, which is the same as equation (6.4) except for the presence of β_0. This is

$$(\beta_0+\tilde{\beta})c\frac{\partial}{\partial z}(n_0+\tilde{n})+\frac{\partial}{\partial t}(n_0+\tilde{n}) = 0 \tag{6.36}$$

which simplifies to

$$(\omega-\beta_0 ck)\tilde{n}-kn_0\tilde{\beta}c = 0 \tag{6.37}$$

when first-order terms only are included.

As before, Maxwell's equations are represented by equation (6.6):

$$\frac{\partial \tilde{E}}{\partial z} = \frac{\tilde{n}q}{\varepsilon_0} \tag{6.6}$$

from which

$$-j\varepsilon_0 k\tilde{E} = q\tilde{n}. \tag{6.7}$$

The dispersion relation may be found in exactly the same way as before; it is

$$(\omega-\beta_0 ck)^2 = \frac{q^2 n_0}{\gamma_0^3 m\varepsilon_0} . \tag{6.38}$$

Two differences from equation (6.7) may be noted. First, the frequency is Doppler shifted by k times the electron velocity; second, there is a γ_0^3 factor on the right-hand side. As already indicated in section 6.2.3 the plasma frequency is defined as that which would be measured in the proper frame of the plasma:

$$\omega_p^2 = \frac{n_0 q^2}{\gamma_0 m_0 \varepsilon_0} . \tag{6.39}$$

In terms of the plasma frequency, the dispersion relation may be written

$$\frac{\omega_p^2}{\gamma_0^2(\omega-\beta_0 ck)^2} = 1 \tag{6.40a}$$

or, alternatively,

$$\beta_w = \frac{\beta_0\omega}{\omega\pm\omega_p/\gamma_0} . \tag{6.40b}$$

Equation (6.40) represents a pair of straight lines (Fig.6.3). The group velocity $\partial\omega/\partial k$ is equal to $\beta_0 c$, the streaming velocity. The phase velocity ω/k, on the other hand, takes all values from $-\infty$ to $+\infty$. For a given value of k there exist two waves with phase velocities ω/k above and below the

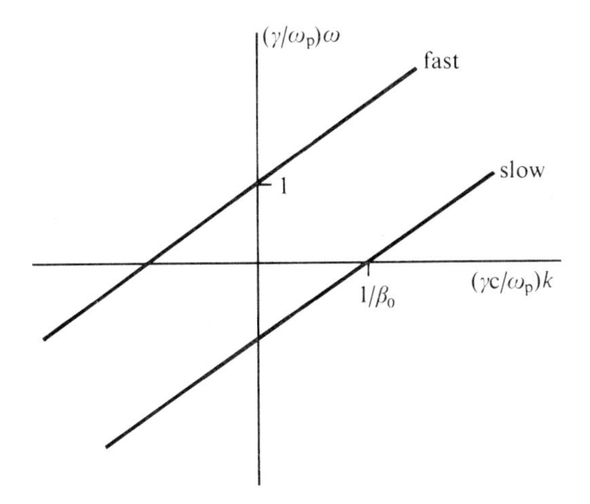

Fig.6.3. Dispersion curve for a plasma drifting through a stationary immobile neutralizing background with velocity $\beta_0 c$, given by equation (6.39). In the first quadrant both waves are forward, one fast and the other slow. In the second and fourth quadrants the phase velocity is negative. In the third quadrant the line marked 'fast' represents a slow wave and *vice versa*. The group velocity for both the fast and slow waves is $\beta_0 c$.

streaming velocity $\beta_0 c$. These are known as the 'fast' and 'slow' waves. When $k < \omega_p/\beta\gamma_0 c$, the slow wave has a negative phase velocity.

An interesting analogy is provided by a moving screw. If the pitch of the thread is p and the screw is rotated with an angular velocity ω, then the phase velocity of the thread in the z-direction is $+p\omega/2\pi$ for a left-hand thread. For ω negative (anticlockwise rotation) the velocity is $-p\omega/2\pi$. If the screw is now moved forwards with velocity v, then the phase velocity of the thread movement will be positive or negative according to whether v is greater or less than $p\omega/2\pi$. This corresponds to $\beta_0 c$ greater or less than $\omega_p/k\gamma_0$ in the plasma.

An alternative method of obtaining equation (6.39) is by means of a Lorentz transformation to a moving frame of the dispersion relation for a plasma (equation (6.8)). If βc is the velocity in the moving frame and subscripts 1 and 2 denote stationary and moving frames, then from the equations

$$\begin{aligned} \omega_{p1}^2 &= \omega_{p2}^2 \\ \omega_2 &= \gamma(\omega_1 - k_1\beta c) \\ k_2 &= \gamma\left(k_1 - \frac{\beta\omega_1}{c}\right) \end{aligned} \tag{6.41}$$

we obtain equation (6.40a) by elimination of ω_2 and k_2. In this particular case k_2 is zero, so that the third equation is not needed. The plasma frequency is unaltered during the transformation. The increase in number density apparent in the moving frame is balanced by the effects of time dilation. If the number density is measured in the stationary frame, as is often the case, then the value of n should be divided by γ to give

$$\omega_p^2 = \frac{n_1 q^2}{\gamma m_0 \varepsilon_0} = \frac{n_2 q^2}{m_0 \varepsilon_0} \tag{6.42}$$

as explained in section 6.2.4.

Stationary plasma is always passive. An imaginary component of k or ω represents a wave which is decaying in a direction away from the source, or in time. A drifting plasma, on the other hand, is essentially an active medium; if it interacts with a stationary background, growing waves can exist in which power is drawn from the kinetic energy of the moving particles.

For the waves considered so far there is a real value of k for each real value of ω and *vice versa*, so that the dispersion relation has no complex roots. When complex roots do occur, the interpretation is by no means always simple. It is not immediately clear which sign should be taken for the imaginary part, not what type of physical behaviour is indicated. Formal techniques for resolving these questions exist, though in many simple cases physical arguments may be used. We return to these questions in section 6.3, but enlarge now in a simple-minded way on the concept of positive and negative energy waves.

If the energy of the moving plasma in the presence of the wave is computed, it will be found to differ from the value in the absence of the wave. For the fast wave the energy is increased by the presence of the wave; for the slow wave it is decreased. Although perhaps at first sight surprising, this result becomes plausible when it is appreciated that the density and velocity modulation of the electrons are correlated in such a way that when a slow wave is excited more particles are slowed down than speeded up, and *vice versa* for the fast wave.

An important consequence of this fact is that such a wave can couple to a positive energy wave in another system in such a way that energy is exchanged between the two to give a growing wave. Such a situation is analysed in the next section. Another important effect is that dissipation causes a negative energy wave to grow rather than damp. As energy is removed from the wave, its amplitude grows. This can be shown very simply by introducing a complex dielectric constant $\varepsilon = \varepsilon_0(1 - j\alpha)$ in equation (6.7), where we take α small compared with unity and of such a sign as to represent dissipation. The plasma frequency then becomes

$$\omega_p \approx \left(\frac{nq^2}{\gamma m\varepsilon_0}\right)^{1/2}(1+\tfrac{1}{2}j\alpha)$$

$$= \omega_{p0}(1+\tfrac{1}{2}j\alpha) \tag{6.43}$$

where ω_{p0} is the value of ω_p when $\alpha = 0$. Equation (6.39) can now be written

$$\omega = \beta_0 ck \pm \left(\frac{\omega_{p0}}{\gamma}\right)(1+\tfrac{1}{2}j\alpha) \tag{6.44}$$

from which it is evident that the fast wave (top sign) is damped, and the slow wave (bottom sign) grows.

We note at this point that it is possible to satisfy equation (6.44) by choosing real k and complex ω or *vice versa*. Which solution is appropriate depends on the boundary conditions of the problem. For an infinite rectilinear system it is usually more convenient to have real ω and complex k, corresponding to a wave growing or damped in space, but with constant amplitude at any point. When the system is closed upon itself, as in an orbital particle accelerator, the complex frequency description is more appropriate. For a finite rectilinear system the choice is not so simple (Rognlien and Self 1972). We shall return to these topics, but meanwhile study systems with two or more interpenetrating plasmas.

6.2.6. Two or more streaming plasmas In the last section the dispersion relation for a cold, drifting, relativistic plasma was found. This is quadratic in form and represents two waves with phase velocities respectively greater and less than the drift velocity. We now enquire into what happens if there are two plasmas streaming through one another in the same or opposite directions.

The dispersion equation for longitudinal waves is calculated precisely as in section 6.2.5, except that there are two components to be considered. Equations (6.35) and (6.36) become two equations, one for each component, and so does equation (6.37). Equation (6.7), however, remains a single equation containing both components. Elimination of $\tilde{n}$ and $\tilde{E}$ as before yields the dispersion relation analogous to equation (6.39):

$$\frac{\omega_{p1}^2}{\gamma_1^2(\omega-\beta_1 ck)^2} + \frac{\omega_{p2}^2}{\gamma_2^2(\omega-\beta_2 ck)^2} = 1. \tag{6.45}$$

A particular case of this dispersion relation relates to a single stream passing through background ions, no longer assumed 'infinitely' heavy;

$$\frac{\omega_{pe}^2}{\gamma_e^2(\omega-\beta_e ck)^2} + \frac{\omega_{pi}^2}{\omega^2} = 1. \tag{6.46}$$

where the subscripts e and i refer to electrons and ions.

Dispersion relations corresponding to the three situations (streams moving in the same direction, one stream stationary, and streams moving in the opposite direction) are sketched in Fig.6.4. The curves are asymptotic to pairs

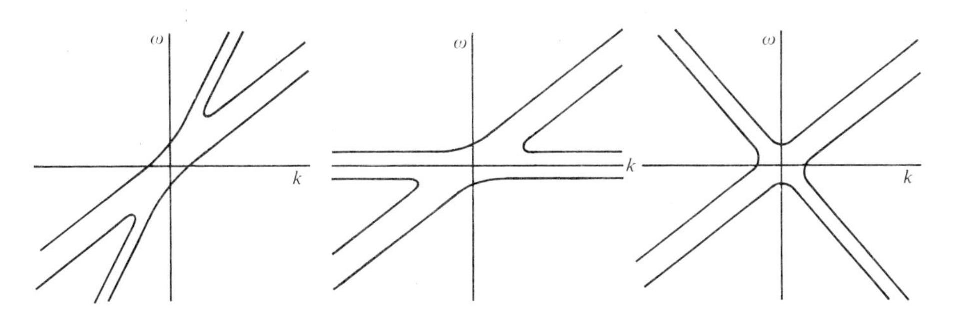

Fig.6.4. Dispersion curves for two-stream instability (equation (6.45)). The velocity of one stream is always positive; the three diagrams correspond to the velocity of the second being positive, zero, and negative.

of straight lines; each pair represents the waves in a single component discussed in the previous section.

Inspection of the dispersion relation shows that for large k there are four real roots, but when k is small enough there are only two roots. For real k, ω is complex, and for real ω, k is complex. We have so far discussed only briefly the problem of which alternative is relevant, and not at all how to decide whether a growing or damped solution is correct. It is possible, however, to advance a plausible physical argument based on the idea of positive and negative energy waves introduced in the last section. First, let us assume that the two streams are moving in the same direction. Waves on the streams can interact if their phase velocities, each in the absence of the other stream, are not too different. We consider first the coupling of a positive energy wave on one stream and a negative energy wave on the other. If a harmonic disturbance is imposed on one (or both) of the streams at some point, a wave is generated which grows in distance along the stream downstream from the disturbance. Energy is exchanged in a straightforward wave between the streams; the disturbance grows with distance, but at any fixed point it is independent of time. The wave number k is complex but ω is real. This phenomenon is known as 'convective' instability. Suppose the waves had both been positive or negative energy waves, it would not then be possible for one to 'feed' the other and there would be no growth; ω and k would be real.

If now the streams move in opposite directions, the group velocities are opposite and the behaviour is different. If one wave has positive energy and the other negative, then instead of merely an interchange of energy between the streams as they move along, energy can flow in a direction opposite to the phase velocity. A disturbance at a point, and in the regions both up and down stream from it, grows with time. The frequency ω is complex, and the instability is known as 'absolute'. Finally, if both waves have either positive or negative energy, the feedback acts in such a way as to reduce the applied disturbance; k is complex, with sign corresponding to attenuation.

This sketch of the types of interaction between two streams is an outline only and contains no formal justification. An extended and more rigorous discussion of the behaviour exhibited by the four cases just considered is given by Sturrock (1958). It is necessary to consider wave packets, rather than a single frequency component as has been done here. For weak coupling the different cases can be readily distinguished by the topology of the curve of the dispersion relation, which is, at the appropriate intersection of coupled waves, hyperbolic.

When the coupling is strong, for example, as it is in the two-stream interaction just discussed, a rigorous determination of the type of instability requires a much more detailed evaluation of the mathematical properties of the dispersion relation. We do not discuss the details of this problem, but references and discussion are given in section 6.R.7. The diagrams in Fig.6.4 represent the dispersion properties of an infinite 'two-stream' medium. It is evident that by changing of a moving co-ordinate system the three situations can be made equivalent, so that whether an instability is convective or absolute depends on the frame of reference in which it is viewed. In the presence of boundaries, or exciters at a fixed point in space, transforming from one situation to another can be complicated;

it is best to work in a frame in which the boundaries are at rest.

The form of the analysis is such that extension to any number of streams is straightforward. The dispersion relation is simply

$$\sum_r \frac{\omega_{pr}^2}{\gamma_r^2(\omega-\beta_r ck)^2} = 1. \tag{6.47}$$

As r increases, the form of the dispersion curve becomes increasingly complicated; it is asymptotic to r pairs of straight lines and has $2r$ roots, many of which may be complex. In the limit as r tends to infinity the system tends to a plasma with a continuous velocity distribution, the shape of which is determined by ω_{pr} as a function of β_r. Proceeding in a straightforward way to the limit $r \to \infty$ in equation (6.47) yields

$$\frac{q^2}{\varepsilon_0 m_0}\int \frac{f(\beta\gamma)\,d(\beta\gamma)}{\gamma^3(\omega-\beta ck)^2} = \int \frac{q^2}{\varepsilon_0 m_0}\frac{f(\beta\gamma)\,d\beta}{(\omega-\beta ck)^2} = 1 \tag{6.48}$$

where n is the density and $f(\beta\gamma)$ is the distribution function of the particles in $\beta\gamma$ space (essentially momentum space) normalized in such a way that $\int f(\beta\gamma)\,d(\beta\gamma) = n$. Integrating by parts yields the alternative form

$$\frac{q^2}{\varepsilon_0 m_0 ck}\int \frac{\partial f(\beta\gamma)/\partial\beta}{\omega-\beta ck}\,d\beta = -1. \tag{6.49}$$

The interpretation of this dispersion integral is made complicated by the presence of the singularity in the denominator of the integrand; it is by no means obvious how to deal with it.

A situation of some practical interest, in which the singularity can be ignored, is a plasma in which the phase velocity of the wave of interest greatly exceeds βc, for all particles. Under these circumstances the integral in equation (6.49) can be written as

$$\frac{q^2}{\varepsilon_0 m_0 \omega^2} \int f(\beta\gamma)\left(1 - \frac{k\beta c}{\omega}\right)^{-2} \mathrm{d}\beta = 1 \qquad (6.50)$$

and expanded by the binomial theorem. Setting $f(\beta)$ appropriate to a non-relativistic distribution with temperature KT and expanding[†]

$$\frac{q^2}{\varepsilon_0 m_0 \omega^2} \frac{m_0 c^2}{2\pi^{1/2} KT} \int_{-\infty}^{\infty} \left\{\exp\left(\frac{-\beta^2 m_0 c^2}{2KT}\right)\right\}\left\{1+3\left(\frac{\beta k c}{\omega}\right)^2\right\}\mathrm{d}\beta = 1, \qquad \text{(N.R.)} \quad (6.51)$$

(The term in the second pair of brackets proportional to β is omitted, since it produces zero contribution to the integral.) Performing the integration and rearranging yields

$$\frac{\omega_p^2}{\omega^2} + \frac{3KT}{m_0}\frac{k^2}{\omega^2} = 1, \text{ (N.R.)} \qquad (6.52)$$

When this approximation is justified the plasma is sometimes referred to as 'warm', as distinct from a 'hot' plasma where there are substantial numbers of particles moving at the wave velocity. The result in equation (6.52) is often obtained by a hydrodynamic argument, by introducing the idea of an electron pressure (see, for example, Clemmow and Dougherty (1969), p.178).

The dispersion relation for a hot plasma, in which the singularity must be resolved, will be discussed in the next section. One method of proceeding is to look very much more carefully at the way in which the limit is arrived at (Dawson 1960). More conventionally, Laplace transform theory can be invoked, as was done originally by Landau (1946).

6.2.7. A continuum of plasma streams; Landau damping In this section we summarize some of the properties of waves in plasma in which the electrons have a continuous velocity distribution. We continue with a discussion of the dispersion integral (equation (4.49)); this might be expected to yield a relation between ω and k, which depends on the distribu-

---

[†]Boltzmann's constant is written as K in this section to avoid confusion with the wavenumber k.

tion function $f(\beta)$. Before we can find this it is necessary to find out how to deal with the singularity which occurs where $\omega = \beta ck$.

It is interesting to note first that, if $\partial f/\partial\beta = 0$ where $\omega = \beta ck$, there is no singularity. This can arise if $f(\beta)$ is of finite width (as it normally is in beams) and ω/ck occurs outside the range of $f(\beta)$. Alternatively $f(\beta)$ may be uniform where $\beta = \omega/ck$. We now calculate the dispersion relation for the latter situation, assuming a distribution centred about $\beta = 0$ of width $2\beta_0$. Setting $f = n/2\beta_0$ and substituting in equation (6.48) yields the dispersion relation, valid when $\omega \neq |\beta_0 ck|$,

$$\omega^2 = \omega_p^2 + \beta_0^2 c^2 k^2, \quad \text{(N.R.)} \tag{6.53}$$

This is quadratic in form, reducing to equation (6.8) when $\beta_0 = 0$, and essentially the same as equation (6.52). It will be seen by inspection that the inequality $\omega \neq |\beta_0 ck|$ is always satisfied.

This condition for the singularity in equation (6.49) may be written $\beta c = \omega/k$, or particle velocity equals wave velocity. Physically this represents a resonant interaction between the particles and the wave. It is not immediately evident what the effect of this resonant interaction will be. The problem was first solved by Landau in 1946 with the aid of Laplace transform theory, and much has been written on the subject since. The mathematical argument is fairly lengthy, and here we merely make a few comments and quote results which will be needed later.

The apparent impasse brought about by the singularity is an indication of the fact that more information is required to solve the problem than is used in deriving equation (6.38). The singularity is associated with resonant behaviour, and the additional information required relates, as might be expected, to the initial conditions ascribed to the initial preparation of the plasma.

Fortunately, for 'well-behaved' distributions (and most practical distributions are of this type) one simple rule enables the integral to be evaluated. The normalized velocity

β is taken as complex, and the integration is taken along the real axis but above the pole at $\beta = \omega/ck$. The real part of the integral, which excludes the region $\omega/ck-\varepsilon$ to $\omega/ck+\varepsilon$, is the Cauchy principal value, denoted by P. The imaginary part is $-\pi j$ times the residue evaluated at ω/ck. The residue of the integral is $-(\partial f/\mathrm{d}\beta)/ck$, so that

$$\frac{q^2}{\varepsilon_0 m_0 ck}\left\{P\int\frac{\partial f\partial\beta}{\omega-\beta ck}\mathrm{d}\beta+\frac{\pi j}{ck}\left(\frac{\partial f}{\partial\beta}\right)_{\omega/ck}\right\} = -1. \tag{6.54}$$

For real value of k this expression yields in general a complex value of ω (with exceptions noted earlier). For all except a few very simple functions equation (6.54) cannot be evaluated explicitly. Although ω as a function of k can only be found by evaluation of equation (6.54), information about the sign of the imaginary part of ω can be obtained more simply.

In the remainder of this section we state some well-established results without proof. A rigorous derivation of them is very lengthy, and may be found in standard texts on plasma physics. The first of these results is that, for a single-humped distribution, $\mathrm{Im}\,\omega$ is always positive, so that waves are damped. Such damping is known as 'Landau damping', after his first analysis of the problem (Landau 1946). It is always possible to choose a frame of reference in which the peak of the distribution has $\beta c = 0$; such a distribution (for example, a Maxwellian, equation (4.54)) has the property that $\partial f/\partial\beta$ always has the opposite sign to that of the velocity of the wave. In order to connect this fact with the phenomenon of damping, it is relevant to enquire into the physical nature of the resonant interaction between the wave and particles when $\beta c \sim \omega/k$.

The theory of particles in travelling fields has been given in section 2.8. Particles moving at nearly the speed of the wave are 'trapped' and have a mean velocity equal to the wave velocity; particles moving substantially faster or more slowly slip continuously in phase. For a wave of constant velocity the mean energy of all particles remains constant. We are interested in the limit where the wave amplitude is vanishingly small (to ensure linearity) and therefore can neglect trapping. Since the energy of any particle remains on the average constant, and there are many particles distributed over all phases, it is not immediately clear where the damping originates. The physical behaviour may be visualized by assuming the rather special initial condition that at $t = 0$ all the particles within a narrow range of velocities about that of the resonant particle are uniformly distributed in phase. What happens is that correlation of particle phase develops, which is of such a nature that particles moving faster than the wave take energy from it, whereas those moving more slowly give up energy. If $\partial f/\partial\beta$ is negative, damping therefore occurs; if it is positive there is growth and instability.

A somewhat different physical picture emerges if we follow the approach of Dawson (1960). This approach, also described in the book by Stix (1962), may be summarized as follows. A system with a very large but finite number N of low-density beams is considered, and the following properties established.

(1) For a given k the dispersion relation yields $2N$ values of ω, some of which are complex; these represent the frequencies of the normal modes of the system.

(2) The phase velocities of these modes lie *between* the beam velocities.

(3) The growth rates of the beam instabilities tend to zero as the number of beams used to approximate the distribution function approaches infinity.

(4) These beam modes form a complete set of normal modes, which are essentially equivalent to those of van Kampen (1955).

A general disturbance may be regarded as a superposition of all the beam modes, and therefore suffers from phase mixing. It is this phase mixing which appears as Landau damping.

It is not immediately evident how to reconcile the two rather different physical viewpoints which have been put forward. The question requires lengthy discussion for a satisfactory resolution, and this will not be attempted here.

The damping is given quantitatively by the expression

$$\operatorname{Im}\omega = -\frac{\pi q^2}{2\varepsilon_0 m_0 k c}\left(\frac{\partial f}{\partial \beta}\right)_{\omega/k}\left(\frac{\omega}{k} - \frac{\partial \omega}{\partial k}\right) \tag{6.55}$$

where ω on the r.h.s. denotes the real part. This formula, valid when $\operatorname{Im}\omega \ll \omega$, was obtained by Jackson (1960) and represents a small correction to the result of Landau. In the limit of very small $\operatorname{Im}\omega$, $\omega \approx \omega_p$, $\partial\omega/\partial k \approx 0$.

For distributions such as the Maxwellian, which are single humped, $(\partial f/\partial\beta)/\beta$ is always negative, so that there is stability; for double-humped distributions, on the other hand, this quantity is positive over a range of β, giving the possibility of unstable growth. Penrose (1960) has shown, using methods closely related to Nyquist-diagram techniques in circuit theory, that the system is unstable if

$$\int \frac{f(\beta\gamma) - f(\beta_0\gamma_0)}{(\beta-\beta_0)^2}\,d\beta > 0 \tag{6.56}$$

where β_0 is the value at the minimum between the two peaks. For well-separated peaks instability is always found; this may be identified with the two-stream instability studied in section 6.2.5. A broad distribution with a small 'dip' of order 10 per cent between peaks, however, is likely to be stable.

In this section we have outlined without detailed proof the main physical characteristics of Landau damping. Further discussion may be found in the references given in section 6.R.2.

<u>6.R.2. Notes and references</u> The properties of plasma waves summarized in section 6.2.2 may be found in many texts on plasma physics. A complete, rather formal, mathematical treatment is given for example by Allis *et al.* (1963) and a more physical discussion, containing all the material used here is given by Clemmow and Dougherty (1969). They also discuss the properties of a cold, relativistic, drifting plasma and analyse the two-stream instability (sections 6.2.5-6).

The 'multiple-stream' approach to the dispersion relation for a plasma with continuous velocity distribution, first employed by Dawson (1960), is presented in the text of Stix (1962). A satisfactory discussion of this problem, and Landau damping (section 6.2.7) is necessarily lengthy and somewhat mathematical. There are many treatments of this important question in various texts and in the literature, with varying styles to suit all tastes. Rather than attempt a summary of the subject, we refer the reader to any modern text on plasma physics. Chen (1974) gives a physical discussion, with references to the classic papers in this field. A more formal mathematical treatment is given by Montgomery and Tidman (1964) and a lengthy discussion by Roos (1969).

A further topic which again needs extended discussion is the distinction between convective and absolute instabilities and the criteria for determining from the dispersion relation which type of instability occurs. The two types of instability have been distinguished in section 6.2.6, and again in section 6.3.7. The concept only applies in a straightforward way to an infinite medium; in a convective instability a disturbance propagates and grows in time, though at any fixed point in the medium the disturbance eventually dies away. In an absolute instability, on the other hand, the disturbance grows continuously at all points. This behaviour is clearly described in the paper by Sturrock (1960) already quoted. Rules for determining the type of instability from causality arguments, making use of Fourier-Laplace transform techniques, were first clearly set out by Briggs (1964). Much has been written on the subject since, and it is treated in various texts such as Clemmow and Dougherty (1967) and Shohet (1971). An extended review by Rognlien and Self (1972) considers in particular the effect of finite boundaries and also contains a complete list of references on this topic. A typical application of these criteria to a practical problem is described by Manickam *et al.* (1975).

The concept of positive and negative energy waves, introduced in section 6.2.5, is referred to again in section 6.3 and section 6.4; a discussion may be found in the text of Steele and Vural (1969).

6.3. Longitudinal waves in beams of finite cross-section

6.3.1. Introduction In section 6.2 an outline was given of some of the wave phenomena encountered in infinite plasmas, both stationary and drifting, with both zero and finite temperatures. Emphasis was on longitudinal waves with electric field in the direction of propagation rather than transverse waves, since in the transition to finite beams the connection between transverse-wave types of interest in the two situations is rather slender, whereas for longitudinal motion it is more direct.

Interaction between waves in different streams of a plasma was discussed in section 6.2.6. In this section the interaction of waves on a beam with the structure of the surrounding wall will also be studied. This topic is very relevant to some forms of microwave tube and to instabilities in particle accelerators.

Attention will be confined in this section to longitudinal waves; transverse waves, where the particle motion in perpendicular to the beam direction, are discussed in section 6.4. We proceed with an analysis of waves on beams surrounded by conducting cylindrical tubes.

6.3.2. Longitudinal waves in a cylindrical beam surrounded by conducting walls The simplest beam configuration is that of a cylinder of moving charge surrounded by free space. In order to prevent beam spreading (section 4.2.5) a strong magnetic field will be postulated. Further, to make the configuration more realistic, the beam, of radius a, will be assumed to be surrounded by a conducting tube of radius b. As will be evident from the discussions in sections 3.2 and 6.2 for very intense beams there will be a large difference of potential between the axis and edge of the beam. In such circumstances a distribution of velocities which is uniform across the beam is unrealistic. Neutralization by massive, immobile, positive ions may then be assumed, since this does not affect time-varying field components.

We now calculate the dispersion relation for longitudinal waves appropriate to the system defined above. This may be done in essentially the same way as in the previous section, except that the transverse variation of field and associated matching complicate the algebra. Such an approach may be found, for example, in the original papers of Hahn (1939) and Ramo (1939) and in the monograph by Beck (1958).

Here we illustrate a slightly more compact approach, making use of the Vlasov equation introduced at the end of the last chapter. The problem studied there was independent of time; here there is a small time-dependent harmonic motion superimposed on steady motion, the characteristics of which are assumed to be known. These two examples illustrate two important uses of the Vlasov equation; to find possible equilibria, and to find time-varying perturbations of such equilibria.

Since the Vlasov formulation includes the distribution of velocities, or more strictly momenta, in a natural way, a continuous distribution will be included from the beginning. For conditions appropriate to our present problem, where the electron motion is constrained to move in the z-direction by a strong magnetic field, the Vlasov equation (4.133) may be written

$$\frac{\partial f}{\partial t} = -\beta c\,\frac{\partial f}{\partial z} - \frac{qE_z}{m_0 c}\,\frac{\partial f}{\partial(\beta\gamma)} \tag{6.57}$$

where $p = \beta\gamma m_0 c$ and f is a function of $\beta\gamma$ rather than of p. The distribution function may be written as $f_0+\tilde{f}$, where $\int f_0\,\mathrm{d}(\beta\gamma) = n$, the particle density. (The equation differs from (4.133) in that velocity is substituted for momentum in the second term and $B_\perp = 0$.) Substituting f into equation (6.57), the steady-state terms in f_0 cancel (since this is a solution with $\partial f/\partial t = 0$); this leaves

$$\frac{\partial f}{\partial t} = -\beta c\frac{\partial\tilde{f}}{\partial z} - \frac{q\tilde{E}_z}{m_0 c}\,\frac{\partial f_0}{\partial(\beta\gamma)} \quad . \tag{6.58}$$

Setting $\tilde{E} = E\exp\mathrm{j}(\omega t-kz)$ and integrating with respect to t yields (dropping the subscript on E)

$$\tilde{f} = \frac{\mathrm{j}q}{m_0 c}\,\frac{\partial f_0}{\partial(\beta\gamma)}\,\frac{\tilde{E}}{\omega-\beta ck} \; . \tag{6.59}$$

This equation, when multiplied by q and integrated over $\beta\gamma$, gives the charge density in terms of $\tilde{E}$. By inserting this in Maxwell's equations, $\tilde{E}$ cancels out to leave the required dispersion relation. The equation for $\tilde{E}$ is found by eliminating $\tilde{B}$ from the two curl equations

$$\left(\nabla^2-\frac{1}{c^2}\frac{\partial^2}{\partial t^2}\right)\tilde{E} = \frac{q}{\varepsilon_0}\nabla\tilde{n}+\mu_0 qc\frac{\partial}{\partial t}(\tilde{n}\beta) . \tag{6.60}$$

(This is a scalar equation, for the z-component of $\underline{E}$. The velocity βc is in the z-direction.) Substituting now the relation

$$\tilde{n} = \int \tilde{f}\mathrm{d}(\beta\gamma) \tag{6.61}$$

equation (6.60) becomes, for a harmonic wave,

$$\left(\nabla_\perp^2-k_z^2+\frac{\omega^2}{c^2}\right)\tilde{E} = \frac{q}{\varepsilon_0}\int\frac{\partial}{\partial z}\tilde{f}\mathrm{d}(\beta\gamma)+q\mu_0 c\int\beta\frac{\partial}{\partial t}\tilde{f}\mathrm{d}(\beta\gamma) . \tag{6.62}$$

Here $\nabla_\perp^2$ is the transverse part of the Laplacian operator and k is the z-component of the wave vector. Substituting for $\tilde{f}$ from equation (6.59) and making use of the fact that $\mathrm{d}(\beta\gamma) = \gamma^3\mathrm{d}\beta$, equation (6.62) becomes

$$\left(\nabla_\perp^2 - k_z^2 + \frac{\omega^2}{c^2}\right)\tilde{E} = \frac{-q^2\tilde{E}}{\varepsilon_0 m_0 c^2}\int\frac{ck_z-\omega\beta}{\omega-\beta ck_z}\frac{\partial f_0}{\partial\beta}\mathrm{d}\beta. \tag{6.63}$$

The integral can be integrated by parts, the first term being zero because f_0 is zero at the limits -1 and 1; dropping the factor $\tilde{E}$ it becomes

$$\left(\Delta^2-k_z^2+\frac{\omega^2}{c^2}\right) = \frac{-q^2}{\varepsilon_0 m_0 c^2}\int\left\{\frac{\partial f_0}{\partial\beta}\mathrm{d}\beta\frac{\partial}{\partial\beta}\left(\frac{ck_z-\omega\beta}{\omega-\beta ck_z}\right)\right\}\mathrm{d}\beta$$

$$= \frac{q^2}{\gamma^3\varepsilon_0 m_0}\left(\frac{\omega^2}{c^2}-k_z\right)\int\frac{f_0\mathrm{d}(\beta\gamma)}{(\omega-\beta ck_z)^2} . \tag{6.64}$$

Introducing γ_w this can be rearranged as

$$\nabla_\perp^2 = -\frac{k^2}{\gamma_w^2}\left(1-\frac{q^2}{\gamma^3 m_0\varepsilon_0}\int\frac{f_0(\beta\gamma)\mathrm{d}(\beta\gamma)}{(\omega-\beta ck)^2}\right) \tag{6.65a}$$

or, in alternative form,

$$\nabla_{\perp}^{2} = -\frac{k^{2}}{\gamma_{w}^{2}}\left[1 - \frac{q^{2}}{\gamma^{3}m_{0}\varepsilon_{0}ck}\int \frac{\partial f_{0}(\beta\gamma)}{\partial\beta}\frac{d(\beta\gamma)}{(\omega-\beta ck)}\right] \qquad (6.65b)$$

where the subscript on k has been dropped. An important approximation, known as the 'quasi-static approximation' is obtained by putting $\gamma_w = 1$. This implies that the phase velocity of the wave is much less than that of light. This is equivalent to putting the velocity of light infinite, or setting curl $\underline{E} = 0$, so that the electric fields are described by Poisson's equation (or its generalization for a tensor permittivity). Referring back to equation (6.60), the left-hand side in this approximation becomes $\nabla^2 E$; by following through the working it is evident that this is equivalent to setting $\gamma_w = 1$ in equation (6.65). The quasi-static approximation should not be confused with the non-relativistic approximation, in which the velocity of the particles is assumed to be small compared with that of light.

Returning now to equation (6.65), to obtain the dispersion relation for a system of finite transverse dimensions we need to find $\nabla_{\perp}$ as a function of the geometry. To do this, we find values of k_r which satisfy the equation

$$(\nabla_{\perp}^{2} + k_{r}^{2})\ \underline{E}, \underline{H} = 0 \qquad (6.66)$$

in cylindrical co-ordinates for a beam of radius a. This is conveniently done in terms of the outward impedance Z_a at $r = a$. For a conducting wall at the beam edge $Z_a = 0$; if the wall is at $r = b$, then its impedance must be transformed to take into account the space between wall and beam. This space can be regarded as a 'radial transmission line'.

In solving equation (6.66) a double infinity of modes is obtained, characterized by azimuthal and radial mode numbers. Although these are of interest in studying waves in plasma columns in beam problems only the lowest radial mode is of general interest in the study of *longitudinal* waves. Rather than write down a general solution, particularizing later, we consider only the solution for the lowest mode, which has axial symmetry and monotonic variation between the centre and wall. In terms of Z_0, the impedance of free space, and γ_w the fields can be written

$$\begin{aligned} E_z &= E_0 J_0(k_r r) \\ E_r &= -j\frac{k_r}{k_z}\gamma_w^2 E_0 J_1(k_r r) \\ Z_0 H_\theta &= -j\frac{\omega k_r}{c k_z^2}\gamma_w^2 E_0 J_1(k_r r)\ . \end{aligned} \tag{6.67}$$

The impedance at $r = a$ is obtained by finding the ratio $-E_z/H_\theta$:

$$Z_a = -\frac{jck_z^2 Z_0}{\gamma_w^2 \omega k_r}\frac{J_0(k_r a)}{J_1(k_r a)} = -\frac{jk_z}{k_r}\frac{Z_0}{\beta_w \gamma_w^2}\frac{J_0(k_r a)}{J_1(k_r a)}\ . \tag{6.68}$$

For a wall at $r = a$, $Z_a = 0$ and k_r is determined by the first root of the equation $J_0\ (k_r a) = 0$. This is $k_r = 2{\cdot}405/a$.

When $b > a$, Z_a must be matched to the wall impedance transformed from $r = b$ to $r = a$. The formula for this transformation in a radial transmission line has been given by Birdsall and Whinnery (1953). If the wall impedance is Z_b and

$$Z_l = \frac{-jk_z}{\omega\varepsilon_0\gamma_w} = \frac{-jZ_0}{\beta_w\gamma_w} \tag{6.69}$$

is the characteristic impedance of the radial transmission line, then

$$\frac{Z_a}{Z_l} = \frac{\text{becosh}\ kx}{\text{Becosh}\ kx}\left(\frac{Z_b + Z_l\ \ \text{Betanh}\ kx}{Z_l + Z_b\ \ \text{betanh}\ kx}\right) \tag{6.70}$$

where $x = (b-a)/\gamma_w$ and the 'Bressel hyperbolic functions' are defined below. For brevity ka/γ_w and kb/γ_w are written as A and B:

$$\text{betanh}\, kx = \frac{K_1(A)I_1(B) - I_1(A)K_1(B)}{K_1(A)I_0(B) + K_0(A)I_1(B)}$$

$$\text{Betanh}\, kx = \frac{K_0(A)I_0(B) - I_0(A)K_0(B)}{K_0(A)I_1(B) + I_0(A)K_1(B)} \tag{6.71}$$

$$\frac{\text{Becosh}\, kx}{\text{becosh}\, kx} = \frac{K_1(A)I_0(B) + I_1(A)K_0(B)}{K_0(A)I_1(B) + I_0(A)K_1(B)}$$

Eliminating Z_a between equations (6.68) and (6.70) determines k_r; the equation so obtained is known as the 'determinantal' equation. For a perfectly conducting wall $Z_b = 0$ and it simplifies to

$$-\frac{\gamma_w k_r J_1(k_r a)}{kJ_0(k_r a)} = \frac{I_1(ka/\gamma_w)K_0(kb/\gamma_w) + I_0(kb/\gamma_w)K_1(ka/\gamma_w)}{I_0(ka/\gamma_w)K_0(kb/\gamma_w) - I_0(kb/\gamma_w)K_0(ka/\gamma_w)} \tag{6.72}$$

(where the subscript on k_z has been dropped). From equation (6.66) $k_r^2 = -\nabla_\perp^2$, so that k_r^2 can now be eliminated from equation (6.65) and the determinantal equation to give the full dispersion relation. In general this is very complicated; the simplification obtained by using the quasi-static approximation $\gamma_w = 1$ is evident.

Even with the quasi-static approximation this expression is somewhat unwieldy. Corresponding expressions for more complicated types of wall, or finite rather than infinite B_z, are much more involved. In subsequent sections, where we are more interested in the physical effects than in mathematical detail, attention will be focused on limiting cases where simplification is possible, rather than the more general situation.

A useful approximation applicable to the study of accelerator instabilities can be made when the wavelength along the beam greatly exceeds the tube diameter divided by γ_w. This allows the Bessel functions in equation (6.72) to be replaced by their asymptotic limits. This is especially helpful, as we see later, when the tube wall is not a smooth perfect conductor so that $Z_b \neq 0$. This is sometimes known as the 'thin-beam' approximation. A further approximation, useful in highly relativistic beams where β_w is small in the frame of reference in which the beam is stationary, is to take $\gamma_w = \gamma$.

In the next few sections we consider several specific situations of pedagogical or practical interest.

6.3.3. Cylindrical beam with a close-fitting conducting tube in the presence of ions When the conducting wall fits closely round the beam, so that $a = b$, the dispersion relation is much simpler than for the more general situation when b exceeds a. The essential physical behaviour is, however, similar; for this reason we examine the simpler case. Setting $a = b$ equation (6.68) reduces to

$$J_0(k_r a) = 0 \tag{6.73}$$

so that to obtain the dispersion relation when $a = b$ it is only necessary to set $k_r a$ equal to the appropriate root of $J_0(x) = 0$. Accordingly, equation (6.65) becomes, for the lowest-order mode where $J_0(2{\cdot}4) = 0$,

$$k_r^2 = \left(\frac{2\cdot 4}{a}\right)^2 = -\frac{k_z^2}{\gamma_w^2}\left(1 - \frac{q^2}{\gamma^3 m_0 \varepsilon_0}\int \frac{f_0(\beta\gamma)\,d(\beta\gamma)}{(\omega-\beta ck)^2}\right) = 0. \tag{6.74}$$

In order to introduce first the simpler features of these waves we replace the continuous distribution of velocities by a δ-function

$$T^2 = -\frac{k_z^2}{\gamma_w^2}\left(1 - \frac{\omega_p^2}{\gamma^2(\omega-\beta ck)^2}\right) \tag{6.75}$$

where T has been written for $2{\cdot}4/a$, the value of k_r which satisfies the determinantal equation. Equation (6.75) may be written in the form

$$1 = \frac{\omega_q^2}{\gamma^2(\omega-\beta ck)^2} \tag{6.76}$$

where, (dropping the subscript on k_z)

$$\frac{\omega_q}{\omega_p} = \frac{k}{(k_z^2+\gamma_w^2T^2)^{1/2}} \tag{6.77}$$

is the 'plasma-frequency reduction factor' F. For many applications, where the wave velocity is much less than that of light, it is possible to use the quasi-static approximation so that $\gamma_w = 1$ and ω_q/ω_p is a function of k and the geometry of the system but not of ω. Curves of F against ka with this approximation are plotted in Fig.6.5. For small

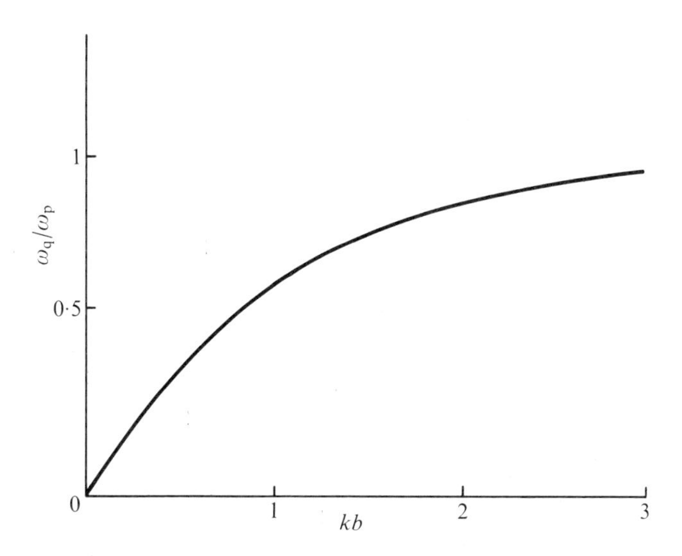

Fig.6.5. Sketch of the plasma frequency reduction factor ω_q/ω_p as a function of ka. The precise form of the curve depends on the geometry; accurately plotted curves with cylindrical geometry are given, for example, by Branch and Mihran (1955).

values of ka, where the wavelength is large compared with the tube diameter, F is small. This may be explained physically by the idea that the lines of electric field are 'shorted out' to the walls, so that their effectiveness in the z-direction is much reduced (Fig.6.6). As ka increases, however, the plasma wavelength ultimately becomes negligible compared with that of the tube and F tends to unity. For more complicated arrangements, where $b \neq a$ or the beam is

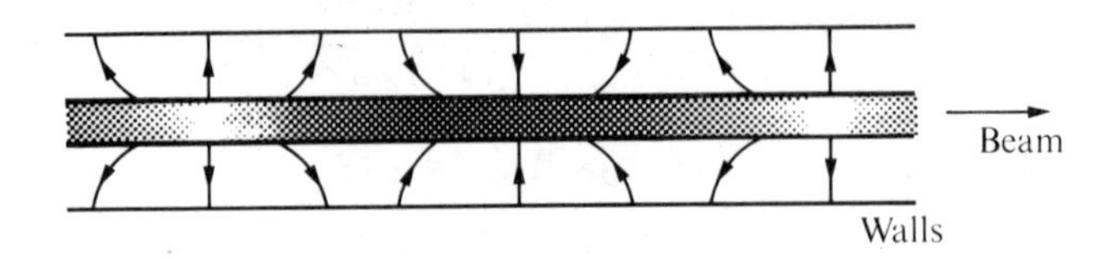

Fig.6.6. Schematic diagram of lines of force associated with a bunched beam in a conducting tube. The signs + and - denote regions of high and low density. The 'short-circuiting' effect of the walls reduces the frequency below the plasma frequency.

hollow, the factor $2{\cdot}4/a$ is replaced by a more complicated function of b and a given by the roots for k_r of the determinantal equation (equation (6.72)) or its analogue for other geometries.

The dispersion curve corresponding to equation (6.75) is sketched in Fig.6.7. It is a quartic with two branches. One corresponds asymptotically to the two free-space waves in an empty guide, and the other to the waves in the beam. For $ka \gg \gamma_w$ the effect of the walls is relatively small. It is near the origin that the behaviour is most modified. The group velocity $\partial\omega/\partial k$ is no longer equal to the beam velocity, not is the phase velocity infinite at zero frequency. Indeed, for a neutralized beam in which the charge density is high the sign of the group velocity can reverse at the origin, giving two backward-wave modes for small ω and k (see Fig.6.7). It is interesting that near the origin for one of these both group and phase velocity are negative. The backward energy associated with the field exceeds the forward energy associated with the particles. Although the frequency of this wave is real, it can produce instability in a finite system where reflections from the ends can produce coupling between the forward and backward waves.

We now calculate the conditions for this wave to appear by finding the condition illustrated in Fig.6.7(b), that $\partial\omega/\partial k = 0$ at the origin. In the region of interest in the analysis (near the origin) ω/k is certainly very much less than c, so that $\gamma_w = 1$ and equation (6.75) may be written

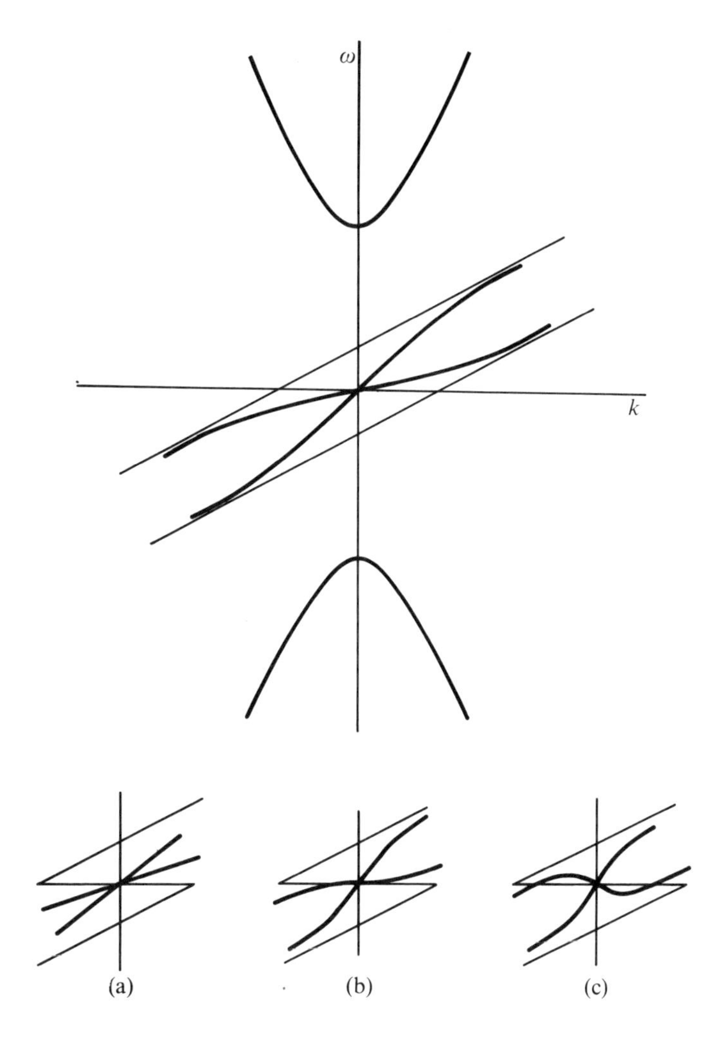

Fig.6.7. Sketch of dispersion relation for cylindrical beam in a tube with conducting walls. The three curves (a) - (c) show conditions at the origin for increasing values of ω_p. Curve (b) illustrates the critical perveance where the group velocity of the slow wave changes sign.

(omitting the subscript on k)

$$\frac{\omega_p^2}{\gamma^2(\omega-\beta ck)^2} = 1 + \frac{T^2}{k^2} . \qquad (6.78)$$

Rearranging,

$$\omega = \beta c k \pm \frac{\omega_p k}{\gamma(k^2+T^2)^{1/2}} \tag{6.79}$$

whence,

$$\frac{\partial\omega}{\partial k} = \beta c \mp \frac{\omega_p T^2}{\gamma(k^2+T^2)^{3/2}} . \tag{6.80}$$

Setting $\partial\omega/\partial k = 0$ at $k = 0$, and $Ta = 2{\cdot}4$ yields the final result. This may be written in various forms, in terms of the plasma frequency, Budker's parameter, or beam current and voltage:

$$\beta c = \omega_p a/2{\cdot}4\gamma \tag{6.81a}$$

$$\nu = \tfrac{1}{2}(2{\cdot}4)^2 \beta^2 \gamma^3 \tag{6.81b}$$

$$I/V^{3/2} = 190 \times 10^{-6}\,\mathrm{A\,V}^{-3/2}\ \text{(N.R.)}. \tag{6.81c}$$

This condition was first identified by Pierce (1944) as the limiting current for a stable beam in the presence of a stationary ion background. He expressed the result as equation (6.81c) and pointed out that it is a few times higher than the limit arising from space-charge depression in an un-neutralized beam (equation (3.13)).

If the ions are not infinitely massive, the system is unstable and represents the two-stream instability in finite geometry. For such a situation the dispersion relation (6.78) contains an extra term ω_{pi}^2/ω^2 on the left-hand side. It can easily be shown that, with singly-charged ions, this results in a factor $\{1 + (m_e/m_i)^{1/3}\}^{3/2}$ on the right-hand side of equation (6.81). This then represents the criterion for the onset of instability; as m_i tends to infinity the growth time increases, and in the limit the system reduces to the stable configuration in which the ions are assumed stationary (Briggs 1964). For a tube in which $b > a$, the analysis is evidently considerably more complicated. For the thin beam approximation, $kb \ll \gamma_w$, the following small argument Bessel function approximations may be used:

$$J_0(z) = 1-\tfrac{1}{4}z^2, \qquad J_1(z) = \tfrac{1}{2}z(1-\tfrac{1}{4}z^2),$$

$$I_0(z) = 1+\tfrac{1}{4}z^2, \qquad I_1(z) = \tfrac{1}{2}z(1+\tfrac{1}{4}z^2),$$

$$K_0(z) = (-\Gamma+\ln\tfrac{1}{2}z), \qquad K_1(z) = -\frac{1}{z}+\tfrac{1}{2}z(\ln\tfrac{1}{2}z+\Gamma)$$

where Γ is Euler's constant, 0·577, usually written as γ. Inserting these in equation (6.60) yields

$$Ta = \{2\ln(b/a)+\tfrac{1}{2}\}^{-1/2} . \tag{6.82}$$

This may be inserted into equation (6.77) and an appropriate value of limiting perveance calculated.

The situation analysed in this section is somewhat academic in the sense that both infinite magnetic fields and totally immobile ions are unrealizable abstractions. Nevertheless, the practical question of the stability of neutralized electron electron beams in conducting tubes has received considerable study, both theoretically and experimentally. The possibility of non-linear relaxation oscillations associated with the formation time of ions complicates the situation, but the main features of the expected theoretical behaviour have been reasonably well verified.

The theory and experiments have been reviewed by Bogdankevich and Rukhadze (1971) in a paper which contains references to earlier work in this field. A relevant paper reporting experiments and giving a general theoretical discussion not included in the above review is that of Frey and Birdsall (1966); further comments, with some corrections to this paper, are given by Faulkener and Ware (1969).

6.3.4. Cylindrical systems with arbitrary wall impedance and with positive or negative mass   Many practical devices, including microwave tubes and particle accelerators, contain beams moving in tubes which do not have zero wall impedance. For smooth walls of thickness large compared with the skin depth the impedance may be expressed in terms of the conductivity σ. For a highly conducting wall.

$$Z_b = (1+j)\left(\frac{\varepsilon_0\omega}{\sigma}\right)^{1/2} Z_0 \, . \qquad (6.83)$$

Walls consisting of a periodic conducting structure are also of interest. Such walls can sometimes be specified by a simple impedance; for example a corrugated wall with deep slots closely spaced compared with the wavelength looks inductive or capacitive according to the depth of slot. In general, however, periodic structures cannot be so simply represented; the mode structure becomes more complicated. To begin with we consider structures in which the wall may be represented by an impedance Z_b. Since Z_b is a function of frequency, the dispersion relation for the whole system includes the dispersive properties of the wall. The thin beam approximation, $kb \ll \gamma_w$, will be used.

The impedance Z_a at $r = a$ can be found from the asymptotic form of the r.h.s. of equation (6.70). Using the small-argument Bessel-function approximations listed in the previous section,

$$Z_a = \frac{a}{b} Z_l \frac{Z_b + Z_l\{kb \ln(b/a)\}/\gamma_w}{Z_l + \frac{1}{2}Z_b k(b+a^2/b)/\gamma_w} \, . \qquad (6.84)$$

Since $kb \ll \gamma_w$, the second term in the denominator may be omitted even if Z_b is not small compared with Z_0, though in practice it often is so. Omitting this term, and substituting for Z_l from equation (6.69), equation (6.84) simplifies to

$$Z_a = \frac{a}{b} Z_b - \frac{jkaZ_0 \ln(b/a)}{\beta_w\gamma_w^2} \, . \qquad (6.85)$$

Equating this to Z_a in equation (6.68) yields the determinantal equation. Inserting the small-argument expansion for the Bessel functions this is found to be

$$1 = \frac{a^2}{b}\frac{\beta_w\gamma_w^2}{2k}\left\{\frac{jZ_b}{Z_0} + \frac{kb}{\beta_w\gamma_w^2}\left[\ln\frac{b}{a} + \frac{1}{4}\right]\right\} k_r^2 \qquad (6.86)$$

Setting $\nabla_\perp^2 = -k_r^2$ in equation (6.65) yields the required dispersion relation. This will now be studied for a number of specific systems.

The very simplest situation occurs when $Z_b = 0$ and $f(\beta\gamma)$ is a δ-function. The dispersion relation is then

$$1 = \frac{a^2}{2}\left(\ln\frac{b}{a} + \tfrac{1}{4}\right)k_r^2$$
$$= \frac{a^2}{2}\left(\ln\frac{b}{a} + \tfrac{1}{4}\right)\left(-\frac{k^2}{\gamma_w^2}\right)\left(1 - \frac{\omega_p^2}{\gamma^2}\frac{1}{(\omega-\beta ck)^2}\right) \quad . \quad (6.87)$$

This may be written as

$$1 = \frac{\omega_q^2}{\gamma^2(\omega-\beta ck)^2} \quad (6.88)$$

where ω_q/ω_p, the plasma-frequency reduction factor is given by

$$\frac{\omega_q^2}{\omega_p^2} = \frac{k^2a^2}{2\gamma_w^2}\left(\ln\frac{b}{a} + \tfrac{1}{4}\right). \quad (6.89)$$

Writing

$$\frac{\gamma_w^2\omega_q^2}{k^2\omega_p^2} = \frac{a^2}{2}\left(\ln\frac{b}{a} + \tfrac{1}{4}\right) = L_0^2 \quad (6.90)$$

where L_0 is a characteristic length which is a function of the geometry; setting $\beta_w = \omega/ck$ equation (6.88) may be written

$$\omega - \beta ck = \pm\frac{L_0 k\omega_p}{\gamma\gamma_w} \quad . \quad (6.91)$$

This represents a pair of straight lines through the origin

$$\omega = \left(\beta c \pm \frac{L_0\omega_p}{\gamma\gamma_w}\right)k. \quad (6.92)$$

These lines are tangents at the origin to the dispersion relation sketched in Fig.(6.7). The 'critical-current' condition when $\partial\omega/\partial k = 0$ for the slow wave discussed in the previous section occurs when the quantity in parentheses is set equal to zero. For a beam filling the tube $L_0 = a/2{\cdot}4$.

If the wall impedance is not zero, then by the same steps as those taken to deduce equation (6.92), it is found that L_0 should be replaced by the complex quantity

$$L_1 = \left\{\frac{a^2}{2}\left[\ln\frac{b}{a}+\frac{1}{4}\right] + \frac{j\beta_w\gamma_w^2 a^2}{2kb}\frac{Z_b}{Z_0}\right\}^{1/2} . \qquad (6.93)$$

For a wall with high conductivity Z_b is given by equation (6.83). In the expression for L_1 the real part is small compared with the first term, and may therefore be omitted, to give

$$L_1 \approx L_0\left\{1 + \frac{j\beta_w\gamma_w^2 a^2}{4kbL_0^2}\left(\frac{\varepsilon_0\omega}{2\sigma}\right)^{1/2}\right\} \qquad (6.94)$$

leading to the dispersion relation

$$\omega = \left(\beta c \pm \frac{L_0\omega_p}{\gamma}\right)k \pm \frac{j\gamma_w^2 a^2\omega_p}{4bL_0\gamma}\left(\frac{\varepsilon_0\omega}{2\sigma}\right)^{1/2} . \qquad (6.95)$$

We now find that for real k, ω is complex and *vice versa*; the top and bottom signs indicate a damped fast wave and a growing slow wave respectively. Real and imaginary parts of ω for real k and vice versa are sketched in Fig.6.8.

If Z_b is purely capacitive, this has the effect of increasing L_0^2. If it is inductive, on the other hand, the value of L_0^2 is decreased. Indeed if Z_b is large enough, the second term in the bracket on the r.h.s. of equation (6.93) exceeds the first, and L_1 becomes pure imaginary. (This implies that the impedance at $r = a$ has now become inductive). Fast and slow waves merge, both having velocity equal to the beam velocity. The beam acquires an increasingly modulated envelope moving with the beam velocity. This is the principle of the 'easitron' microwave tube (Pierce 1954c). This same physical effect is observed under 'negative-mass' conditions in the presence of zero-impedance walls. In section 2.7.4 it was shown that in circular machines above transition energy, $\gamma > \gamma_t$, the effective mass

$$m^* = \frac{\gamma m_0}{(1/\gamma^2)-\alpha} = \frac{\gamma m_0}{(1/\gamma^2)-(1/\gamma_t^2)} \qquad (2.144)$$

becomes negative. For a betatron field, with field index $1 > n > 0$, $\gamma_t^2 = 1/(1-n)$ and m^* is always negative. In a betatron the beam is not constrained by a longitudinal magnetic field in the way we have assumed, but nevertheless it

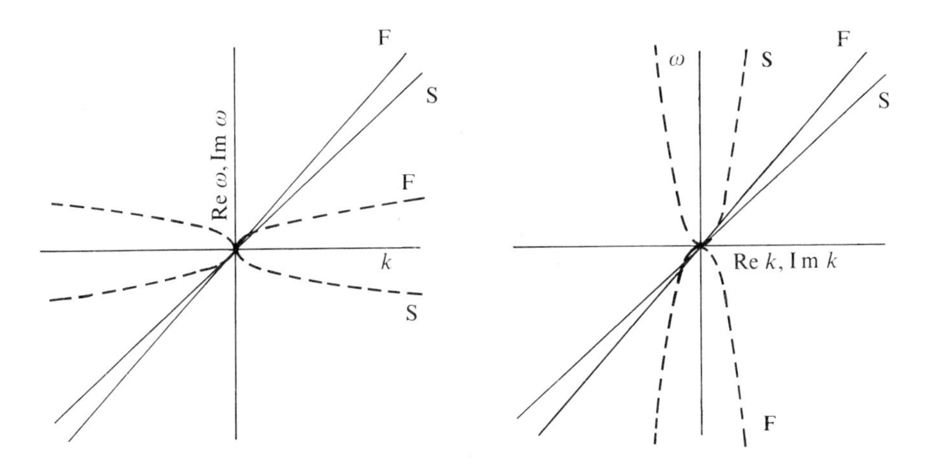

Fig.6.8. Dispersion diagram for a beam in a cylindrical tube with a resistive wall when $L_0\omega_p/\gamma \ll \beta$ (equation 6.95) and $kb \ll \gamma_w$. The wall impedance is low, so that the real frequency shift associated with the imaginary part of the impedance may be neglected. The first diagram shows real (solid) and imaginary (broken) parts of ω for real k, and the second shows real and imaginary parts of k for real ω. The imaginary parts associated with the fast and slow waves (F and S) respectively give damping and growth.

occupies a well-defined region and for the particular case of $n = 1/\surd 2$ the focusing in vertical and horizontal planes is equal, so that it is realistic to consider a beam of radius a in a tube of radius b. Under these circumstances equation (6.79) holds, except that ω_p^2 is replaced by $\omega_p^2 m_0/m^*$, a negative quantity. Again, therefore, the modulation of the beam increases with time.

This problem has been analysed in terms of accelerator parameters by Neil and Sessler (1965). Two differences are evident. In the first place the closed-orbit geometry only allows quantized values of k, so that results are more conveniently expressed in terms of $n\theta$, where n is the azimuthal mode number, rather than kz. A second, very minor, difference is that their logarithmic factor is $(\ln a/b+\frac{1}{2})$ in place of $(\ln a/b+\frac{1}{4})$. The reason for this may be found in slightly different assumptions in the derivation. There is in addition the minor difference in beam characteristics arising from the accelerator focusing system rather than a confining B_z field. Particles do not stay at a fixed radius in the beam, but swing across it as they execute betatron oscillations.

In the same way that an inductive impedance at $r = a$ makes a positive-mass beam unstable, it is now found that such an impedance tends to stabilize a negative-mass beam, whereas a capacitive impedance tends to make it more unstable. This fact was pointed out by Briggs and Neill (1967). The beam characteristics obtained in the present section are summarized in Table 6.1. In the negative-mass regime it is

TABLE 6.1

Characteristics of longitudinal wave motion on beams in terms of the impedance of the surrounding wall

Form of Z_a	Sign of mass	Motion
$-jX$ (capacitive)	+	Stable
jX (inductive)	-	Unstable
$-jX$	-	Unstable
jX	+	Stable
$\pm jX + R$	+ or -	Unstable

the slow wave which carries positive energy and is damped and the fast wave which grows.

In this section the approximation $kb \ll \gamma_w$ has been used throughout. In the more general (though non-relativistic) paper by Birdsall and Whinnery (1953) numerical mappings of relevant complex functions are presented. In the next section we consider continuous velocity distributions and the role of Landau damping in stabilizing the situations listed as unstable in Table 6.1.

6.3.5. Note on accelerator notation Many of the applications of the ideas in the last section are to cyclic particle accelerators. The approach to longitudinal instabilities adopted here was laid down in the paper by Neil and Sessler (1965); later contributions by Keil and Schnell (1969) and others reformulated it somewhat in terms of readily measurable impedances of elements of the vacuum chamber.

Similar effects of course occur in microwave tubes, but these are best studied after introducing coupled-mode theory (section 6.3.7). Here we introduce for convenience parameters used in accelerator theory and relate them to the notation of the last section. Instead of z and $\beta\gamma m_0 c$, canonical variables θ and W are taken. For a circular orbit $W = 2\pi(P_\theta - P_0)$ where P_0 is the canonical angular momentum of a particle on the equilibrium orbit. For a non-circular orbit a more general definition may be used, $W = m_0 c^2 \int \mathrm{d}\gamma/\omega_0$ where ω_0 is the frequency on the equilibrium orbit. The limits of the integral are taken such that W is zero on the equilibrium orbit. Groupings of constants analogous to L_0 and L_1 of equations (6.93) and (6.94) are included in two parameters U and V. Indeed, a complete translation of Neil and Sessler's theory into our notation can be made with the aid of the 'dictionary' shown as Table 6.2.

The parameters U and V, which have the dimensions of energy, are defined as

$$\begin{aligned} U &= \frac{N_0 q^2 n\{2\ln(b/a)+1\}}{2\gamma_{\mathrm{w}}^2 C\varepsilon_0} \\ V &= \frac{q^2 N_0 \beta_{\mathrm{w}}}{2\pi\varepsilon_0 b}\left(\frac{\varepsilon_0\omega}{2\sigma}\right)^{1/2} \end{aligned} \qquad (6.96)$$

As in equation (6.95) the real part of the wall impedance Z_b is neglected. In accelerator notation equation (6.78) becomes

$$\begin{aligned} \omega - n\omega_0 &= \pm\,(nk_0)^{1/2}\,(U+\mathrm{j}V)^{1/2} \\ &\approx (\pm\, nk_0 U)^{1/2}(1+\tfrac{1}{2}\mathrm{j}V/U). \end{aligned} \qquad (6.97)$$

In subsequent work the quantities U and V will be used when comparison with accelerator papers is of interest.

TABLE 6.2

'Dictionary' giving notation in accelerator papers in terms of that used here

Neil and Sessler†	Present notation
R	$C/2\pi$
$R\dot{\theta}$	$\beta_{\mathrm{w}}c$
$R\omega_0$	βc
n/R	k
N	N_0, $2\pi^2 Ra^2 n_0$
k_0	$1/2\pi R^2 m^*$
$\eta Nq^2/2\pi^2 Ra^2\gamma m_0$	$-\omega_{\mathrm{p}}^2/\gamma$
$\mathcal{R}$	$(\varepsilon_0\omega/2\sigma)^{1/2}$

† The quantity n refers to the azimuthal mode number; n_0 is the number density of particles.

6.3.6. Cylindrical systems with continuous velocity distribution and Landau damping In the previous section it was shown that for a variety of wall conditions longitudinal waves in a mono-energetic beam are unstable. If an energy spread is introduced, Landau damping occurs. When this is sufficient the system is stabilized. We now look for the conditions for this in the positive- and negative-mass regimes where Z_b (or V) is not zero.

The appropriate dispersion relation, for a wall of arbitrary impedance, may immediately be found by inserting $k_r = -\nabla_\perp^2$ from equation (6.65b) into equation (6.86), with $kb \ll \gamma_{\mathrm{w}}$:

$$1 = \left\{\frac{\mathrm{j}Z_b}{Z_0} + \frac{kb}{\beta_{\mathrm{w}}\gamma_{\mathrm{w}}^2}\left(\ln\frac{b}{a}+\tfrac{1}{4}\right)\right\}\frac{q^2a^2\beta_{\mathrm{w}}}{2\gamma^3 m_0\varepsilon_0 bc}\int\frac{\partial f_0(\beta\gamma)}{\partial\beta}\frac{\mathrm{d}(\beta\gamma)}{(\omega-\beta ck)} \qquad (6.98)$$

where, as before, $\int f_0(\beta\gamma)\mathrm{d}(\beta\gamma) = n$. The corresponding integral in accelerator notation is

$$1 = -(U+jV)\int \frac{\partial F}{\partial W}\frac{dW}{\omega - n\dot{\theta}}$$
$$= -\frac{U+jV}{nk_0}\int \frac{\partial F}{\partial W}\frac{dW}{W-W_1} \quad . \tag{6.99}$$

All the symbols are defined in the previous section, except for W_1 which is equal to $(\omega-n\omega_0)/nk_0$; $F(W)$ is normalized so that its integral over W is unity. It is important to distinguish between ω, the frequency of the disturbance, ω_0, the angular velocity of a particle on the equilibrium orbit, and $\dot{\theta}$, the angular velocity of the wave.

Equation (6.99) is in convenient form for a discussion of the balance between a tendency to growth and Landau damping; we follow the treatment in the paper by Neil and Sessler, relevant to cyclic accelerators, though of course a parallel argument could be applied to the straight beams analysed earlier. Equation (6.99) is not in general tractable analytically. An exception is the physically unrealistic 'resonance' distribution

$$F_0(W) = \frac{\Gamma}{\pi(W^2+\Gamma^2)} \tag{6.100}$$

for which equation (6.99) is readily integrable to yield

$$\omega = n\omega_0 \pm (nk_0U)^{1/2}\{1+\tfrac{1}{2}j(V/U)\}+jnk_0\Gamma . \tag{6.101}$$

For a mono-energetic beam $\Gamma = 0$, and there is a growing wave corresponding to the negative sign. When $\Gamma \neq 0$ this growth is just cancelled when

$$nk_0\Gamma = \frac{1}{2}\left(\frac{nk_0V^2}{U}\right)^{1/2} = \frac{1}{\tau_0} \tag{6.102}$$

where τ_0 is the e-folding growth time for the growing wave when $\Gamma = 0$. Because of the long 'tails' on the resonance distribution this result turns out to be much less stringent than that obtained from a more realistic distribution.

Following this somewhat academic example we now examine

the problem of an arbitrary distribution, allowing k_0 to be positive or negative corresponding to both positive- and negative- mass regimes. It is important first to realize that in a cyclic accelerator arguments based on distribution functions extending to infinity must be used with caution. Only a finite range of energies can be held within the finite aperture, and furthermore there will be some correlation between energy and radial position in the beam. In the limit of a machine with zero amplitude of betatron oscillation, the energy is related directly to the position of the particle in the beam cross-section; in practice, however, there will be fairly thorough mixing and the correlation between energy and position will be ignored. There will nevertheless be a fairly well-defined 'cut-off' at both sides of the distribution function beyond which the amplitude is zero. For waves which have an angular phase velocity outside the range of angular velocities of the particles, there can of course be no damping to counteract the growing wave. Neil and Sessler discuss this point and show that in most practical situations the sufficient condition for a growing wave noted above is almost a necessary one; relatively few particles at the wave velocity are needed to produce sufficient damping to counteract the growth.

The behaviour of any distribution function $F(W)$ in the presence of a wall with impedance Z_b can be found by solving equation (6.99). This is usually tedious; often, however, we are interested only in whether unstable modes exist, not in their growth times or frequencies. This information can be obtained by a technique which we now describe. Its application to plasmas has been described by Penrose (1960); we outline here his method, but retain the notation of equation (6.99) which we write in the form

$$\frac{nk_0(U-\mathrm{j}V)}{U^2+V^2} = \int \frac{\partial F}{\partial W}\frac{\mathrm{d}W}{W-W_1} \tag{6.103}$$

where $W_1 = (\omega - n\omega_0)/nk_0$ is a complex quantity. For positive n, growing waves occur if there is a solution of this equation for which $\mathrm{Im}\, W_1/k_0$ is negative. We now define the function

$$Z(W_1) = \int \frac{\partial F}{\partial W} \frac{dW}{W-W_1} \tag{6.104}$$

If F is 'well behaved', satisfying certain smoothness and integrability conditions, then it can be shown that, (c.f. equation (6.54))

$$Z(\mathrm{Re}\, W_1 - j0) = \lim_{\mathrm{Im} W_1 \to 0} \int_{-\infty}^{\infty} \frac{W - \mathrm{Re}\, W_1 - \mathrm{Im}\, W_1}{(W - \mathrm{Re}\, W_1)^2 + (\mathrm{Im}\, W_1)^2} \frac{\partial F}{\partial W}\, dW$$

$$= \int \frac{\partial F}{\partial W} \frac{dW}{W-W_1} - j\pi \left(\frac{\partial F}{\partial W}\right)_{W_1} . \tag{6.105}$$

For a well-behaved distribution function the map of this path from $-\infty$ to $+\infty$ forms a closed path on the Z-plane which starts from and ends at the origin. This path evidently crosses the real axis a number of times equal to the number of points at which $\partial F/\partial W$ is zero; that is, the number of maxima and minima in the distribution curve.

The lower half of the W-plane maps into the interior of the curve, so that values of Z in this region correspond to roots of the equation in the lower half-plane and thus to instability. (The usual conventions as to what is meant by 'inside' a multiply-looped curve apply; if a point is encircled an odd number of times in traversing the curve it is 'inside'.) Points outside the curve correspond to stability. The map for the gaussian distribution, from Penrose's paper (with change of notation), is reproduced as Fig.6.9. The situation is unstable for small negative values of Z; this is the region of the negative-mass instability. For larger values of $-Z$ stability is restored by Landau damping. (This procedure is analogous to the Nyquist-diagram technique used in circuit stability theory.) Making the correspondence between U and Z from equations (6.103) and (6.104), the stability criterion for a negative mass instability, when $V = 0$, is found to be

$$-Z > -2 = -\left(\frac{2n\sigma^2|k_0|}{U}\right)^{1/2} \tag{6.106}$$

or

$$\sigma = \left(\frac{2U}{n|k_0|}\right)^{1/2} \tag{6.107}$$

where σ is the r.m.s. width of the gaussian distribution. In the non-relativistic limit this can be put in an interesting form. Setting $\gamma = \gamma_w = 1$, $\sigma^2/2\pi^2 = kT/m_0^2R^2$, then in terms of Budker's parameter ν and the effective mass m^*

$$\frac{kT}{m^*c^2} > 2(2\ln\frac{b}{a}+1)\nu \ . \tag{6.108}$$

A very similar technique has been used by Ruggiero

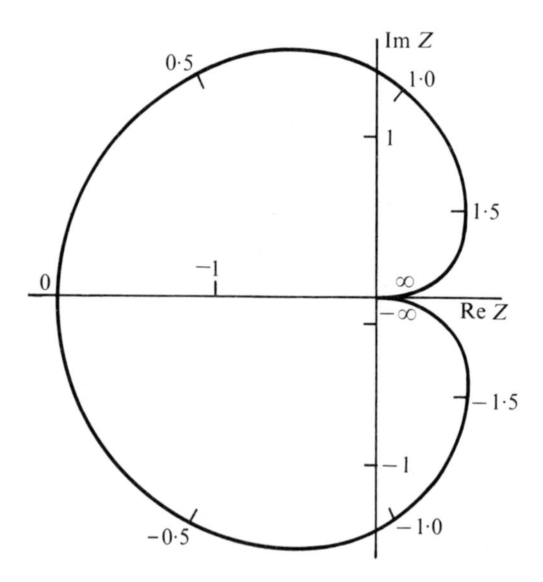

Fig.6.9. Map on the Z plane of the function given by the r.h.s. of equation (6.105), where W is the gaussian function

$$F(W) = (W_0/\sqrt{\pi})\exp(-W^2/W_0^2)$$

and the integral is performed just below the real W-axis. Values of W/W_0 are marked on the curve. For values of Z lying inside the curve ω has an imaginary part corresponding to growing waves.

and Vaccaro (1968) in connection with calculations for the CERN intersecting storage rings. They, however, map on to the $(U-jV)/n\delta_0 k_0$ plane, so that the diagrams are 'inside out'. Their results, for a wide range of distributions, are reproduced in Fig.6.10. The value of δ_0 is defined as the normalized half-width of the distribution curve at half the maximum value. The curves suggest a general stability criterion for most 'reasonable' distributions:

$$|U-jV| < 0{\cdot}7n\delta_0^2 k_0. \tag{6.109}$$

Since $\delta_0^2 = \sigma^2 \ln 4$, this is seen to be in agreement with the more special relation (6.107).

This criterion has been put in the more practical form by Keil and Schnell (1969), starting with the impedance seen by a beam in a closed-orbit accelerator

$$Z_{\parallel} = \oint_{\text{orbit}} \frac{E_s}{I}\,ds = \frac{RE_z}{aH_\phi} = -\frac{R}{a} Z_a \tag{6.110}$$

where I is the alternating component of the current. Making use of equations (6.85) and (6.96),

$$Z_{\parallel} = \frac{j2\pi R}{q^2 N\beta_w c}(U-jV). \tag{6.111}$$

Expressing the momentum spread in the beam as $\Delta p/m_0 c$, where m_0 is the rest mass rather than the effective mass m^*, the criterion (6.109) may be written

$$\frac{Z_{\parallel}}{n} < \frac{\Delta p^2}{m^* I_0 q} \tag{6.112}$$

where I_0 is the circulating current; the constant $0{\cdot}7\pi/2 \approx 1$ multiplying the right-hand side has been omitted and β_w has been set equal to β, a good approximation in relativistic beams. An alternative form for equation (6.112), in terms of Budker's parameter ν, is

$$\frac{Z_{\parallel}}{n} < \left(\frac{\Delta p}{m_0 c}\right)^2 \frac{m_0}{m^*} \frac{Z_0}{4\pi\nu\beta}. \tag{6.113}$$

Up to now a uniform wall impedance has been considered. This equation applies equally well to a situation with localized wall discontinuities; $Z_{\|}$ now represents the sum of series impedances in series round the circumference.

Not all discontinuities can be represented as simple

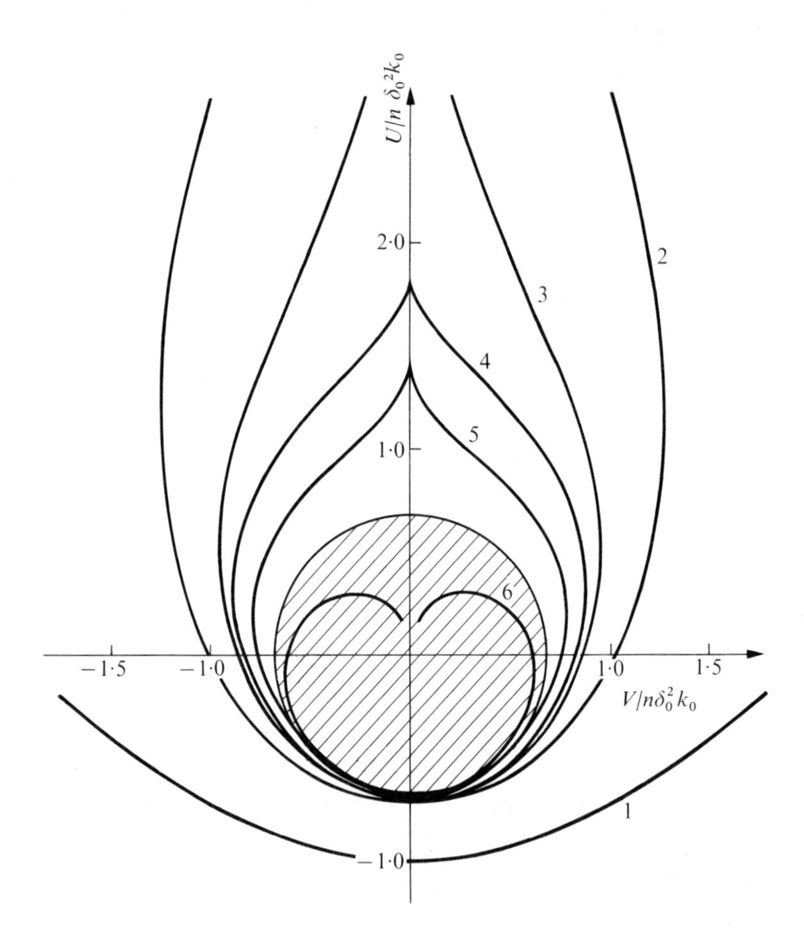

Fig.6.10. Boundaries between stable and unstable regions in the plane $(U/n\delta_0^2k_0,\ V/n\delta_0^2k_0)$ for various distribution functions $F(W)$ with normalized half-width δ_0. In contrast to the curve in Fig.6.9, the points *outside* the curve correspond to instability. The distribution functions are of the form given below; x is proportional to W/δ, but for simplicity they are not normalized:

cont.....

Fig.6.10 continued...

(1) $(1+x^2)^{-1}$
(2) $\exp(-x^2)$
(3) $(1-x^2)^4$, $(|x|<1)$)
(4) $\cos^2 x$, $(|x|<\pi/2)$
(5) $(1-x^2)^2$, $(|x|<1)$)
(6) $\cos x$, $(|x|<\pi/2)$.

All except the somewhat unrealistic cosine distribution enclose the circular shaded area, of radius 0·7. This fact is used in specifying the stability criterion (equation (6.109)).

series impedances; an example is provided by a clearing electrode consisting of a plate mounted between the wall and the beam, connected to the tube at one point (Sessler and Vaccaro 1967). Under these circumstances the propagation characteristics of the structure become important. This happens when the phase velocity of the wave on the structure (or one of its space harmonics) is close to the phase velocity of the wave on the beam. In the travelling-wave tube and backward-wave oscillator this coupling of waves on the structure and on the beam is an essential feature. The problem is most elegantly treated by coupled-mode analysis, in which equations characterizing the dynamics of the beam and the dispersion characteristics of the structure together with an appropriate coupling constant are put into normal-mode form. In the next section coupled-mode theory is examined as a preliminary to the study of microwave tubes.

<u>6.3.7. Cylindrical beam weakly coupled to a propagating structure; normal modes</u> When the walls surrounding a beam can support a travelling wave which has a phase velocity close to that of a wave on the beam, resonant interaction may occur. This interaction, which may be between the beam wave and the fundamental wave on the structure, or with one of the space harmonics, is exploited in microwave tubes such as the travelling-wave (TWT) tube and backward-wave oscillator (BWO). A complete treatment of the problem requires that the

beam and structure be treated as a single system, which can then be described in terms of its normal modes. Actual problems often display considerable complexity, and can be simplified greatly if it is legitimate to assume that the coupling between beam and structure is weak. In this section coupled-mode theory, devised to exploit this approach in its simplest form, will be developed.

Before tackling the problem in a formal manner, we discuss in a descriptive way the behaviour of the ω-k dispersion diagram as the coupling is gradually increased. In the limiting case of no coupling the combined dispersion curve may be represented as that of the beam and that of the structure superposed. For very weak coupling a gap appears at any point where the two curves intersect; the line pair through this point is replaced by a hyperbola which joins on to the original curve. This type of behaviour may be seen in Fig.6.12. As the coupling increases the distance between the two branches of the curve increases, and the hyperbolic approximation deteriorates. Ultimately the 'hyperbolae' associated with neighbouring crossing points of the two original curves begin to overlap and interfere; this is the point where the weak-coupling approximation breaks down entirely and more than two waves need to be considered. This is the situation encountered in the two-stream instability analysed in section 6.2.6 and illustrated in Fig.6.3, where the branch inside the central parallelogram has disappeared.

In the theory of the travelling-wave tube and backward-wave oscillator the coupling of three waves, involving three branches of the combined dispersion curve, needs to be considered. The essential physical behaviour can, however, be seen in a simpler model in which only two are included.

We shall now look in a general way at the weak coupling between two modes, and examine the physical significance of the four essentially different types of hyperbolic configuration which may occur at the coupling point.

We first study the normal-mode description of waves in the absence of coupling. These are represented by four first-order differential equations for amplitudes which are linear combinations of the variables characterizing the wave.

Thus, for electromagnetic plane waves propagating in free space there are two such variables, $\underline{E}$ and $\underline{H}$. The equations are necessarily in terms of complex variables, and, for a practical system which can be described in terms of real variables, the amplitudes occur in complex-conjugate pairs. This form is convenient when considering coupling of waves on two systems, such as a beam and a structure. For weak coupling only one of the four waves in each set needs to be considered. The complex amplitudes a satisfy the equations

$$\begin{aligned} \left(\frac{\partial}{\partial z} + jk\right) a^{+} &= 0 \\ \left(\frac{\partial}{\partial z} - jk\right) a^{+*} &= 0 \\ \left(\frac{\partial}{\partial z} - jk\right) a^{-} &= 0 \\ \left(\frac{\partial}{\partial z} + jk\right) a^{-*} &= 0 \end{aligned} \tag{6.114}$$

where the four factors $\exp j(\pm\omega t \mp kz)$ and $\exp j(\pm\omega t \pm kz)$ respectively are understood. For a stationary medium the value of k is the same for all four modes; the first two travel forwards, whereas the second two travel with equal velocity in the opposite direction. For a moving medium, on the other hand, $k^{\pm}$ is of the form $(\omega \pm \omega_1)/V$. When $\omega_1 > \omega$ all waves travel forward and represent the fast and slow waves discussed in section 6.2.4. The combination of variables making up the normal-mode amplitude can be chosen and normalized in such a way that the power carried by the two pairs in the forward and backward directions is $2|a^{+}|^2$ and $2|a^{-}|^2$ respectively. If all the modes are forward, this represents forward transport of *negative* energy by the *slow* modes.

Returning to the very simple example of a wave in free space, a combination of both sets of waves gives rise at some value of z to fields $\underline{E}$ and $Z_0\underline{H}$, which are not in general equal in amplitude or in phase. In terms of the transverse components of these fields a straightforward, though detailed, argument shows that the power flow is

$$\overline{W} = \frac{(E+Z_0H)^2}{8Z_0} \tag{6.115}$$

and the appropriate normal-mode amplitudes are

$$a^{\pm} = \frac{E \pm Z_0 H}{4Z_0^{1/2}} \tag{6.116}$$

It is evident that analogous equations with V and I instead of E and H apply to waves on transmission lines. In this example, the power flow, group velocity, and phase velocity are all in the same direction. In a positive energy wave the power flow is in the direction of the group velocity, and for this reason can be negative in a dispersive medium. In a negative-energy wave, however, the group and power-flow velocities are opposite, so that a negative power flow occurs when the group velocity is positive. These possibilities give rise to a variety of types of behaviour when two waves interact. This we now discuss.

For two weakly-coupled systems there are eight equations corresponding to equations (6.114). The waves interact when the phase velocities in the two systems are nearly equal. We consider now the coupling of two forward waves, one from each set of four: k_1 is positive and k_2 could be k^+ in the first equation or $-k^-$ in the third. It is shown in appendix 5 that, writing a for a^+ and introducing a coupling constant K, the coupled-mode equations may be written

$$\begin{aligned} \left(\frac{\partial}{\partial z} + jk_1\right) a_1 - Ka_2 &= 0 \\ \left(\frac{\partial}{\partial z} + jk_2\right) a_2 \pm K^* a_1 &= 0 \end{aligned} \tag{6.117}$$

where the top and bottom signs refer to power flow in the same or opposite directions in the two systems. By eliminating k_1, k_2, K from these equations and their complex conjugates we obtain

$$\frac{\partial}{\partial z}(|a_1|^2 \pm |a_2|^2) = 0 \tag{6.118}$$

with the same sign convention as above. This confirms that the sum or difference of the power flows on the two component systems is conserved, depending on whether the directions of flow are the same or opposite.

Replacing $\partial/\partial z$ by $-jk$ in equation (6.117) and eliminating a_1 and a_2 yields the dispersion relation for the system:

$$\pm KK^* = \pm|K|^2 = \{k-k_1(\omega)\}\{k-k_2(\omega)\} . \tag{6.119}$$

For very weak coupling the dispersion relation is only modified from the form

$$\begin{aligned} k - k_1(\omega) &= 0 \\ k - k_2(\omega) &= 0 \end{aligned} \tag{6.120}$$

when $k - k_1(\omega)$ and $k - k_2(\omega)$ are small. Writing $\Delta\omega = \omega - \omega_0$, $\Delta k = k - k_0$, where k_0 and ω_0 are the values of k and ω for which $k_1 = k_2$, the dispersion relation may be written as

$$\pm|K|^2 = \left(\Delta k - \frac{\Delta\omega}{v_1}\right)\left(\Delta k - \frac{\Delta\omega}{v_2}\right) \tag{6.121}$$

where v_1 and v_2 are the group velocities $\partial\omega/\partial k$ of the two waves at (ω_0, k_0) when $K = 0$.

Because of the two possible signs for $|K|^2$, representing energy flow in the same or opposite directions in the two systems, and for v_2, the group velocity in the second system, the configuration of the hyperbola in equation (6.122) can take four forms. These are exhibited in Table 6.3, together with the signs of the terms in the equation and of the relevant physical quantities. In all except the first case illustrated complex values of one of the variables $\Delta\omega$ or Δk exist for real values of the other, so that either evanescence, convective instability, or absolute instability can occur. Which of these is relevant can be argued physically by the method used in section 6.2.6. Evanescence results when the waves both have positive energy but the group velocities are opposite. This may be regarded as providing negative feedback. If, on the other hand, one of the waves

TABLE 6.3

Four types of coupling between two beams (1 and 2) and summary of their physical characteristics

Signs in eqn (6.121)	$\pm\lvert k^2\rvert$	+	+	-	-
	V_2	+	-	+	-
Sign of energy	1	+	+	-	-
	2	+	+	+	+
Sign of group velocity	1	+	+	+	+
	2	+	-	+	-
Physical behaviour		Beating	Evanescence	Convective instability	Absolute instability
Local shape of the dispersion curve in the neighbourhood of resonant interaction. (ω versus k).			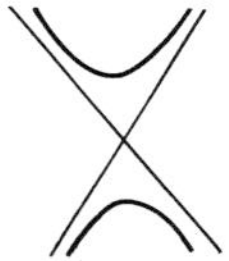		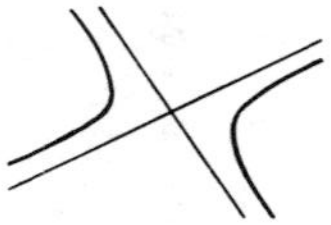

has negative energy, the feedback becomes positive and the instability is absolute. The remaining possibility, both waves with positive group velocity but different signs for the energy, gives rise to convective instability. Energy passes from one wave to the other continuously as they propagate; both increase in amplitude with distance but not in time at a given position.

These results have been arrived at by simple physical argument. When the coupling is no longer weak, so that the topology of the dispersion curves is no longer hyperbolic (as in Fig.6.4 for example), then the criterion for distinguishing the different types of instability becomes much less simple. References to this more general problem were given earlier (section 6.R.2). Extended tables, including growth rates and other information not in Table 6.3 have been given by Barnes (1964).

We now return to the problem of interaction between a beam and a circuit. To proceed it is necessary to put the equation for the waves on the beam in normal-mode form. This is not quite so simple as for the circuit, since we need to be clear what is understood by the notion of power on the beam. Even with no wave present the beam carries 'd.c. power' associated with the kinetic energy of the charges; it is necessary to sort out the steady and alternating components in a consistent way. In the next section this is done for a system with longitudinal waves; this is relatively straightforward, though, as we see later, the subject becomes complicated in the more general situation where transverse motion is possible.

6.3.8. Kinetic power theorem for confined flow In a plane electromagnetic wave in free space the power flow may be specified in terms of the fields $\underline{E}$ and $\underline{H}$. In this section we determine the appropriate quantities for specifying the power flow in a beam. In order to do this we make use of the kinetic power theorem of Chu (1951).

We start by considering power flow in a drifting medium in an infinite B_z field so that all motion is in the z-direction, and then particularize to a beam. As in section

6.2.5, all field quantities, currents, and velocities are written as the sum of a constant and a harmonically varying part:

$$A = A_0 + \tilde{A} \tag{6.122}$$

where $|\tilde{A}| \ll A_0$.

Maxwell's equations for the alternating components of the fields and currents may be written

$$\begin{aligned} \text{curl}\,\underline{\tilde{H}} &= \underline{\tilde{i}} + j\omega\varepsilon_0\underline{\tilde{E}} \\ \text{curl}\,\underline{\tilde{E}} &= -j\omega\mu_0\underline{\tilde{H}} \end{aligned} \quad . \tag{6.123}$$

If now the complex conjugate of the first of these equations is scalar multiplied by $\underline{\tilde{E}}$ and the second by $\underline{\tilde{H}}^*$, then, by subtraction and use of the identity $\text{div}(\underline{A}\times\underline{B}) = \underline{B}.\text{curl}\underline{A}-\underline{A}.\text{curl}\underline{B}$, we find

$$\text{div}\,(\underline{\tilde{E}}\times\underline{\tilde{H}}^*) = -\underline{\tilde{i}}^*.\underline{\tilde{E}} + j\omega\,(\varepsilon_0|\underline{\tilde{E}}|^2-\mu_0|\underline{\tilde{H}}|^2)\ . \tag{6.124}$$

This is the small-signal Poynting theorem, valid when $\tilde{\beta} \ll \beta$. The presence of the first term on the r.h.s. indicates energy interchange between the field and the charges. To proceed we require the equations of motion and continuity; these have already been used in section 6.2.5, and are

$$q\tilde{E}_z = j\gamma^3 m_0\tilde{\beta}c(\omega - \beta_0 ck) \tag{6.35}$$

$$(\omega - \beta_0 ck)\tilde{n} - kn_0\tilde{\beta}c = 0. \tag{6.37}$$

Introducing now the current density

$$\tilde{i} = qc\,(n_0\tilde{\beta}+\beta_0\tilde{n}) \tag{6.125}$$

and defining a new quantity the *kinetic voltage* V_k

$$-q\tilde{V}_k = \gamma^3\beta_0\tilde{\beta}m_0c^2 \tag{6.126}$$

these four equations can be manipulated to yield

$$\tilde{i}^*\tilde{E}_z = j\gamma_0^3 m_0 \omega n_0 |\tilde{\beta}|^2 c^2 - \quad (\tilde{V}_k \tilde{i}^*) \ . \qquad (6.127)$$

From the real parts of equations (6.124) and (6.127) it follows from integration over a volume with element $d\tau$ that

$$\mathrm{Re} \int_{\mathrm{vol}} \mathrm{div}\,(\underline{E}\times\underline{H}^* - \tilde{V}_k \tilde{i}^*)\,d\tau = 0 \ . \qquad (6.128)$$

For a beam, it is convenient to convert this into a surface integral and replace the current density by the total current I. With the notation indicated in Fig.6.11 this becomes

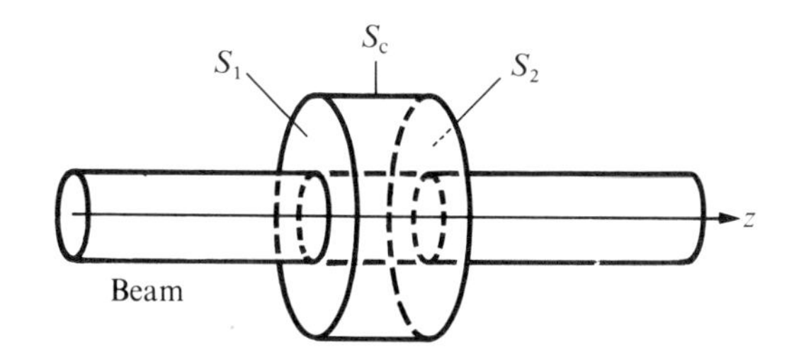

Fig.6.11. Surface of integration for equation (6.129).

$$\mathrm{Re}\int_{S_1+S_2+S_c} \tfrac{1}{2}\underline{E}\times\underline{H}^* dS = \left[\mathrm{Re}\int_{S_1} - \mathrm{Re}\int_{S_2}\right] \tfrac{1}{2}(V_k i^*)\,ds$$
$$= \mathrm{Re}\left(\tfrac{1}{2}V_k I|_{S_1} - \tfrac{1}{2}V_k I|_{S_2}\right) \ . \qquad (6.129)$$

This is the kinetic power theorem for current constrained to flow in the z-direction. The theorem states that the electromagnetic power flowing out of the volume bounded by the surface $S_1+S_2+S_c$ is equal to the difference between the kinetic power flowing in through S_1 and that leaving through S_2.

It is instructive to calculate the kinetic power associated with fast and slow waves in the infinite medium analysed in section 6.2.5. From equations (6.125) and (6.126), making use of equations (6.35) and (6.37), the kinetic power

$$W_{\mathrm{k}} = \mathrm{Re}(\tfrac{1}{2}\tilde{V}_{\mathrm{k}}\tilde{i}^*) = \mathrm{Re}(\tfrac{1}{2}\beta_0\tilde{\beta}\tilde{i}^*\gamma^3 m_0 c^2/q) \qquad (6.130)$$

can be expressed in the form

$$W_{\mathrm{k}} = \frac{\beta_0 c\omega\omega_{\mathrm{p}}^2\varepsilon_0|\tilde{E}_z|^2}{2\gamma^2(\omega-\beta_0 ck)^3}\,. \qquad (6.131)$$

This can be simplified by making use of equation (6.29), the dispersion relation for waves in a moving plasma; it becomes

$$W_{\mathrm{k}} = \frac{\tfrac{1}{2}\beta_0 c\varepsilon_0 E^2}{1-\beta_0/\beta_{\mathrm{w}}}\,. \qquad (6.132)$$

This changes from positive to negative as the beam velocity $\beta_0 c$ exceeds the phase velocity of the wave, ω/k. Thus, since the group velocity is positive, the energy associated with the fast and slow waves is positive and negative respectively. As β tends to zero the power flow also tends to zero also, but the energy density $W_{\mathrm{k}}/\beta_0 c$ becomes $\tfrac{1}{2}\varepsilon_0 E^2$; this is the energy density associated with plasma oscillations in a stationary medium.

The same result is obtained by using the expression for the field energy density in a dispersive medium, given for example by Landau and Lifshitz (1960). This may be written

$$U = \tfrac{1}{4}\{\frac{\partial}{\partial\omega}(\omega\kappa)\varepsilon_0\underline{\tilde{E}}.\underline{\tilde{E}}^* + \frac{\partial}{\partial\omega}(\omega\mu)\mu_0\underline{\tilde{H}}.\underline{\tilde{H}}^*\} \qquad (6.133)$$

where κ and μ are the dielectric and permittivity functions. The dielectric function, from which the dispersion relation may be obtained by putting $\kappa = 0$, is, for a moving medium

$$\kappa = 1 - \frac{\omega_{\mathrm{p}}^2}{\gamma^2(\omega-\beta_0 ck)^2}\,. \qquad (6.134)$$

The permeability μ is unity; substituting for κ and μ in equation (6.133) and multiplying by the group velocity $\beta_0 c$ yields the power flow W_{k} given by equation (6.133).

It must be emphasized that the energy associated with a wave in a medium is only valid for perturbations sufficiently small that higher-order terms in equations involving products of harmonic qualities can be ignored. It may readily be verified that the permitted field amplitude vanishes as $\omega-\beta ck$ tends to zero. In this resonant situation a small field produces a very large velocity perturbation.

A knowledge of the energy of the wave defined above and the mean velocity of the beam particles to first order is not sufficient to enable the actual energy of the whole system to be calculated. If v_0 is the mean velocity of the particles and U is the wave energy, then it is not possible to assert that (for a non-relativistic situation) the total energy is $U+\frac{1}{2}m_0v_0^2$. This may be seen by considering a beam with mean velocity v_0 carrying a small-amplitude wave which gives rise to velocity modulation $\Delta\tilde{v}_0$. Thus, if v_0 is known to first order, but is uncertain by an amount εv_0 to second order, this gives an uncertainty of energy $m_0\varepsilon v_0$, which is of the same order as $\frac{1}{2}m_0\Delta v_0^2$, a quantity of the same order as the wave energy. The significance of this point will become clearer after the discussion on the energy of transverse waves in section 6.4.3.

If the power flow is to a resistive wall or propagating circuit surrounding the beam, then the power flow into the wall or circuit is

$$W = \mathrm{Re}\int_{S_\mathrm{c}} \tfrac{1}{2}\underline{\mathrm{E}}\times\underline{\mathrm{H}}^*\,\mathrm{d}s. \tag{6.135}$$

For a propagating circuit (transmission line) with $\omega/k \ll c$ and a thin beam, most of the electromagnetic energy is concentrated near the circuit and the kinetic energy is confined to the beam. Expressing the power on the transmission line in terms of the line voltage $\tilde{V}_\mathrm{C}$ and its characteristic impedance Z_C, the kinetic power theorem reduces to the simple form

$$\frac{\partial}{\partial z}\left\{\tfrac{1}{2}\mathrm{Re}(\tilde{V}_\mathrm{C}^2/Z_\mathrm{C}-\tilde{V}_\mathrm{k}\tilde{I}^*)\right\} = 0. \tag{6.136}$$

We can now express the waves on a beam in the same normal-mode form as for the waves on a transmission line, and treat such devices as the travelling-wave tube by the coupled-mode theory developed in the last section. This will be done in the section which follows.

6.3.9. Travelling-wave tube and backward-wave oscillator

The essential physical mode of operation of these two devices has already been described; energy is continuously coupled from a growing negative-energy slow wave on the beam to a positive-energy wave on the circuit. If the group velocity of both waves is forward, this results in a convectively growing wave and forms the basis of an amplifier. If the group and phase velocity are oppositely directed, this results in instability and the build-up of oscillation.

These two types of interaction are illustrated in Fig. 6.12. The beam wave can be 'tuned' by adjusting the voltage

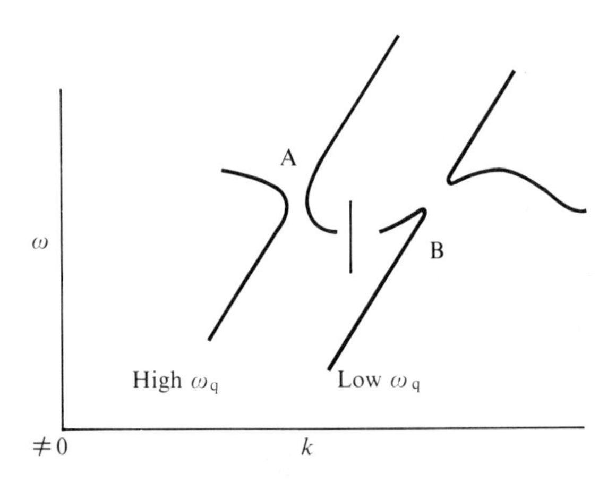

Fig.6.12. Coupling between a periodic structure and the slow wave on a beam. The dispersion relation of the structure is represented by the wavy curve; the slanting lines correspond to the slow wave on a beam. Two cases are shown, with different values of ω_q, chosen so that they intersect the dispersion curve of the structure where the slope is negative and positive respectively. The intersections at A and B give rise to absolute and convective instability, corresponding to the operation of backward-wave oscillator and travelling-wave tube. (The diagram is schematic only, and the axes do not pass through the origin.)

and current to give a line with varying slope and position. Tuning for operation as a travelling-wave tube (TWT) and backward-wave oscillator is indicated. By coupling the fast wave to the circuit, attenuation or beating can be achieved. By feeding a positive-energy wave on the circuit, and adjusting the beam so that the amplitude of the circuit wave output has dropped to zero at some point along the tube, the conditions for a quarter-wavelength of the 'beat wave-length' can be determined experimentally. This enables the coupling parameters of the tube to be determined, and is known as the 'Kompfner dip' condition. The topology of the dispersion curves at the intersection point for these three types of interaction corresponds respectively to the diagrams in the third, fourth, and first columns of Table 6.3.

The situation is actually rather more complicated than indicated so far. In the first place, the assumption of only two interacting waves is often rather poor; secondly, actual devices have ends, and the detailed behaviour requires consideration of the appropriate boundary conditions. It is found, for example, that there is a minimum length for a backward-wave oscillator.

A proper consideration of the theory of the travelling-wave tube and backward-wave oscillator requires lengthy discussion. Here we merely discuss the coupling between the waves on a beam and those on a periodic circuit, making use of the expressions for the normal-mode amplitudes obtained in the previous sections, and comment in a general way on the application to actual devices.

For a start we consider longitudinal waves in an infinite drifting medium. Writing $\partial/\partial z$ instead of $-jk$, equations (6.35) and (6.37) become

$$\begin{aligned} \left(\frac{\partial}{\partial z} + \frac{j\omega}{\beta_0 c}\right)\tilde{\beta} &= \frac{jq}{\gamma^3 m_0 \omega_0 \beta_0 c^2}\,\tilde{i} \\ \left(\frac{\partial}{\partial z} + \frac{j\omega}{\beta_0 c}\right)\tilde{i} &= \frac{j\omega n_0 q}{\beta_0}\,\tilde{\beta} \end{aligned} \qquad (6.137)$$

where use is made of equation (6.125) in relating current to velocity and $\underline{E}$ is eliminated from equation (6.35) by means of equation (6.123) with curl $\tilde{\underline{H}} = 0$. The curl of $\tilde{\underline{H}}$ vanishes because, as explained in section 6.2.4, $\tilde{\underline{H}}$ is zero and the total (convection plus displacement) current is zero. Introducing now the kinetic voltage $\tilde{V}_k$ (equation 6.126) and the relativistically corrected voltage $-\beta^2\gamma^2 m_0 c^2/q$ (section 2.2.2), denoted as V_B, equation (6.137) may be written in the form

$$\left(\frac{\partial}{\partial z} + \frac{j\omega}{\beta_0 c}\right)\tilde{V}_k = \frac{jcZ_0}{\omega}\tilde{i}$$
$$\left(\frac{\partial}{\partial z} + \frac{j\omega}{\beta_0 c}\right)\tilde{i} = -\frac{j\omega n_0 q^2 \tilde{V}_k}{2\gamma V_B} . \tag{6.138}$$

We now wish to write these in normal-mode form. This involves finding new variables, which are a linear combination of $\tilde{i}$ and $\tilde{V}_k$. The combination must be chosen such that equation (6.138) takes the form of equation (6.114), and the new variables normalized such that the power flow equation, equation (6.116), in terms of V_k and i rather than E and H, is satisfied. At the same time we apply the result to a beam of finite cross-section, and work in terms of the current I_B instead of the current density i. Introducing the d.c. beam impedance

$$Z_B = V_B/I_B = V_B/2\pi\int ir\mathrm{d}r \tag{6.139}$$

where I_B is the time-averaged beam current, equations (6.138) take the form

$$\left\{\frac{\partial}{\partial z} + \frac{j}{\beta_0 c}(\omega-\omega_q)\right\}\tilde{a}^+ = 0$$
$$\left\{\frac{\partial}{\partial z} + \frac{j}{\beta_0 c}(\omega+\omega_q)\right\}\tilde{a}^- = 0 \tag{6.140}$$

where

$$\tilde{a}^{\pm} = \tfrac{1}{4}\left(\frac{\gamma^2\omega}{2\omega_q Z_B}\right)^{1/2}\left(\tilde{V}_k \pm \frac{2\omega_q Z_B \tilde{I}_B}{\gamma^2\omega}\right) \tag{6.141}$$

and the reduced plasma frequency ω_p, appropriate to waves on a finite beam, has been substituted for the plasma frequency ω_p. It may readily be verified that substitution of $\tilde{a}^{\pm}$ in equation (6.140), together with equation (6.139), yields equation (6.138) and furthermore the product $2\tilde{a}^{+}\tilde{a}^{+}*$ gives the kinetic power flow $\frac{1}{2}\tilde{V}_k\tilde{I}*$.

We have now obtained normal-mode equations for the beam (equations (6.140) and (6.141)); those for the circuit to which the beam is to be coupled can be obtained by writing V_C, I_C, Z_C in place of E, H, and Z_0 in equations (6.115) and (6.116). The subscript C denotes 'circuit', and as before the subscript B will be used to denote quantities pertaining to the beam, so that henceforth the kinetic voltage $\tilde{V}_k$ will be written as $\tilde{V}_B$. The next step is to consider a composite system of beam and circuit, and to find a suitable form for the coupling constant between the two sub-systems.

In order to ascertain the form of this coupling term, it is necessary to consider the physical nature of the interaction between beam and transmission line. The most commonly used circuit is a helix, which carries a wave with phase velocity of approximately $c \sin \psi$ where ψ is the angle of the helix. The theory of the helix is covered in many texts (for example Collin 1966); here we merely note that it can be represented at low frequencies by a loaded line. This is shown in Fig.6.13.

The beam is coupled capacitively to the circuit and induces a current in it of $\partial\tilde{I}_B/\partial z$ per unit length. In addition, the particles in the beam experience a force arising from the electric field associated with the circuit. These coupling terms may be included in a straightforward way in the un-normalized transmission-line equations and beam equations. The transmission-line equations are based on equations (6.116) and (6.117); for the beam equation we use equation (6.138), expressed in terms of I_B, the total beam current, rather than the current density i. The equations are

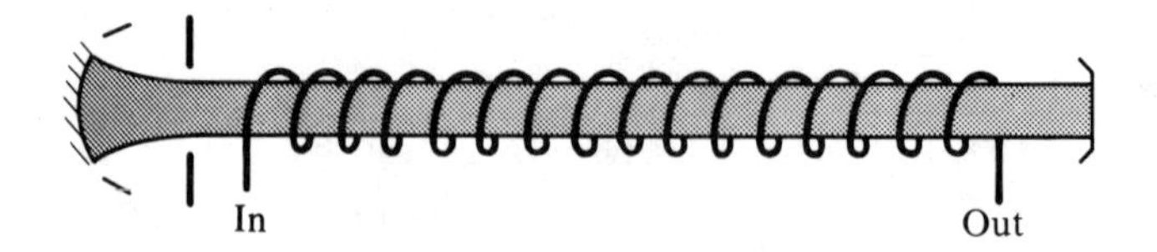

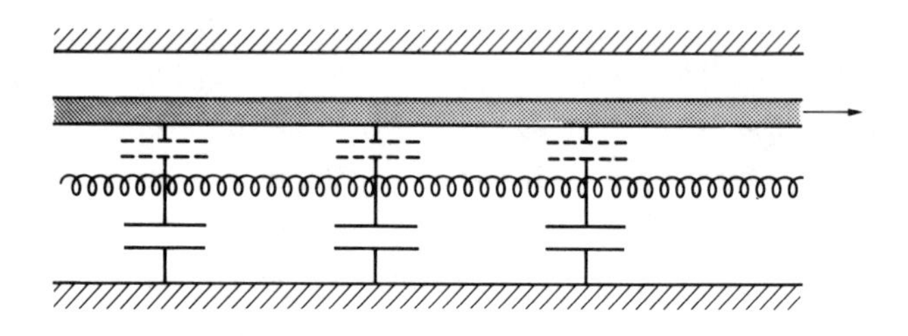

Fig.6.13. Schematic diagram of travelling-wave tube or backward-wave oscillator and its equivalent circuit.

$$
\begin{aligned}
\frac{\partial \tilde{V}_C}{\partial z} &= -j\omega L \tilde{I}_C \\
\frac{\partial \tilde{I}_C}{\partial z} &= -j\omega C \tilde{V}_C - \frac{\partial \tilde{I}_C}{\partial z} \\
\left(\frac{\partial}{\partial z} + \frac{j\omega}{\beta_0 c}\right) \tilde{V}_B &= \frac{jcZ_0}{\omega} \tilde{I} + \frac{\partial \tilde{V}_C}{\partial z} \\
\left(\frac{\partial}{\partial z} + \frac{j\omega}{\beta_0 c}\right) \tilde{I}_B &= \frac{j\omega n_0 q^2}{2\gamma V_B} \tilde{V}_B \, .
\end{aligned}
\tag{6.142}
$$

These may now be put into normal-mode form, using the definitions of $a_B^{\pm}$ and Z_B in equation (6.141) and those of $a_C^{\pm}$ in equation (6.116), but with V and I in place of E and H, and writing $1/\surd(LC) = \beta_0 c$:

$$
\begin{aligned}
\left(\frac{\partial}{\partial z} \pm \frac{j\omega}{\beta_C c}\right) \tilde{a}_C^{\pm} &= \pm \left(\frac{\omega_q Z_C}{8\omega Z_B}\right)^{1/2} \frac{\partial}{\partial z} (\tilde{a}_B^{+} - \tilde{a}_B^{-}) \\
\left\{\frac{\partial}{\partial z} + \frac{j}{\beta_0 c} (\omega \mp \omega_q)\right\} a_B^{\pm} &= \left(\frac{\omega_q Z_C}{8\omega Z_B}\right)^{1/2} \frac{\partial}{\partial z} (\tilde{a}_C^{+} + \tilde{a}_C^{-}) .
\end{aligned}
\tag{6.143}
$$

These equations are in slightly different form from those already studied in section 6.3.7. Four waves rather than two are coupled, namely $\tilde{a}_B^{\pm}$ and $\tilde{a}_C^{\pm}$, and there are derivatives on the r.h.s. of the equation. Since, however, the coupling is weak and the mode amplitudes change slowly with z, it is a fair approximation to include only the harmonic variation and replace $\partial/\partial z$ by $-j\omega/\beta_0 c$. Making also the replacement

$$c_{CB} = \frac{j\omega}{2\beta_0 c}\left(\frac{\omega_q Z_C}{2\omega Z_B}\right)^{1/2} \tag{6.144}$$

where c_{CB} is the coupling constant between beam and circuit, equation (6.143) can be written

$$\left(\frac{\partial}{\partial z} \pm \frac{j\omega}{\beta_C c}\right)\tilde{a}_C^{\pm} = \pm\, c_{CB}\,(\tilde{a}_B^{+} - \tilde{a}_B^{-})$$

$$\left\{\frac{\partial}{\partial z} + \frac{j}{\beta_0 c}\,(\omega \mp \omega_q)\right\}\tilde{a}_B^{\pm} = c_{CB}\,(\tilde{a}_C^{+} + \tilde{a}_C^{-})\;. \tag{6.145}$$

Two parameters used in travelling-wave tube theory are related to the constants used here in the following way. The 'gain parameter' is given by

$$C = \tfrac{1}{2}\left(\frac{\omega Z_C}{\omega_q Z_B}\right)^{1/3} \tag{6.146}$$

and the 'space-charge parameter' $2(QC)^{1/2}$ is defined by

$$2C(QC)^{1/2} = \omega/\omega_q\,. \tag{6.147}$$

In practical situations $C \simeq 0{\cdot}01$ and $2(QC)^{1/2}$ is of order unity.

Equations (6.145) represent four linear coupled equations. Eliminating the amplitudes as in section 6.3.7 produces a quartic equation, which is analytically intractable. For very weak coupling it is permissible to include only the two waves of interest at the point where their velocities are very nearly synchronous. Introducing a dimensionless parameter b the approximate equality between $\omega/\beta_0 c$ and k_c can be expressed as

$$\frac{\beta_0 c k_{\mathrm{C}}}{\omega} = 1 + Cb. \tag{6.148}$$

Considering only the forward circuit wave and the slow beam wave $\tilde{a}_{\mathrm{C}}^{+}$ and $\tilde{a}_{\mathrm{B}}^{-}$, equation (6.143) with the notation introduced in equations (6.146) and (6.147) reduces to

$$\begin{aligned} \left\{\frac{\partial}{\partial z} + \frac{j\omega}{\beta_0 c}(1+Cb)\right\}\tilde{a}_{\mathrm{C}}^{+} &= c_{\mathrm{CB}}\tilde{a}_{\mathrm{B}}^{-} \\ \left[\frac{\partial}{\partial z} + \frac{j\omega}{\beta_0 c}\{1+2(CQ^3)^{1/2}\}\right]\tilde{a}_{\mathrm{B}}^{-} &= c_{\mathrm{CB}}\tilde{a}_{\mathrm{C}}^{+} \end{aligned} \tag{6.149}$$

This is of the general form of equation (6.116), discussed in section 6.3.7, and it can be used to demonstrate the essential features of TWT and BWO behaviour in terms of the parameters b, C, and Q. In practice this approximation is too crude, and the fast beam wave $\tilde{a}_{\mathrm{B}}^{+}$ needs to be included. This leads to a cubic dispersion relation of form sketched in Fig.6.14.

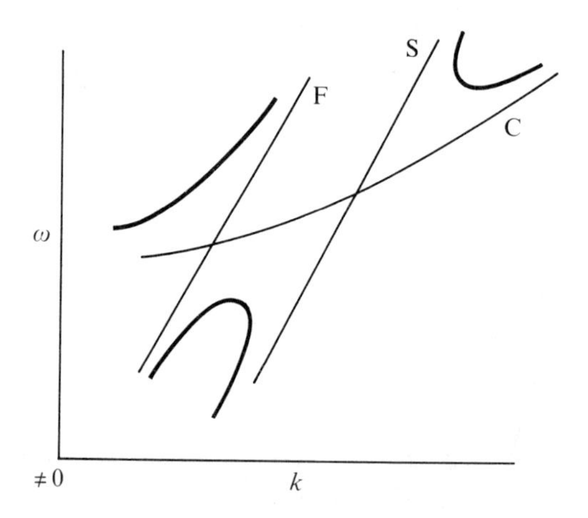

Fig.6.14. Sketch showing topology of the dispersion curve for a travelling-wave tube. The lines marked F, S, and C represent the fast, slow, and circuit waves on an uncoupled system.

The analysis in this section is condensed from that in Louisell's (1960) book, with some changes of notation, and convention (we have $2\omega_q Z_{\mathrm{B}}/\omega$ where he has Z_b). He continues with a detailed analysis of the three-wave system and a more complete discussion of the TWT than that sketched here.

The essential physical features of TWT and BWO operation have now been outlined. For practical design very much more detailed theories exist, taking into account the precise nature of the fields associated with the helix and other structures. This is supplemented by much experimentally determined information. An interesting example of the approach using a full field analysis may be found in the paper of Chu and Jackson (1948). This is essentially the same as that used in section 6.3.4, with the complications that k as well as ω needs to be considered for the structure and also that hybrid waves, requiring consideration of modes with azimuthal symmetry, need to be taken into account. A simpler treatment along the same lines is given by Collin (1966).

In this section we have described only the basic beam-structure interaction in beam-wave tubes. For a description of complete systems, and a discussion of such important factors as gain and bandwidth characteristics, efficiencies, etc. the texts referred to in section 6.R.3 should be consulted.

6.3.10. Longitudinal beam-plasma interaction Beam-plasma interaction in an infinite cold plasma with massive ions has already been studied in section 6.2.6. It is a particular case of the two-stream instability when the velocity of one of the streams is zero, giving a dispersion diagram as shown in Fig.6.3(b). This diagram is replotted in Fig.6.15, showing real and imaginary parts of k for real ω, and *vice versa*, for the region $\omega < \omega_p$.

Below ω_p the beam is essentially traversing an 'inductive' medium, and the dispersion relation may be written as equation (6.38) with ε_0 replaced by $\varepsilon_0(1-\omega_p^2/\omega^2)$. This is the same as equation (6.45) with $\beta_2 = 0$, but brings out the analogy with equation (6.93), which represents (with a suitable value for L_0) the 'easitron' oscillator or negative-mass instability.

Beam-plasma interaction in finite transverse geometry

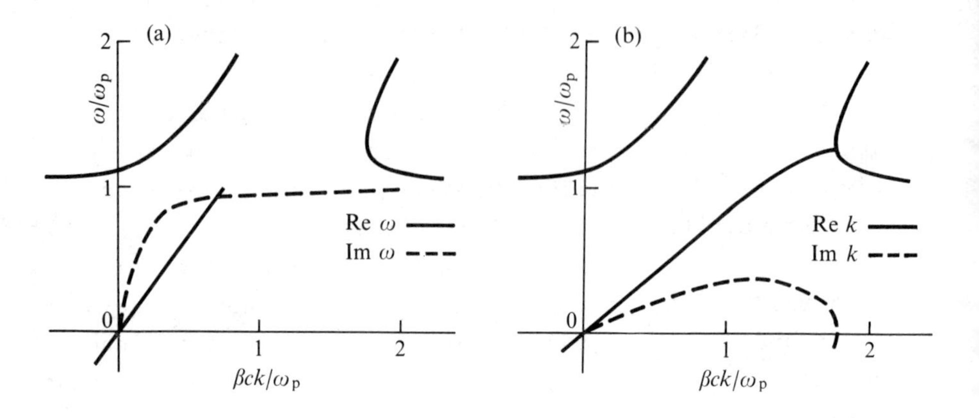

Fig.6.15. Sketch showing topology of dispersion curve for infinite beam-plasma system: (a) real k and complex ω, (b) real ω and complex k.

was mentioned briefly in section 6.3.3; the plasma consisted of heavy background ions through which an electron beam passed. For the idealization in which the ions are regarded as immobile, two-stream instability can be expected if a beam passes through a stationary plasma, or if two beams, in the same or opposite directions, pass through one another. Unlike the situation for an infinite beam, however, there is a threshold for the effect. This can be seen by considering the dispersion relation corresponding to the presence of either one stream or the other. The two parallel pairs of straight lines corresponding to unbounded plasma necessarily intersect at four points, but two pairs of curved lines of the form shown in Fig.6.7 only intersect at the origin. Only if the beam intensity is high enough can complex roots occur. This may be seen in Fig.6.16 where the dispersion curve near the origin is sketched for the interaction of a beam with a stationary plasma in a cylindrical conducting tube. The ions are considered to be immobile. The effect of increasing the beam density and the consequent change of topology may be seen. In the first diagram there are four real values of ω for each value of k. In the third diagram complex ω is possible for a range

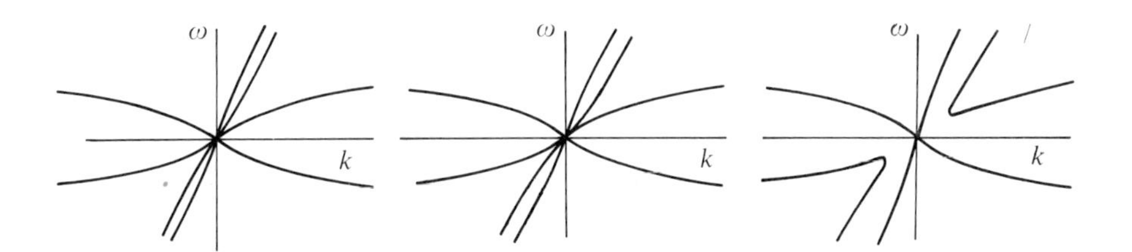

Fig.6.16. Dispersion curve near the origin for beam-plasma interaction in finite cylindrical geometry with massive ions. Conditions in the three diagrams are the same, except that the intensity of the beam is progressively increased through the threshold for instability.

of k. The central diagram represents the transition, which, for given geometry and beam velocity, occurs when the densities of beams and plasma are sufficiently high.

This is an idealization of the actual situation, where the ions have finite mass and the magnetic field is not infinite. Considerable experimental study of this type of situation has been made; in such studies other features need to be included, and we return to a more general discussion of beam-plasma interaction in section 6.4.6. after the introduction of transverse oscillations.

6.R.3. Notes and references The material of this section is of importance in the theory of microwave tubes and accelerators. Microwave applications are now well documented in a number of books. In the accelerator field only the text of Bruck (1966) deals with beam instabilities; many developments have occurred, however, since his book was published.

Treatments in a microwave context similar in spirit to that of the present section may be found in the texts of Johnson (1965) and Collin (1966). More specialized books on the theory of microwave tubes which also cover many aspects not dealt with here are those of Pierce (1950), Beck (1958), Hutter and Harrison (1960), Chodorow and Susskind (1964), Gewartowski and Watson (1965), Sims and Stephenson (1963), and Shevchik, Shvedov and Soboleva (1966). All deal at length with waves on beams, and many contain material on focusing and space-charge effects to be found in earlier chapters. A particularly illuminating early review of beam-wave concepts in microwave tubes is that of Pierce (1954c). Many other texts of a more specialized nature exist, as well as more popular accounts of microwave tubes. There is also a great deal of material in electronic engineering journals up to the early 1960's; the importance of many of these tubes has declined in recent years, however, with the advent of solid-state microwave devices.

References to accelerator applications will be made later (6.R.5), after transverse instabilities and bunched beams have been studied. Of the more general concepts treated here, coupled mode theory applied to microwave tubes form the subject of a book by Louisell (1960); it was first introduced by Pierce (1954b) in the context of microwave tubes, and an early paper giving a clear physical account of its application to the TWT and BWO is that of Gould (1955). Further references to the kinetic power theorem and wave-energy concept are given in section 6.R.4 after the introduction of transverse waves.

6.4. Transverse waves

6.4.1. Introduction So far, only longitudinal waves have been studied. Although the electric and magnetic fields associated with the waves have components in all three directions, the particle motion associated with the wave has been restricted to the direction of propagation. We now introduce transverse waves in which there are components of particle velocity perpendicular to the direction of propagation. The analysis of transverse motion is in general more complicated than that of longitudinal motion. A sideways displacement of a beam in cylindrical geometry necessarily breaks the axial symmetry, so that the field equations are more awkward to solve. In some waves both transverse and longitudinal velocity components occur.

Various geometrical configurations can be characterized

by considering the azimuthal displacement of the edge of the beam. If this is expressed in the form

$$r = a_0 + \delta a_0 \cos m\theta \tag{6.150}$$

then for $m = 0$ axial symmetry is maintained and area of the beam varies. If $m = 1$, then this describes a sideways movement of the beam without change of area ('kinking' or 'snaking'). Higher modes imply distortion; $m = 2$ describes the quadrupole mode in which the beam cross-section is elliptical. These configurations are shown in Fig.6.17. The azi-

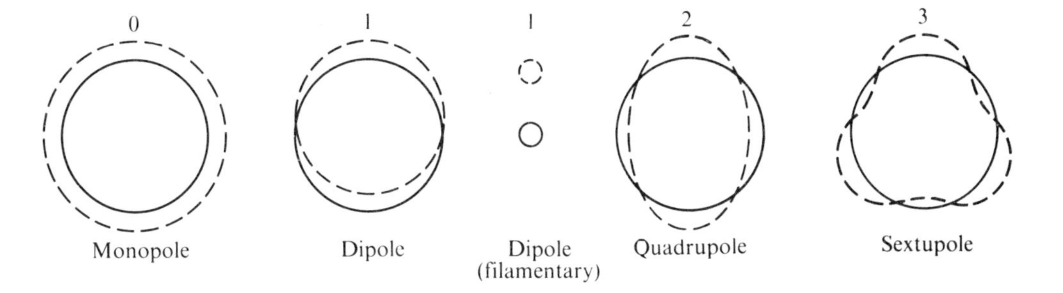

Fig.6.17. Types of transverse displacement which can occur on beams.

muthal variation described so far is of standing-wave form. The perturbation δa is of form $\delta a_0 \cos \omega t$; if $m = 1$, for example, the beam at a point during a half-cycle of oscillation appears elliptical, then circular, then elliptical with major and minor axes reversed. A travelling-wave form, where the argument of the cosine in equation (6.150) is $(m\theta - \omega t)$ so that a fixed spatial pattern rotates, is of course possible.

Since $a \gg \delta a_0$ the motion can be described by linear equations. For the dipole perturbation, where the beam cross-section remains constant, it is possible to preserve linearity for larger displacements provided that the beam radius is small compared with that of the surrounding tube. An approximation which is frequently useful is that of a 'filamentary' beam, where the radius is small compared with the displacement which is again small compared with the tube diameter. At a given value of z and t all particles have essentially the same value of x and y.

We start by studying waves on filamentary beams in the absence of walls, and then look at the effect of walls of various types. We then examine beams in which self-fields cannot be neglected surrounded by simple walls. The situation in which both space charge is important and walls are complex is somewhat intractable.

6.4.2. Filamentary beams, coherent betatron oscillations, and cyclotron waves A particularly simple idealization which we study first is the filamentary beam, in which the diameter is assumed small compared with a typical displacement. If there are linear focusing forces towards the axis, then a characteristic frequency exists which, for a low-intensity beam, is independent of beam density. This contrasts with the situation for longitudinal waves, where the plasma frequency tends to zero as the beam density decreases.

We analyse first motion in a system with two planes of symmetry, the xz and yz planes, where as usual the z-axis is the axis of the beam. The only force acting on the beam particles has two components, one parallel to the xz plane proportional to x and the other in a perpendicular direction proportional to y. The force may be specified in terms of the frequency of oscillation of a single particle, measured in the laboratory frame. The oscillation frequencies in the two planes are denoted by ω_x and ω_y. Since the motion of individual particles is uncorrelated, the value of k is arbitrary. The frequency in the x-direction of the wave, observed at a fixed point, is Doppler shifted from ω_x by an amount βck to $\beta ck \pm \omega_x$, depending on whether the wave seen in the frame of reference moving with the particle moves forwards or backwards. Setting this equal to ω and re-arranging

$$(\omega-\beta ck) = \pm\omega_x \tag{6.151}$$

or, in conventional form, writing $\omega_x = Q\omega_0$ for reasons explained below

$$\frac{Q_x^2\omega_0^2}{(\omega-\beta ck)^2} = 1 \ . \tag{6.152}$$

This is of the same form as equation (6.39) for an infinite plasma. It does not depend on the beam intensity, and should be contrasted with the corresponding equation for longitudinal oscillations in a finite beam, where the relation depends on intensity through ω_p and the lines come together at the origin where $\omega = 0$, $k = 0$.

A similar relation exists for the yz plane, giving four waves in all, one fast and one slow in either plane. The notation has been chosen to accord with that of section 2.5.3 for particle accelerators, if z is replaced by the distance measured along the orbit and ω_0 by $2\pi\beta c/C$, where C is the circumference of the orbit. In this situation ω_0 is a function of β. Waves of the kind just discussed are normally described in terms of 'coherent betatron oscillations'. We shall refer to them as betatron waves.

When the frequencies in the two planes are equal, it is possible to specify the motion in terms of complex conjugate variables $\xi = x+jy$, $\xi^* = x-jy$ as was done in section 2.2.2. (see Table 2.1). These correspond to left-handed and right-handed waves in which every particle moves in a helix. At a given time the particles lie on helices which may be right or left handed. Furthermore, the particles can be rotating in a positive or negative sense. This gives four possibilities, corresponding to the four types of wave. For a helix with pitch $2\pi/k = 2\pi\beta c/\omega$ these are shown in Table 6.4. These four waves form a Larmor-frame description of cyclotron waves in a uniform magnetic field in the z-direction, where $Q\omega_0 = \Omega_L = \frac{1}{2}\omega_c$. Two of them consist of particles rotating in the field with frequency $+\omega_c$; the other two consist of particles in a helical configuration but moving along the lines of force. These 'synchronous' waves are similar to those obtainable from a garden hose by rotating the nozzle either clockwise or anticlockwise at constant frequency. For the cyclotron waves we have a corresponding table,(6.5). It is interesting to note that of the two synchronous waves, both of which move at

TABLE 6.4

Transverse filamentary waves on a focused beam

Direction of helix	+	+	-	-
Rotation frequency	$\omega_0 Q$	$-\omega_0 Q$	$\omega_0 Q$	$-\omega_0 Q$
Wave velocity	$(\omega+\omega_0 Q)/k$	$(\omega-\omega_0 Q)/k$	$(\omega-\omega_0 Q)/k$	$(\omega+\omega_0 Q)/k$
Type of wave	Fast	Slow	Slow	Fast

	$\omega_0 Q$	$-\omega_0 Q$	$\omega_0 Q$	$-\omega_0 Q$

Angles with the axis are exaggerated; the horizontal lines represent a hypothetical opaque cylinder inserted to give three-dimensional information.

TABLE 6.5

Transverse filamentary waves in an axial magnetic field

Direction of helix	+	+	-	-
Rotation frequency	ω_c	0	ω_c	0
Wave velocity	$(\omega+\omega_c)/k$	ω/k	$(\omega-\omega_c)/k$	ω/k
Type of wave	Fast	Slow	Slow	Fast
	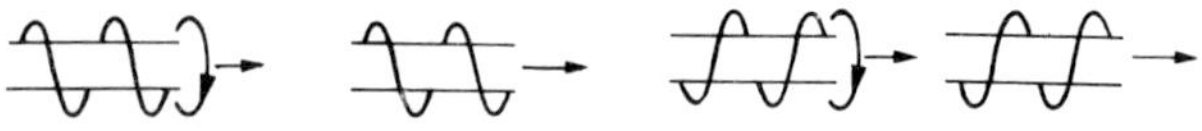			
	ω_c		ω_c	

velocity ω/k, one is slow and the other fast.

<u>6.4.3. Behaviour of filamentary waves in a resistive environment</u> The way in which a filamentary wave is influenced by a resistive environment may be seen by a simple physical argument. We consider a transverse filamentary beam sym-

metrically placed between plates perpendicular to the plane of the oscillations as shown in Fig.6.18. Initially the

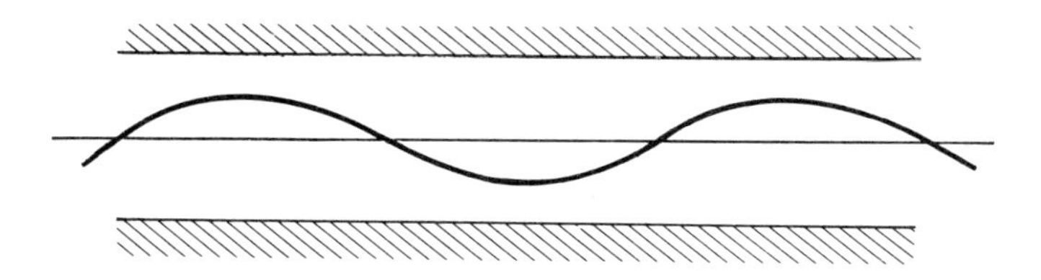

Fig.6.18. Filamentary current wave between infinite conducting plates. If the beam flows to the right, there is an equal current in the plates which flows to the left. For a vertical displacement of the beam additional currents are induced which exert a force towards the axis; if the beam is un-neutralized the image charges produce an even stronger force in the opposite direction.

plates will be assumed to be perfect conductors. The particles induce in the plates image charges of opposite sign and image currents flowing in the opposite direction to the actual current. These charges and currents repel and attract respectively the beam particles with a force which, for small displacements, is proportional to x, the distance from the axis. In an un-neutralized beam the effect of the charges predominates, the net focusing is strengthened and the frequency of the wave undergoes a 'coherent Q-shift' to a lower value.

If now some dissipation is introduced in the plates, the behaviour is slightly modified. The image force at any point is not simply proportional to the beam displacement at that point; a step displacement of part of the beam does not give rise to a step force, but rather to a force which rises sharply to the same value that it would have for a perfect conductor, and then decays. A pulse of displacement gives rise to a force of the form shown in Fig.6.19. A moving step gives rise to a decaying 'wake' behind it which exerts a force on subsequent particles. If the displacement of subsequent particles is *greater* than that of the original

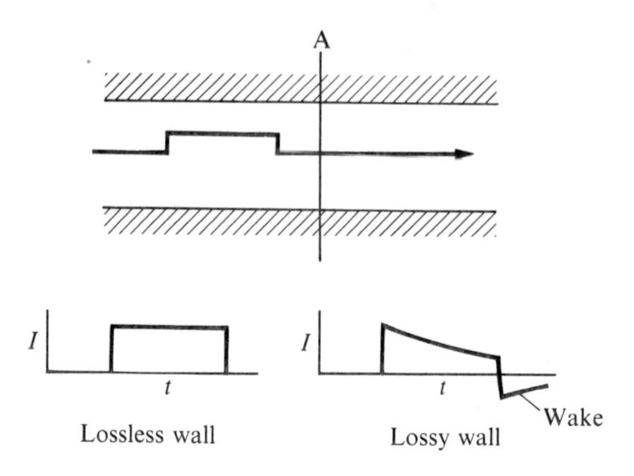

Fig.6.19. A beam with a 'step' representing a sideways displacement gives rise to a pulse of current at some point A in the surrounding walls. For a lossless wall and pulse length large compared with the tube radius divided by γ the pulse shape is as shown in the first diagram; when the wall is lossy, it leaves a 'wake' as shown in the second diagram.

particles, the net force on them will be *reduced*, and *vice versa*. This force at some amplitude x is clearly proportional to $\partial x/\partial t$ for the wave. This means that in a slightly lossy medium a sinusoidal wave experiences an additional inward force, phase-shifted by $\pi/2$, of such a sign as to produce an extra inward force at places where the wave is moving outwards, and *vice versa*. Now although the wave may be moving outwards, it does not follow that the particles which compose it are also moving outward. Indeed, if the wave is moving more slowly than the beam, an observer moving with the beam sees the particles move inward as the wave moves outward; if, on the other hand, it moves faster, then the particles move outward. This is illustrated in Fig.6.20. For a slow wave, therefore, the particles are accelerated outward as they are moving outward and the wave grows. For a fast wave the force is opposite to the motion and the wave is damped. As with longitudinal waves, dissipation produces growth in slow waves and damping in fast waves.

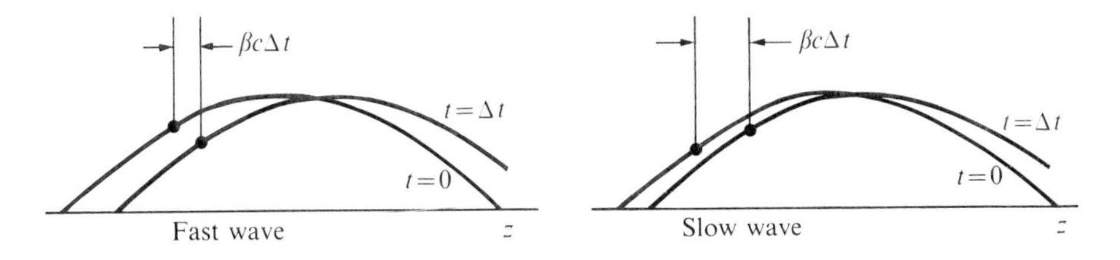

Fig.6.20. Diagram to illustrate the motion of particles and waves in fast and slow transverse waves. For both fast and slow waves the displacement is plotted against position at two times $t = 0$ and $t = \Delta t$. For a fast wave an observer moving with the beam velocity sees both wave and particles moving inward together; for a slow wave, on the other hand, he sees the wave and particle moving in opposite directions.

Similar arguments may be applied to the cyclotron waves. At first sight the two synchronous waves might appear to be symmetrical, since in neither does the beam interact with the magnetic field. Table 6.5, however, shows that one is fast and the other slow. In the presence of walls electric fields with an azimuthal component are set up. The direction of this field, and hence the radial velocity arising from the product $E_\theta B_z$, depends on the sense of the helix. A full treatment of cyclotron waves is given by Louisell (1960). These can be coupled to transverse field components on a structure in a manner that corresponds to the coupling between longitudinal field components and waves studied in section 6.3.9.

It is pertinent to enquire in what sense these transverse waves may be said to possess negative energy. Clearly, if every particle in the beam is given an appropriate transverse velocity, without changing the longitudinal velocity, the system with a wave present will have more energy than in the absence of a wave, whether it is slow or fast. If, however, realistic methods of exciting the wave are considered, it is found that a second-order change in longitudinal velocity occurs which is sufficient, for a first-order transverse velocity, to give a total energy change of the whole system which is positive for a fast wave and negative for a slow wave. A specific simple example is analysed in appendix 6.

If the longitudinal and transverse velocities are known only to first order, then as indicated in section 6.3.8, it is not possible to determine in an unambiguous way the energy associated with the wave. This point caused a certain amount of confusion during attempts to generalize the kinetic power theorem to include transverse motions. The formal difficulty was circumvented by Sturrock (1960, 1962), who introduced a quantity which he termed the pseudo-energy, which is uniquely defined by the first-order terms and therefore not necessarily to be identified with the actual energy in the beam. The highly formal procedure starts from postulating suitable Lagrangian density and leads to relations between the pseudo-energy and pseudo-momentum of the waves of the form

$$\begin{aligned} E &= J\omega \\ p &= Jk \end{aligned} \tag{6.153}$$

where J is an action density. The suggestion of negative frequency associated with negative energy appears in an interesting way in the example discussed in appendix 6. Parallels with quantum phenomena, where 'absorption of a negative-energy quantum' excites a negative-energy wave, are discussed by Musha (1964).

An interesting physical discussion of the meaning of negative-energy waves, contrasting with the formal treatment of Sturrock, is given by Pierce (1961). Extension to relativistic energies has been made by Briggs (1964b). The more general form of the kinetic power theorem, including a reference to earlier papers and discussion of some of the difficulties of interpretation, is reviewed by Bobroff, Haus, and Klüver (1962).

6.4.4. Transverse instability in accelerators The physical nature of filamentary waves was explored in the previous section and the interaction with walls discussed in a qualitative manner. Some similarities and differences compared with longitudinal waves were found. For both classes of wave conducting walls near the beam shift the oscillation frequencies, and resistive walls damp fast waves but cause growth of slow waves. A significant difference, however, is that for longitudinal waves the frequency tends to zero as the density decreases, whereas for transverse waves it remains constant (apart from a small Q-shift).

Although transverse cyclotron waves have been studied in connection with transverse travelling-wave tubes and the 'Adler tube', a transverse-wave parametric amplifier described, for example, by Sims and Stephenson (1963), these devices have not come into general use. The principal interest at the time of writing is in the field of particle accelerators, where the aim is suppression rather than excitation.

The method of study parallels that developed for longitudinal oscillations. Quantities $U_{\perp}$ and $V_{\perp}$ are defined and practical criteria expressed in terms of a suitably defined impedance. The definitive paper is by Laslett, Neil, and Sessler (1965). Here we use a slightly different approach, and lead on to the impedance formalism developed at CERN. The waves are not filamentary in this treatment; the beam is assumed to retain its circular cross-section and to be displaced an amount small compared with the radius. Transverse waves present the additional complication that not only is there a distribution of particle velocity, but also a distribution in oscillation amplitude of particles within the beam. To preserve simplicity some approximations are required, but nevertheless a useful theory of practical value emerges.

A physical description of the nature of the force which acts on the displaced beam was given in the previous section; this force, acting on a sinusoidal wave, has a component proportional to the displacement which produces a shift in oscillation frequency and a component proportional to velocity which can produce growth or decay. We now examine the magnitude of this force, which is a function of beam geometry and intensity, and of wall geometry and resistivity. This has been calculated by Laslett *et al.* (1965) for rectangular and cylindrical geometries; here we consider only the latter, illustrated in Fig.6.21. In the derivation the thin beam

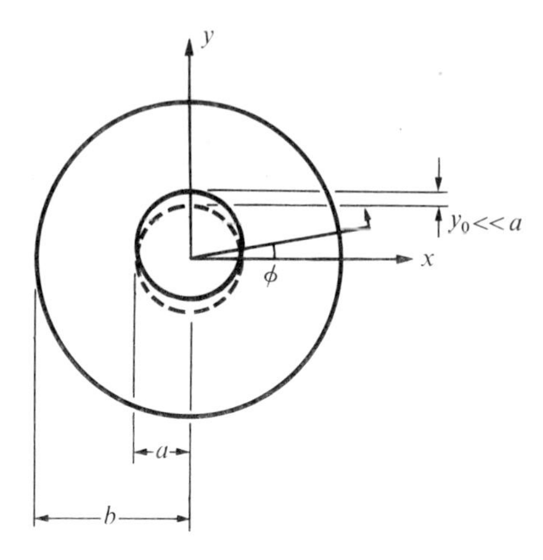

Fig.6.21. Geometry and notation for analysis of transverse instability.

approximation that the perturbation wavelength is long compared with b/γ_w, will be used. This means that in calculating the transverse fields arising from a displacement of the beam the problem reduces to a two-dimensional one, and k_z can be assumed zero except in the exponential factor. The beam-density is n_0, and the geometry and displacement are defined in Fig.6.21. Motion is assumed in the yz plane, corresponding to vertical motion in accelerators.

The sources for the fields are

$$n_1 q = n_0 q y_0 \sin\phi\,\delta(r-a)\,\exp j(\omega t - kz)$$

$$i_{1z} = n_1 q\beta c\ . \tag{6.154}$$

We now evaluate the fields inside and outside the beam, assuming first that the walls are perfect conductors. Inside the beam the fields are uniform, whereas outside this is not so. It is convenient to use cylindrical co-ordinates and match the fields at the inner and outer boundaries, and we follow closely the working in the paper of Laslett *et al.* (1965). The fields are expressed first in terms of unknowns E_{1y}, H_{1y}, E_{20} and H_{20}, which are then found by matching to the source specified in equation (6.154). The fields are, inside the beam

$$\begin{aligned} \underline{\mathrm{E}}_1 &= E_{1y}(\sin\phi\hat{r}+\cos\phi\hat{\phi}) \\ \underline{\mathrm{H}}_1 &= H_{1x}(\cos\phi\hat{r}-\sin\phi\hat{\phi}) \end{aligned} \tag{6.155}$$

and outside

$$\begin{aligned} \underline{\mathrm{E}}_2 &= E_{20}[\{(b/r)^2+1\}\sin\phi\hat{r}-\{(b/r)^2-1\}\cos\phi\hat{\phi}] \\ \underline{\mathrm{H}}_2 &= H_{20}[\{(b/r)^2-1\}\cos\phi\hat{r}+\{(b/r)^2+1\}\sin\phi\hat{\phi}]. \end{aligned} \tag{6.156}$$

Matching boundary conditions at the source and wall yields for the unknown fields inside the beam

$$E_{1y} = \frac{-\frac{1}{2}n_0 q y_0\{1-(a/b)^2\}}{\varepsilon_0} \tag{6.157}$$

$$Z_0 H_{1x} = -\beta E_{1y}$$

and outside

$$\begin{aligned} E_{20} &= \frac{n_0 q y_0 (a^2/b^2)}{\varepsilon_0} \\ Z_0 H_{20} &= -\beta E_{20} \end{aligned} \tag{6.158}$$

If now the pipe is not quite a perfect conductor, but has a wall conductivity σ, this can be included (to first order) by adding small additional field components which satisfy the new boundary conditions at the wall. The condition to be satisfied is

$$E'_{1z} = -(1+j)Z_0\left(\frac{\varepsilon_0\omega}{2\sigma}\right)^{1/2} H_{1\phi}. \tag{6.159}$$

From this we wish to determine the additional components of

E_y and H_x at $r \leq a + y_0$. Inserting $H_{1\phi}$ from equations (6.157) and (6.158), equation (6.159) becomes, (omitting the exponential factor)

$$E'_{1z} = (1+j)n_0 q\beta y_0 \frac{a^2 r}{b^3}\left(\frac{\varepsilon_0\omega}{2\sigma}\right)^{1/2} \sin\theta \tag{6.160}$$

which is valid for all values of r. The magnetic field may be found from the relation $\mu_0 \operatorname{curl} \underline{H} = -\partial\underline{E}/\partial t$ as

$$\underline{H}'_1 = -(1-j)n_0 q\beta y_0 c\frac{a^2}{b^3}\left(\frac{1}{2\sigma\omega\varepsilon_0}\right)^{1/2}(\cos\phi\hat{r} - \sin\phi\hat{\phi}). \tag{6.161}$$

This expression is independent of r, and does not fall to zero at $r = b$. For this reason equation (6.159) is satisfied to first order only.

We can now write down the force on a particle in the beam arising from a displacement y_0 as

$$\begin{aligned} F_y &= (E_y + \beta c B_x)q \\ &= -\tfrac{1}{2}n_0 q^2 y_0 \left\{\frac{(a^2-b^2)}{b^2\varepsilon_0\gamma^2} - 2(1-j)\frac{\beta^2 a^2 c}{b^3}\left(\frac{1}{2\sigma\omega\varepsilon_0}\right)^{1/2}\right\}. \end{aligned} \tag{6.162}$$

Following Laslett *et al.* we write this as

$$F_y = 2\gamma m_0\omega Q_y\{U_\perp + (1-j)V_\perp\}y_0 \tag{6.163}$$

where

$$\begin{aligned} U_\perp &= \frac{-\omega_p^2\{1-(a/b)^2\}}{4Q_y\omega_0\gamma^2} \\ V_\perp &= \frac{\omega_p^2 a^2\beta c}{2\omega\omega_0 b^3 Qy}\left(\frac{\varepsilon_0\omega}{2\sigma}\right)^{1/2} \end{aligned}. \tag{6.164}$$

(An alternative procedure would be to write F_x directly in terms of the impedance $Z_\perp$ defined later in equation (6.172).) These quantities have the dimensions of frequency. The parameter $U_\perp$ depends essentially on the beam density ω_p^2, but V depends on $a^2\omega_p^2$, which is proportional to the total number of particles on the beam. In contrast to section 6.3.4 the plasma frequency ω_p^2 contains the transverse mass γm_0 rather than m^*. It is evident from equation (6.163) that the image forces represent a shift in the value of Q_y^2 given by (dropping the subscript)

$$\Delta Q^2 = \frac{-2Q\{U_\perp+(1-j)V_\perp\}}{\omega_0} \qquad (6.165)$$

so that the dispersion relation for transverse waves, equation (6.152), becomes

$$\frac{\omega_0^2Q^2-2\omega_0Q(U_\perp+V_\perp)+2j\omega_0QV}{(\omega-\beta ck)^2} = 1. \qquad (6.166)$$

In a cyclic accelerator ω_0 is the orbital frequency and βck is replaced by $n\theta$, where n is integral.

Setting $V_\perp \ll \omega_0Q$, a condition which holds in practical situations, equation (6.165) can be written, in terms of n rather than k,

$$\omega = (n\pm Q)\omega_0 \mp (U_\perp+V_\perp-jV_\perp). \qquad (6.167)$$

When $n > Q$ the lower signs represent a slow wave, which grows with e-folding time $V_\perp^{-1}$ inversely proportional to the number of particles per unit length of beam. When $n < Q$, the equation represents a fast wave with negative phase velocity, which is damped. The terms $U_\perp+V_\perp$ decrease the effective Q-value by $(U_\perp+V_\perp)/\omega_0$; in addition to the danger of beam loss associated with the growing wave, there exists the possibility that the Q will be shifted to a resonant value to produce loss from the mechanisms described in section 2.7.3.

As with longitudinal waves, the result of equation (6.167) can readily be generalized to include velocity spread in the beam; there will in addition be a spread of oscillation amplitude y_0. Integrating over the appropriate distribution functions as before, and retaining the accelerator notation (section 6.3.5), the dispersion equation becomes

$$1 = Q\omega_0\{U_\perp+(1-j)V_\perp\}\int\frac{(\partial h/\partial y_0)(y_0)y_0^2dy_0f(W)\,dW}{(\omega-n\dot{\theta})^2-Q^2\dot{\theta}^2} \qquad (6.168)$$

where $y_0 h(y_0)\mathrm{d}y_0$ is the fraction of particles in the interval $\mathrm{d}y_0$ and $\mathrm{d}W$ is the fraction in the interval $\mathrm{d}W$ at $W = 2\pi(P_\theta - P_0)$, so that

$$\int (\partial h/\partial y_0)(y_0)y_0\mathrm{d}y_0 = 1, \quad \int f(W)\mathrm{d}W = 1. \quad (6.169)$$

The angular velocity $\dot{\theta}$ is $\omega_0 - k_0 W$, as in equation (6.99). The dispersion relation (6.168) is given in the original paper by Laslett *et al.*, but is there derived with the aid of the Vlasov equation.

The form of equation (6.168) differs from that of equation (6.99), but by making an approximation it is possible to cast it in the same form, and then to proceed in a similar manner. First, we note that the integration over y_0 is straightforward, unless it is necessary to take into account the non-linear dependence of Q on y_0. Second, since we are only interested in growing waves, it is sufficient to consider values of ω near $(n-Q)\omega_0$. Separating the integrand of equation (6.168) into partial fractions,

$$\{(\omega - n\dot{\theta})^2 - Q^2\dot{\theta}^2\}^{-1} = (2Q\dot{\theta})^{-1}[\{\omega-(n+Q)\dot{\theta}\}^{-1} - \{\omega-(n-Q)\dot{\theta}\}^{-1}] . \quad (6.170)$$

The singularity in the first term on the r.h.s. lies outside the range of interest, so that this term will have negligible effect on the result. The approximate integral then becomes

$$1 \approx -\tfrac{1}{2}(U_\perp + V_\perp - \mathrm{j}V_\perp) \int \frac{h'(y_0)y_0^2\mathrm{d}y_0 f(W)\mathrm{d}W}{\{\omega-(n-Q)\dot{\theta}\}} \quad (6.171)$$

with $\dot{\theta} = \omega_0 - kW$.

It is now possible to proceed as in section 6.3.6 to determine stability thresholds in the presence of Landau damping. We do not follow this topic here, but this development is continued in the original paper and followed as for longitudinal oscillations by detailed calculations (Hübner, Ruggiero, and Vaccaro 1969).

As before, it is possible to express the term containing $U_\perp$ and $V_\perp$ as an impedance and find a stability condition in terms of the energy spread. A convenient definition of this transverse impedance is

$$Z_{\perp} = \frac{\oint (E_y + \beta c B_x)\,\mathrm{d}s}{I y_0 \beta} \tag{6.172}$$

where qE_y/y_0 and $q\beta c B_x/y_0$ are the electric and magnetic forces per unit displacement of the beams integrated round the orbit of the machine and I is the current. This impedance has dimensions Z_0 divided by length. For this reason a better convention might have been to multiply the r.h.s. of equation (6.172) by the machine circumference.

Although introduced here in connection with instabilities in circular accelerators, it is interesting to note that the impedance has also been used in the analysis of r.f. beam separators (Larsen, Altenmueller, and Loew 1963). In such a device beams of particles with the same momentum are separated by a travelling-wave field with a transverse component of electric field. It is arranged that the phase velocity of this field should match the velocity of the wanted (or unwanted) component and thus deflect it preferentially.

Proceeding in a manner analogous to that described in section 6.3.6 it is again possible to find a stability criterion valid for 'reasonable' distribution functions. This time it takes the form

$$Z_{\perp} \leq \frac{\pi \beta m_0 c \omega_0 Q}{qI} \left| (n-Q)\eta + \frac{\partial Q}{\partial(\Delta p/p)} \right| \frac{\Delta p}{m_0 c} \tag{6.173}$$

where $\eta = \gamma_t^{-2} - \gamma^{-2} = -\gamma m_0/m^*$. Substituting $I = Nq\beta c$, $\nu = Nq^2/4\pi\varepsilon_0 m_0 c^2$, $\partial Q/(\Delta p/p) = Q\xi$ (equation 2.139) this becomes

$$\frac{\nu}{\gamma} Z_{\perp} \leq \frac{\omega_0 Q Z_0}{4c} \left| (n-Q)\eta + Q\xi \right| \frac{\Delta p}{m_0 c} \ . \tag{6.174}$$

A numerical discussion, indicating the extent to which wall effects limit the performance of the CERN intersecting storage ring, is given by Keil (1972b). Steps must be taken to keep both the longitudinal and transverse impedance of the vacuum chamber as small as possible.

Electron storage rings, and all cyclic accelerators with the exception of betatrons, operate with bunched beams.

The calculation of stability conditions in such beams is more complicated, as will be seen in section 6.5.

6.4.5. Transverse interaction between beams and travelling waves; 'beam break-up' In the previous section the interaction of transverse waves with a smooth wall characterized by finite resistivity was analysed. As with longitudinal waves, interaction with a structure which supports propagating waves is also possible. A coupled-mode analysis on the lines of that for longitudinal waves set out in section 6.3.7 has been given by Louisell (1960), and there is a chapter on transverse travelling-wave tubes in the book by Pierce (1950).

In electron linear accelerators transverse instability limits the current obtainable. The behaviour corresponds exactly to that in a transverse backward-wave oscillator; the disc-loaded accelerating structure can propagate many modes, among them one with transverse electric field on the axis which has negative group velocity. This travels backwards down the waveguide and feeds back energy in the way described in section 6.3.7. This effect is analyzed in detail in review articles by Helm, Loew, and Panofsky (1968) and by Helm and Loew (1970), and a broad review has been given by Loew (1967). The external focusing is not always included, so that the fast and slow waves on the beam cannot then be distinguished.

An instability of similar, but not identical, nature is also found when a beam passes through a series of cavities, or groups of cavities, which are not connected, so that no backward wave is propagated. This is analogous to the inductive-wall longitudinal instability (easitron) discussed in section 6.3.4. Energy is fed continuously into the cavities or wall material until the growth is limited by dissipation or non-linearity (such as the beam striking the wall). This is also analysed in the reviews referred to above.

6.4.6. Transverse two-stream instability In particle accelerators the transverse two-stream instability is sometimes troublesome. In unbunched beams ions of opposite charge produced by ionization of the residual gas are held within the potential well of the circulating beam. The beam of moving particles experiences restoring forces arising from the accelerator focusing field and the trapped particles; the trapped particles, on the other hand, only experience the force arising from the moving particles. Since we are only interested in calculating the threshold and growth rate, it is sufficient to consider small amplitudes, where the system is necessarily linear. Effects associated with modified beam shape (monopole, quadrupole, and higher-order modes) will not be considered. To be specific, we consider a beam in which the beam particles are electrons and the stationary particles positive ions. The transverse distribution of both types of particle is assumed uniform, with radius a, and the density ratio of (singly charged) ions to electrons is f. Let x_e and x_i be the displacements of electrons and ions from the axis, and X_e, X_i the displacements of the centres of the beams (Fig.6.22).

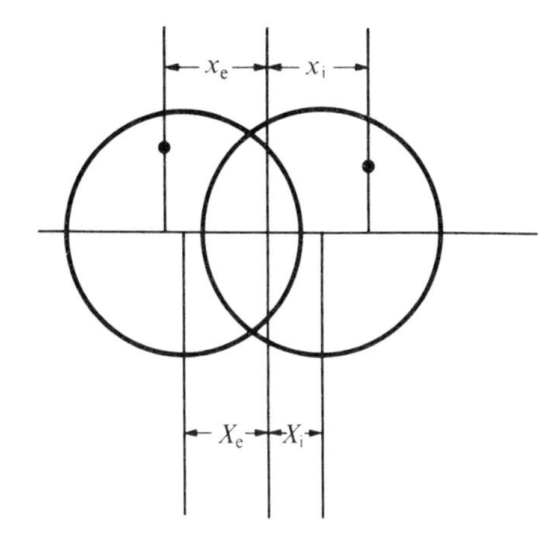

Fig.6.22. Notation for transverse two-stream instability. The co-ordinates of the beam centre, and of individual particles, are denoted by X and x respectively.

Three frequencies enter into the problem. The frequency of electrons in the focusing field of the machine is $Q\omega_0$; the frequencies of the ions in the field of the electrons is $Q_i\omega_0$ and the frequency which the electrons would have in the field of the ions if the focusing associated with the machine, $Q\omega_0$, were zero is $Q_i\omega_0$. By considering the space-charge forces as-

associated with the species of opposite sign, it is readily shown that these frequencies are

$$\begin{aligned} Q_e^2\omega_0^2 &= \tfrac{1}{2}f\omega_{pe}^2 \\ Q_i^2\omega_0^2 &= \omega_{pi}^2/2f \end{aligned} \tag{6.175}$$

where ω_{pe} and ω_{pi} are the plasma frequencies of the electrons and ions. The motion of the two beams is now given by the coupled equations

$$\begin{aligned} \ddot{x}_e + Q^2\omega_0^2 x_e &= Q_e^2\omega_0^2(x_e - X_i) \\ \ddot{x}_i &= Q_i^2\omega_0^2(x_i - X_e) \end{aligned} \tag{6.176}$$

If we now assume that the values of Q are the same for all particles of each species, we can sum over all the ions and electrons to find the equations of motion of the beam centres. These are the same as (6.176), but with X written in place of x. Substituting a solution of the form

$$X = X_0 \exp j\{(\omega - \beta ck)t - kz\} \tag{6.177}$$

which represents travelling transverse waves in equation (6.176) with X in place of x yields

$$\begin{aligned} \{-(\omega-\beta ck)^2 + \omega_0^2Q^2\}X_e &= Q_e^2\omega_0^2(X_e - X_i) \\ -\omega^2 X_i &= Q_i^2\omega_0^2(X_i - X_e) \end{aligned} \tag{6.178}$$

where βc, the beam velocity, is equal to $C\omega_0/2\pi$. Eliminating X_i/X_e yields the dispersion relation

$$(\omega^2 - \omega_0^2 Q^2)\{(\omega-\beta ck)^2 - \omega_0^2(Q^2 + Q_e^2)\} = \omega_0^4 Q_e^2 Q_i^2 \,. \tag{6.179}$$

This can be written in the alternative form

$$\frac{\omega_0 Q_i^2}{\omega^2} + \frac{\omega_0^2 Q_e^2}{(\omega-\beta ck)^2 - \omega_0^2 Q^2} = 1 \tag{6.180}$$

which is similar, but not identical, in form to the dispersion relation for the longitudinal two-stream instability described by equation (6.46). The difference lies in the term $\omega_0^2 Q^2$ in the second denominator which has the effect of making the coupling weaker. A loop appears inside the central parallelogram formed by the asymptotes, which in Fig.6.3 is empty.

A detailed analysis of the dispersion relation has been made by Koshkarev and Zenkevich (1972), who plot areas of stability in the $Q_e Q_i$ plane for various values of Q. The quadrupole mode is also analysed, and the effect of Landau damping considered. The effect of wall currents as well as Landau damping are included in a study by Laslett, Sessler, and Möhl (1972). Landau damping, as with other forms of instability, gives rise to a threshold, but neither damping nor the effect of images has a great influence on the limitations imposed by the instability. Conversely, however, the introduction of ions into an otherwise ion-free electron beam does considerably lower the transverse instability threshhold.

In the electron ring accelerator this instability forms an important constraint on the allowable parameters (Möhl *et al.*1973). The reverse situation of the instability arising from electrons in a proton beam has been observed in a proton accelerator (Grunder and Lambertson 1971) and the CERN proton storage rings (Keil 1972b). Although sometimes troublesome, such oscillations are less destructive than in an electron ring.

In machines in which the beam is bunched, ions formed in the bunches escape before the next bunch arrives, and so the effect is of no importance.

6.R.4.1-6. Notes and references Transverse waves on beams were originally studied in connection with microwave tubes. Siegman (1960) described filamentary cyclotron waves and discussed the sign of the energy. Since such tubes have not, however, been found to be useful, interest waned. The subject again became of interest in connection with particle accelerators; instabilities were observed in experimental machines built by the 'MURA' (Midwestern Universities Research Association) group in the early 1960's. The paper by Laslett *et al.* (1965), discussed in section 6.4.4, contains comparisons between the theory and these experimental results.

Although earlier work used the U and V notation, it is customary now to work directly in terms of the transverse impedance; the conceptual simplification in doing this is emphasized in connection with bunched beams by Sacherer, as will be described in section 6.5.

A further discussion of the literature on accelerator instabilities is given after the section on bunched beams in section 6.R.5.

6.4.7. Axially symmetrical transverse waves in a laminar paraxial beam The study of small-amplitude transverse waves on a beam in paraxial approximation is relatively straightforward. In this approximation longitudinal components of the fields arising from displacements are neglected. Surface waves of the type studied in section 6.2.3, which involve both transverse and longitudinal motion of the charges, will be considered in the next section. The subject of 'rippling' on magnetically confined beams has already been discussed in section 3.3.1. The present section is essentially on extension of this, with emphasis on arriving at a dispersion equation which includes travelling as well as stationary waves.

We start by considering an electron beam in which the flow is laminar, and, in the absence of wave motion, the space-charge repulsion of the electrons is annulled by a massive immobile ion background which extends beyond the edge of the beam. This background produces an inward force on an electron proportional to its distance from the axis, and in the discussion which follows we regard the ions as forming an external focusing system, so that the force on an electron displaced radially outward is proportional to its distance from the axis. Later we introduce a magnetic field; as noted earlier the two systems are essentially equivalent, since one can be transformed into the other by a Larmor transformation.

The paraxial envelope equation (3.57) (which is here identical to the equation of motion of the outermost electron, since the flow is laminar) is of the form

$$a'' + \kappa_e a - \frac{K}{a} - \left(\frac{P_{\theta a}}{\beta\gamma m_0 c}\right)^2 \frac{1}{a^3} = 0. \qquad (6.181)$$

The external focusing *including the ion background* is represented by κ_e and K is the generalized perveance (section 3.2.5). Since $B_z = 0$, $P_{\theta a} = p_{\theta a}$. The subscript a, which denotes that this is the value at the edge of the beam, will be omitted.

We proceed as in section 3.3 and linearize by setting $a = a_0 + \rho$, where $\rho \ll a$. The perveance is given by equation (3.30):

$$K = \frac{2\nu}{\beta^2\gamma^3} = \frac{a_0^2\omega_p^2}{2\beta^2\gamma^2 c^2} . \qquad (6.182)$$

(As explained above, this is the form appropriate to an unneutralized beam, since the ion background is considered as a contribution to the external focusing.) The angular momentum is $\gamma m_0 \dot{\theta} a_0^2$, so that the last term in equation (6.181) becomes $a_0^4\dot{\theta}^2/a^3\beta^2c^2$. Substituting $a_0+\rho$ for a and linearizing yields the two relations for the equilibrium and perturbed motion respectively:

$$\kappa_e - \frac{\omega_p^2}{2\beta^2\gamma^2c^2} - \frac{\dot{\theta}^2}{\beta^2c^2} = 0 \qquad (6.183)$$

and

$$\rho'' + \left(\kappa_e + \frac{\omega_p^2}{2\beta^2\gamma^2c^2} + \frac{3\dot{\theta}^2}{\beta^2c^2}\right)\rho = 0. \qquad (6.184)$$

Setting $d^2\rho/dt^2 = \beta^2c^2\rho''$, and substituting for κ_e from equation (6.183) in (6.184), gives

$$\frac{d^2\rho}{dt^2} + \left(\frac{\omega_p^2}{\gamma^2} + 4\dot{\theta}^2\right)\rho = 0. \qquad (6.185)$$

For $\omega_p = 0$, this gives an effective value of $Q_r = 2$ for a radial field increasing linearly with r as expected from section 2.5.3. Projections of individual orbits on the xy plane are ellipses. From equation (6.185)

$$\omega^2 = \left(\frac{\omega_p}{\gamma}\right)^2 + 4\dot{\theta}^2. \qquad (6.186)$$

Proceeding now as in section 6.4.2, and writing the frequency as seen in the stationary frame, $\omega-\beta ck$ for ω yields the dispersion relation

$$(\omega-\beta ck)^2 = \left(\frac{\omega_p}{\gamma}\right)^2 + (2\dot{\theta})^2 . \tag{6.187}$$

When $\dot{\theta}=0$, this is the same as for longitudinal waves in an infinite plasma. It is important to note that this relation implies a focusing system which keeps the beam in equilibrium.

Although it is straightforward, by moving into rotating co-ordinates, to find the properties of waves in a longitudinal magnetic field, it is perhaps more instructive to calculate the behaviour directly, and this is what we do. Except under conditions of Brillouin flow, the particles in such a beam possess a finite canonical angular momentum. For a particle at the edge of the beam

$$P_\theta = \gamma m_0 a_0^2(\dot{\theta}-\tfrac{1}{2}\omega_c) = -\tfrac{1}{2}\gamma m_0 a_0^2\omega_v \tag{6.188}$$

where ω_v is the vortex frequency defined in section 3.6. The appropriate form of paraxial equation is then

$$a'' + \frac{\omega_c^2}{4\beta^2c^2}a - \left(\frac{P_\theta}{\beta\gamma m_0 c}\right)^2\frac{1}{a^3} - \frac{K}{a} = 0. \tag{4.48}$$

Substituting for P_θ and writing $\mathrm{d}^2a/\mathrm{d}t^2 = \beta^2c^2a''$ it becomes

$$\frac{\mathrm{d}^2a}{\mathrm{d}t^2} + \frac{\omega_c^2a}{4} - \frac{(\omega_c-2\dot{\theta})^2a_0^4}{4a^3} - \frac{\beta^2c^2K}{a} = 0. \tag{6.189}$$

The condition for equilibrium is found as before by setting the first term equal to zero and setting $a = a_0$. Using equation (6.182)

$$\tfrac{1}{4}\omega_c^2 - \tfrac{1}{4}(\omega_c-2\dot{\theta})^2 - \tfrac{1}{2}\omega_p^2/\gamma^2 = 0 \tag{6.190}$$

which reduces to

$$\dot{\theta}^2 - \omega_c\dot{\theta} + \omega_p^2/2\gamma^2 = 0. \tag{6.191}$$

This is the same as equation (3.42) except for the γ^2 term. (Equation (3.42) was derived without paraxial assumptions, but also non-relativistically. When the motion is relativistic the force component (3) in the list at the beginning of section 3.2.8 must be taken into account. It acts to reduce the outward space-charge force in a way analogous to the corresponding force in the beam studied in section 3.2.6.)

Expanding equation (6.189) and setting $a = a_0+\rho$ results in the equation corresponding to (6.186):

$$\frac{d^2\rho}{dt^2} + \left\{\tfrac{1}{4}\omega_c^2 + \tfrac{3}{4}(\omega_c-2\dot{\theta})^2 + \frac{\omega_p^2}{2\gamma^2}\right\}\rho = 0. \qquad (6.192)$$

For Brillouin flow $\omega_c = 2\dot{\theta}$, $\omega_c^2 = 2\omega_p^2/\gamma^2$, and equation (6.192) reduces to equation (6.185) with $\dot{\theta} = 0$ as expected. Eliminating $\tfrac{1}{4}\omega_c^2$ between the curly bracket in equation (6.192) and equation (6.191), we find the dispersion relation by analogy with the derivation of equation (6.187) to be

$$\begin{aligned}(\omega-\beta ck)^2 &= \left(\frac{\omega_p}{\gamma}\right)^2 + (\omega_c-2\dot{\theta})^2\\ &= \left(\frac{\omega_p}{\gamma}\right)^2 + \omega_v^2 .\end{aligned} \qquad (6.193)$$

The situations described by equations (6.191) and (6.193) are completely equivalent. It is often more convenient to work in the simpler system with no magnetic field, since there is symmetry between $-\dot{\theta}$ and $+\dot{\theta}$. Properties of an unneutralized beam in a magnetic field can then be derived by the simple transformation $\dot{\theta} \to \omega_v$.

The results in this section have been confined entirely to paraxial beams, with the additional assumption that longitudinal components of the alternating field are not important. Such fields are of second order in smallness for realistic values of k. This implies that the fields outside the beam remain unchanged and that there is no interaction with the walls. Their presence has no effect. In longitudinal waves, however, there is a first-order change in the charge per unit length of the beam, giving rise to first-order longitudinal fields. Such waves will be discussed in the next section. The analysis has been confined so far to laminar, or cold beams. Equations (6.187) and (6.193) may be readily generalized to include the effect of a finite emittance arising from a K-V distribution (equation 4.26). As explained in section 4.3.5 the emittance arising from finite canonical angular momentum P_θ in a laminar beam is equivalent to that of a K-V distribution with emittance ε, when $P_\theta/\beta\gamma m_0 c$ is set equal to ε. For a system without a magnetic field $P_\theta/\beta\gamma m_0 c$ is equal to $a^2\dot{\theta}/\beta c$, so that in equation (6.187) the term $(2\dot{\theta})^2$ can be replaced by $4\beta^2 c^2\varepsilon^2/a^4$.

For a beam with a Maxwellian transverse-velocity distribution the dispersion relations (6.187) or (6.193) are not applicable, since such beams necessarily have a non-uniform radial density profile and ω_p is not constant. One might expect, however, that the dispersion relation would be valid if some average value were used. Under these circumstances the emittance is given by

$$\varepsilon \approx 2a\left(\frac{kT}{\gamma m_0\beta^2 c^2}\right)^{1/2} \tag{4.42}$$

and the second term on the r.h.s. of equation (6.187) by

$$\frac{4\beta^2 c^2\varepsilon^2}{a^4} = \frac{4kT}{\gamma m_0 a^2}\,. \tag{6.194}$$

The condition that this should have a small effect is

$$\frac{4kT}{\gamma m_0 a^2} \ll \frac{\omega_p'^2}{\gamma^2} \tag{6.195}$$

or

$$a \gg \lambda_D \tag{6.196}$$

where λ_D is the Debye length. This condition is met in plasma columns and in carefully designed Brillouin flow beams. When $\lambda_D \gg a$ the dispersion relation requires a different interpretation. The emittance can be expressed in terms of the focusing strength, and the dispersion relation becomes equivalent to equation (6.151) for a filamentary beam.

6.4.8. Surface waves on non-vortical beams In section 6.2.3 the properties of a surface wave on a semi-infinite plasma medium were calculated. These were obtained in a straightforward way by matching impedance and phase velocity across the boundary; the plasma is simply a dielectric with negative κ, and the currents near the surface flow in a similar way to those in a dielectric, except that the normal component of $\underline{E}$ reverses at the surface. As pointed out earlier, in planar geometry this gives rise to a negative group velocity. In these waves the longitudinal electric field has a first-order effect; there is a longitudinal electric field and a magnetic field parallel to the surface. The magnetic field is in phase with the normal component of electric field and out of phase with the component in the direction of propagation.

We tackle the problem now for cylindrical geometry in a frame of reference in which the beam is at rest. The situation to be analysed is simply that of a dielectric rod of radius a with dielectric constant $\kappa = 1 - \omega_p^2/\omega^2$ in a perfectly conducting waveguide. This is straightforward, and has been done, for example, by Trivelpiece and Gould (1959). They obtain the dispersion relation which, in our notation, may be written (for modes without angular variation)

$$\left(\frac{\omega_p^2}{\omega^2} - 1\right) \frac{I_1(k_r a)}{k_r a \; I_0(k_r a)}$$

$$= \frac{\gamma_w}{ka} \frac{I_1(ka/\gamma_w)K_0(kb/\gamma_w) + K_1(ka/\gamma_w)I_0(kb/\gamma_w)}{I_0(ka/\gamma_w)K_0(kb/\gamma_w) - K_0(ka/\gamma_w)I_0(kb/\gamma_w)} \tag{6.197}$$

where

$$k_r^2 = \frac{k^2}{\gamma_w^2} + \frac{\omega_p^2}{c^2} \tag{6.198}$$

and the subscript on k_z has been omitted. For very large values of k, this has the same characteristics as the surface wave studied in section 6.2.3.

When $b = \infty$ the r.h.s. of equation (6.197) simplifies considerably to give, for a stationary beam in free space,

$$\left(\frac{\omega_p^2}{\omega^2} - 1\right) \frac{I_1(k_r a)}{k_r a I_0(k_r a)} = \frac{\gamma_w}{ka} \frac{K_1(ka/\gamma_w)}{K_0(ka/\gamma_w)} . \tag{6.199}$$

Curves of ω against k have been plotted by Trivelpiece and Gould. For small values of ω and k the curves rise rapidly with slope corresponding to the velocity of light, but for large k they become asymptotic to a value of $\omega_p/\surd 2$ pected from equation (6.27). The behaviour for large k is not greatly affected by the value of b; for small k and ω, however, there are differences. To study the case of small k we make the 'thin-beam' approximation ka/γ_w and $kb/\gamma_w \ll 1$, so that the Bessel-function expansions listed in section 6.3.3 may be used.

We consider first a stationary beam in free space with $b = \infty$. Equation (6.199) then becomes

$$\left(\frac{\omega_p^2}{\omega^2} - 1\right) = -\frac{\gamma_w^2}{k^2a^2} \frac{1}{\ln(ka/2\gamma_w) - \Gamma} \tag{6.200}$$

where Γ is Euler's constant. For $\omega \ll \omega_p$, this may be written

$$\frac{\omega^2}{\omega_p^2} = -\frac{k^2a^2}{\gamma_w^2}\left(\ln\frac{ka}{2\gamma_w} - \Gamma\right) \tag{6.201}$$

or, alternatively,

$$\beta_w^2\gamma_w^2 = \frac{\omega_p^2a^2}{c^2}\left[\ln\frac{2\beta_w\gamma_w c}{\omega a} + \Gamma\right]. \tag{6.202}$$

As $k \to 0$, $\gamma_w \to \infty$, so that for small values of ω and k the phase velocity is almost equal to that of light. Substituting for β_w and γ_w in terms of ω, c, and k (except in the logarithmic term) enables equation (6.202) to be written

$$c^2k^2 - \omega^2 = \frac{\omega c^2/\omega_p^2 a^2}{\frac{1}{2}\ln(\beta_w^2\gamma_w^2c^2/\omega^2a^2) + \Gamma} \tag{6.203}$$

or, in terms of Budker's parameter (equation (3.2)), with $\gamma = 1$

$$c^2k^2 = \omega^2\left\{1 + \frac{1}{2\nu\ln(\beta_w^2\gamma_w^2\omega_p^2/\omega^2\nu) + \Gamma}\right\} \tag{6.204}$$

Since the argument of the logarithm is large, this may be expanded to give the approximate relation

$$\frac{\omega}{k} = c\left\{1 - \frac{1}{2\nu \ln(\beta_w \gamma_w \omega_p / \omega \nu^{1/2})}\right\}, \quad \frac{\omega}{\omega_p} \ll 1. \tag{6.205}$$

For a beam this can readily be transformed to a frame moving with velocity βc by means of equations (6.41). This leads to a somewhat inelegant relation which we do not write down explicitly. As might be expected, the condition that $\beta_w \approx 1$ for small ω/ω_p remains.

Outside the beam the fields fall off with the characteristic decay distance in free space, $x = \gamma_w ƛ$. As ω tends to zero, x tends to infinity, giving a $1/r$ field dependence. Inside the beam, when γ_w tends to infinity, we see from equation (6.198) that $x \approx c/\omega_p$, the collisionless skin depth. From the relation (equation (3.2)) $c^2/\omega_p^2 = a^2\gamma/4\nu$, it follows that, for paraxial beams in which $\nu/\gamma \ll 1$, we have $x \gg a$. This implies that E_z is essentially uniform across the beam, whereas E_r and H_ϕ, both of which are $\pi/2$ out of phase with E_z, are proportional to r. This type of wave results in both longitudinal and radial motion of the charges.

We now examine the behaviour when b is finite and b/a is not large. For this situation the second factor on the r.h.s. of equation (6.200) is modified, and the equation becomes

$$\frac{\omega_p^2}{\omega^2} - 1 = \frac{2\gamma_w^2}{k^2 a^2} \frac{1}{\ln(b/a)}. \tag{6.206}$$

It is interesting to see what happens in the limit $b/a \to 1$. This may be done by writing $b = a + \Delta a$. Substituting for b and for γ_w this becomes

$$\left(\frac{\omega_p^2}{\omega^2} - 1\right)\left(k^2 - \frac{\omega^2}{c^2}\right) = \frac{2}{a\Delta a}. \tag{6.207}$$

As $\Delta a \to 0$, $\omega \to 0$ for all values of k.

In section 6.4.7 using paraxial and thin-beam approximations we considered waves for the general situation of finite or zero vorticity. In section 6.4.8 a different type of wave was discussed but only for the special case of zero magnetic field and a neutral laminar beam with zero vorticity and zero temperature.

The more general situation of a surface wave with $B_z \neq 0$, so that ω_v is finite, is complex and difficult to analyse rigorously. In the quasi-static approximation the problem is more tractable, at least in the limiting case of a beam filling a conducting tube, $b = a$. We now analyse these waves and find two branches to the dispersion curve. As ω_v tends to zero, one of these branches disappears when $b = a$. When $a \neq b$, however, both remain, and they correspond to the two types of wave discussed in sections 6.4.7 and 6.4.8. The correspondence is, however, somewhat obscured by the different approximations in the two treatments.

In the description of these beams two lines of work come together; we shall use the analysis of Trivelpiece and Gould (1959) of waves in plasma columns, and make the appropriate transformation into translating and rotating coordinates to obtain the corresponding results for magnetically focused beams. This synthesis was first achieved in a definitive paper by Pötzl (1965), where he made the connection between the theory of Trivelpiece and Gould and a group of papers studying waves on magnetic beams. These latter papers are not all correct, and gave rise to a certain amount of controversy finally resolved by Potzl, whose paper should be consulted for details and early references.

Following Trivelpiece and Gould we now find the dispersion relation in a manner analogous to that of section 6.3.2. Because of the possibility of transverse currents, the solution of Maxwell's equation is more complicated; accordingly we simplify the problem by using the quasi-static approximation $\gamma_w = 1$. Since this implies that the velocity of light is effectively infinite the electric fields are derivable from a potential $\underline{E} = -\text{grad}\,\phi$. Regarding the plasma as a dielectric

$$\text{div}\,\underline{D} = \text{div}\,(\underline{\underline{\kappa}}.\underline{E}) = 0. \tag{6.208}$$

Using the tensor dielectric constant (equations (6.15) and (6.16)) the equation for the potential in cylindrical co-ordinates is

$$\kappa_\perp \left\{\frac{1}{r}\frac{\partial}{\partial r}\left(r\frac{\partial\phi}{\partial r}\right) + \frac{1}{r^2}\frac{\partial^2\phi}{\partial\theta^2}\right\} + \kappa_\parallel \frac{\partial^2\phi}{\partial z^2} = 0. \tag{6.209}$$

This has a suitable solution, valid inside the plasma,

$$\phi = A_1 J_n(k_r r)\exp\{-j(m\theta + k_z z)\} \tag{6.210}$$

where

$$k_r^2 = -\frac{k_z^2\kappa_\parallel}{\kappa_\perp} = -k_z^2\left\{\frac{(\omega^2-\omega_p^2)(\omega^2-\omega_c^2)}{\omega^2(\omega^2-\omega_p^2-\omega_c^2)}\right\} \tag{6.211}$$

Between the plasma, of radius a, and the conducting wall, of radius b, a solution which vanishes at $r = b$ is required;

$$\phi = A_2\{I_n(k_z r)K_n(k_z b) - I_n(k_z b)K_n(k_z r)\ \exp\{-j(m\theta + k_z z)\}. \tag{6.212}$$

At the plasma boundary, $r = a$, the normal displacement and tangential electric field must be continuous. This leads to two relations between the constants A_1 and A_2. As before, we particularize to the lowest radial mode with no azimuthal variation, though the general case is not much more complicated. The necessary relation between A_1 and A_2 in this particular case leads to the determinantal equation

$$\kappa_\perp k_r \frac{J_1(k_r a)}{J_0(k_r a)} = \kappa_1 k \frac{I_1(ka)K_0(kb) + I_0(kb)K_1(ka)}{I_0(ka)K_0(kb) - I_0(kb)K_0(ka)} \tag{6.213}$$

where the subscript has been omitted from k_z and κ_1 represents the dielectric constant of the material between the plasma and the wall. This is analogous to equation (6.72) except that it is in quasi-static approximation. In the limit of $\omega_c = \infty$, $\kappa = 1$, equation (6.213) in identical to equation (6.72) with $\gamma_w = 1$.

When the plasma fills the tube, $a = b$ and equation

(6.213) simplifies considerably to give $J_0(k_r a) = 0$. Denoting as in section 6.3.3 the appropriate value of k as $T = 2{\cdot}4/a$, from equation (6.209) the dispersion relation then becomes

$$k_z^2 = T^2\left\{\frac{-\omega^2(\omega^2-\omega_p^2-\omega_c^2)}{(\omega^2-\omega_p^2)(\omega^2-\omega_c^2)}\right\} . \tag{6.214}$$

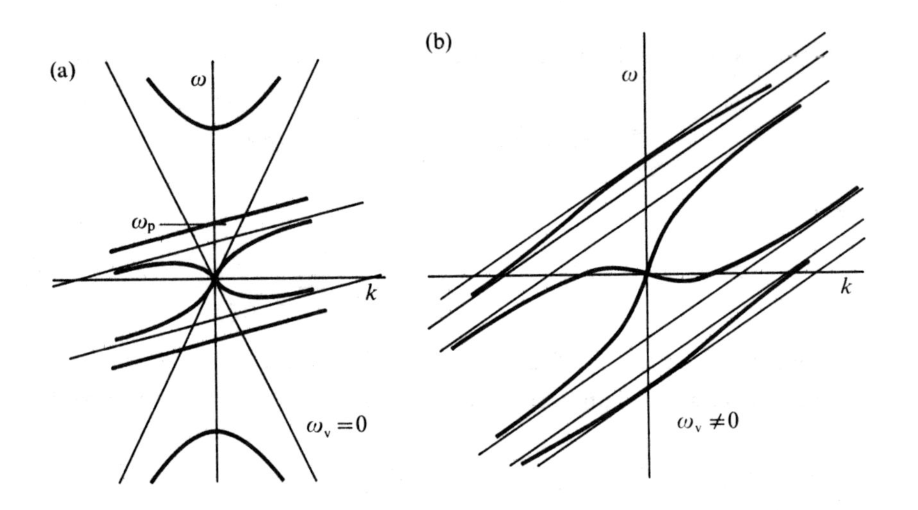

Fig.6.23. Sketches of dispersion relations for some of the waves discussed in sections 6.4.7 - 6.4.9. Diagrams refer to the lowest mode only, on a beam rather than a stationary plasma.
(a) Waves on electron beam, in tube such that $b > a$. The three branches correspond to the wave described in equations (6.191) with $\omega_v = 0$, equation (6.203) modified to apply to a moving beam, and the high-frequency 'waveguide' mode. As $b/a \to 1$, the central branch tends to the line $\omega - \beta c k = 0$.
(b) Waves on a beam with finite vorticity, in quasi-static approximation. This is obtained by replacing ω by $\omega - \beta c k$ in equation (6.213) or, when $a = b$, equation (6.214).

When $a = b$ the points at which the asymptotes and top curve cross the frequency axis are respectively

Fig.6.23. cont.

$\pm\omega_p$, $\pm\omega_v$, and $\pm(\omega_p^2+\omega_v^2)^{1/2}$, or, when $\omega_p > \omega_c$, the same but with ω_p and ω_c reversed. When either frequency is zero, only the straight lines $\omega=\beta ck \pm\omega_p$ and 0 or $\omega = \beta ck \pm\omega_v$ and 0 remain. When $b > a$, on the other hand, the topology remains the same unless $\omega_v = 0$, in which case the central branch does not disappear and the curve becomes similar to that of the first figure, without the outer branches. The slope at the origin becomes infinite, rather than $\omega/k = c$ as in the accurate calculation. If $\omega_p = 0$, then the curve becomes three straight lines at $\omega = \beta ck$ and $\omega = \beta ck \pm\omega_c$. In the second curve ω_v can have either sign, whereas in the curves in Fig.6.1(a), for example, only the positive value of ω_c is taken. This is because in the bounded system the wave is elliptically polarized, and this implies the simultaneous presence of both signs of the frequency.

For an infinite radius $T = 0$, and this yields $\omega = \omega_p$ or $\omega = \omega_c$ as expected in quasi-static approximation. For a plasma of finite radius with $\omega_c < \omega_p$ there are two real values of ω for each value of k, in the ranges $0 < \omega < \omega_p$ and $\omega_c < \omega < (\omega_c^2+\omega_p^2)^{1/2}$. When $\omega_p > \omega_c$, the terms ω_p and ω_c in these two inequalities are interchanged.

For a beam all that is necessary is to transform to moving co-ordinates. In quasi-static approximation this is achieved by replacing ω by $\omega-\beta ck$. (Some idea of the complexity involved in not making the quasi-static approximation may be obtained from the paper of Groendijk, Vlaadingerbroek, and Weimer (1965), equation (22).) To apply these results to a charged beam in a magnetic field it is only necessary to find the appropriate value of $\dot{\theta}$ which, with the same values of ω_p and ω_c, yields a rotating equilibrium beam in the absence of ions, and then find the apparent magnetic field and cyclotron frequency as seen in this frame, ω_{c2}. This value of ω_{c2} is then inserted into the dispersion relation. From equation (2.21), setting $\omega_f = \dot{\theta}$, it is evident that the required relation is $\omega_{c2} = \omega_{c1} - 2\omega_f = \omega_{c1} - 2\dot{\theta}$, so that $\omega_c \to \omega_v$, as found in section 6.4.7.

It is relevant at this point to enquire into the relation between the waves just analysed and those studied in sections 6.4.7 and 6.4.8. In section 6.4.7 the assumption was made that only transverse fields are important. This implies that k is small, and it is seen that the two dispersion relations agree for $k = 0$. When $b \neq a$, the dispersion relation is considerably more complex than equation (6.213); the general form is the same, however, except that the bottom branch (in the frame in which the beam is at rest, $\beta = 0$) becomes asymptotic for large k to a value somewhat higher than ω_v. In particular, when $\omega_v = 0$ the bottom branch has not disappeared. In this case it is equivalent to the surface wave considered in section 6.4.8, and the top branch is equivalent to the wave studied in section 6.4.7. Dispersion relations corresponding to waves discussed in this and the previous two sections are sketched and commented upon in Fig.6.23.

6.4.10. Multipolar and higher-order transverse waves Multipolar transverse perturbations of beams with circular cross-section were described briefly in section 6.4.1 and sketched in Fig.6.17. In the presence of space charge and walls they behave in a generally similar way to the types of wave described so far, though the analysis is somewhat more complicated (Lee, Morton, and Mills 1967). Such waves occur in fast and slow pairs, and show the phenomena of growth, decay, beating, etc., analysed in earlier sections when coupled to each other and to the environment. There is evidently scope for parametric coupling to the accelerator focusing lattice, or between modes, since the contribution of the beam self-fields to focusing is a function of the transverse-oscillation amplitude. Some of these effects have been studied by Smith (1963) and by Sacherer (1968) who has calculated the normal modes of a uniformly focused uniform beam in terms of azimuthal and radial mode numbers. Higher radial mode numbers imply a modulation of beam density with radius, a situation not likely to occur in practice.

One interesting question arises, as to whether in a beam in which the external focusing is small compared with the self-field effects unstable modes of this type can exist in the presence of smooth walls. Analysis by Gluckstern (1970) suggests that for a sharp-edged beam with a K-V distribution, the azimuthally symmetrical mode with radial mode number 2, becomes unstable if $\omega_0 Q/\omega_p < \frac{1}{4}$. This, however, is a somewhat academic result, since the distribution is far from realistic. When the distribution function decreases monotonically with velocity and radius, the beam appears to be stable.

These multipolar and radial waves clearly introduce many degrees of freedom into the behaviour of beams in the presence of focusing lattices and of image fields in non-uniform walls. Many of these detailed problems have been studied in depth in the design of high-energy, high-current accelerators and storage rings.

6.4.11. Transverse beam-plasma interaction in an unbounded plasma

In this section we examine first the unbounded two-stream system in a magnetic field. This is followed in the next section by consideration of a finite beam in an infinite or finite plasma.

Before considering a drifting plasma, we discuss the dispersion relation for a stationary plasma quoted earlier:

$$1 - \frac{c^2k^2}{\omega^2} = \frac{\omega_p^2}{\omega(\omega \pm \omega_c)} \tag{6.20}$$

or

$$1 = \frac{-\beta_w^2\gamma_w^2\omega_p^2}{\omega(\omega \pm \omega_c)} \tag{6.22}$$

where, as before, $\beta_w = \omega/ck$, $\gamma_w^2 = (1-\beta_w^2)^{-1}$. The two signs refer to waves with right- and left-handed polarization. When $\omega = \omega_c$, resonance occurs for the left-handed polarization and the phase velocity becomes infinite. The dispersion relation for various values of ω_p/ω_c has already been sketched in Fig.6.1. In the quasi-static approximation $\gamma_w = 1$ and $\beta_w = 0$; under these circumstances the dispersion relation takes the simple form

$$\omega^2 = \omega_c^2 \tag{6.215}$$

which is independent of ω_p. If ω_p is zero then equation (6.215) is exact. It is of the same form as for filamentary waves, discussed earlier, with $\beta = 0$. In these waves it may readily be verified that the electrons move in circular orbits; as ω approaches ω_c the radii of the circles increase, until $\omega = \omega_c$ where resonance occurs and the orbits become spirals. These are finally limited by relativistic mass increase as in a cyclotron. For a finite plasma, boundary conditions at the edge clearly need careful consideration.

For a plasma drifting with velocity βc parallel to the field the Lorentz transformation (6.41) may be applied, with the result

$$1 - \frac{c^2k^2}{\omega^2} = \frac{1}{\beta_w^2\gamma_w^2} = \frac{\omega_p^2(\omega-\beta ck)}{\omega^2(\omega\pm\omega_c-\beta ck)} . \tag{6.216}$$

During the transformation ω_p remains invariant, since it is defined always in the proper frame; ω_c, on the other hand, is decreased in the moving frame by a factor γ. This may be considered to arise from the increase of transverse mass. The dispersion relation is sketched in Fig.6.24. It represents a generalization of Fig.6.1 for βc, the drift velocity, not equal to zero.

If now we have both a drifting and a stationary plasma, interactions occur where the individual dispersion curves for the drifting and stationary plasmas intersect. Obviously there are many possibilities depending on the relative values of the parameters ω_c, ω_{p1}, ω_{p2}, and β. Some of these are explored in the book by Briggs (1964a), including the presence of ions and a finite temperature in the stationary plasma.

6.4.12. Further discussion of beam-plasma interaction In earlier sections of this chapter many forms of wave on charged-particle beams have been identified, and for some of the simpler of these interaction with walls, either with

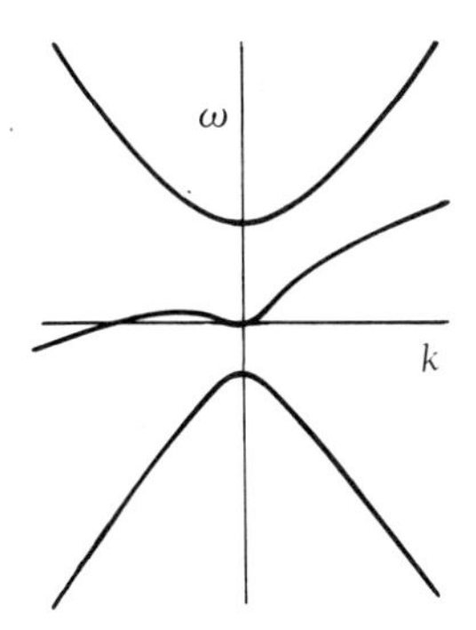

Fig.6.24. Sketch of the dispersion relation for waves in an infinite plasma drifting along the direction of a finite magnetic field. The central branch crosses the k-axis, and the phase velocity becomes zero and reverses at this point. When ω_c/ω_p is small or β is large, however, the curve crosses only at the origin and the phase velocity is everywhere positive.

finite impedance or with propagating characteristics of their own, have also been considered. Many classes of wave can be propagated in plasma; these are now well understood and have been described, for example, by Allis *et al.* (1963). For beams passing through infinite plasma or bounded columns there are accordingly very many possibilities for coupling. All beam-plasma interactions nevertheless represent two-stream instabilities, the basic features of which have already been investigated.

In practice, both beam and plasma have finite transverse dimensions. Typically the plasma is contained in a dielectric (glass) tube, which is surrounded by a metal tube. Down the axis of the tube a beam is propagated. Many such configurations have been investigated in connection with particular experiments. If the plasma is produced by a conventional gas discharge tube, both collision loss and finite temperature effects, and possibly density variations across the column, need to be taken into account. The effects described introduce intolerable complications if all are included together; many studies include some, but not all, of them. The complications introduced do not give rise to any physical features essentially different from those studied already. Finite plasma temperature, as discussed already in section 6.2.7, tends to suppress growing waves or reduce the growth rate.

Collisions between plasma particles introduce dissipation, which damps positive energy waves in the beam and enables negative energy waves to grow. These are broad observations, and it is necessary to analyse individual situations in detail. Interactions of many kinds are described in the text by Briggs (1964a), which contains a good bibliography of work up to 1963. A strong motivation for work in this field was the (largely unfulfilled) hope of application to microwave devices, with the plasma column taking the place of the structure in TWT- and BWO-type devices. These have the apparent advantages of simplicity, with a straightforward uniform medium rather than intricate and at times very small periodic structures. The basic difficulty, however, turns out to be that of producing a suitably behaved quiescent plasma. A review by Crawford (1971), covering the physical principles of the beam-plasma interaction, discusses the application in the microwave field and contains a comprehensive list of references. Another reason for studying these instabilities is in connection with possible schemes for heating plasma in fusion experiments.

In addition to this practical motivation, much of the enquiry was conducted to test general agreement between theory, which is always to some extent idealized, and 'practical' plasmas, which in detail are extremely complicated. Much has been learnt in this respect, contributing to confidence in the application of theory to experimental situations. In general, 'reasonable' agreement is found. A good example of this type of work may be found in the pair of papers by Self (1969), which contain a careful discussion of the factors to be taken into account in comparing experimental measurements with theoretical expectations.

Most work has been done on rectilinear systems; toroidal geometry has, however, been studied in connection with plasma betatrons and suggestions for fusion reactors. Rotating sheet beams in plasma background have also been investigated for this latter application. It is found that, as might be expected, in the negative-mass regime the inductive nature of the plasma medium at frequencies below the plasma frequency tends to stabilize the instability (Lau and Briggs (1971). In the presence of a dissipative medium, however, growth of betatron oscillation occurs, again as expected. This has been analysed by Furth (1965) in a broader plasma context. In linear geometry a similar effect appears as the 'hose instability' studied, for example, by Kino and Gerchberg (1963).

Now that the linear behaviour of beam-plasma systems is fairly well established, it is possible to look in a systematic way at non-linear phenomena, such as three-wave parametric effects and limitations to amplitude growth in unstable systems. This important but complicated field is, however, outside the scope of our present discussion.

6.4.13. Instabilities arising from shear in laminar beams

The instabilities studied so far in this chapter have been associated with beams which are essentially homogeneous over the cross-section. An important class of instabilities exist in inhomogeneous beams when there is shear between the layers of flow which is laminar or approximately so. Several examples of motion of this type have been studied in Chapter 3 without consideration of whether or not they are stable. The motion can be parallel to an external (or self) magnetic field or at an angle to it. (Flow cannot be parallel to the magnetic field in an unneutralized beam.) When the velocity, electric field, and magnetic field are mutually perpendicular, the instability is sometimes known as the diocotron instability.

Crossed-field flow, described in sections 3.4.1 and 4.5.6, is utilized in oscillators and amplifiers related to the magnetron. In these devices the frequency of the instability is determined by the characteristics of the structure placed in the neighbourhood of the beam. Very high efficiencies of conversion from beam energy to high-frequency power can be achieved; under these circumstances the motion is highly non-linear, and a good theoretical description of the behaviour is notoriously difficult to achieve.

The wide range of parameters required to describe adequately the various situations in which shear occurs makes a general discussion difficult. Even a well-defined configuration, such as a beam of uniform charge density in a two-dimensional planar geometry, leads to equations which are difficult to solve; indeed, the general solution of this problem is not known. Different aspects of the motion have, however, been studied over certain ranges of the parameters. The problem is discussed, for example, by Buneman (1961) and Knauer (1966); earlier references may be found in these papers. Cylindrical geometry with inner and outer conducting cylinders has been studied by Levy (1965) for $\omega_c \gg \omega_p$, and appropriate stability criteria found. A lengthy paper by Buneman, Levy, and Linson (1966) contains detailed physical discussion and analysis and draws attention to the role of the parameter $q = \omega_p^2/\omega_c^2$ which can vary from zero to unity. Different types of behaviour are discussed in terms of the parameters q and ka, as qa varies from being small to large compared with unity. The conclusions in these papers are not easy to summarize succinctly. We do not attempt this, but as an illustration, give a simplified analysis of the effect in a thin sheet beam.

In the limiting case of a beam of zero density, the equilibrium configuration is that of rectilinear motion in crossed fields B_x and $E_y = \dot{z}B_x$. It is convenient to work in a frame of reference with respect to which the beam is at rest. In such a frame B_x remains, but E_y is zero. Such a system can support 'rippled-sheet' waves, analogous to filamentary waves in a thread beam, with frequency $\omega = \omega_c$. This can be achieved by setting all the particles into rotation with a successive phase delay. Unlike the waves discussed in section 6.6.2 this wave has longitudinal density modulation in addition to transverse motion. There is a further type of wave which can exist which is trivial in the zero-density limit. The charges can be stationary, but arranged in a sinusoidal curve. The 'wave' has zero frequency, but all values of k are possible.

We now study what happens when the beam has a finite charge, but very small thickness, so that the perturbed state is an oscillating rippled sheet where the lateral perturbation is to be considered large compared with the width of the sheet. This configuration is shown in Fig.6.25. In

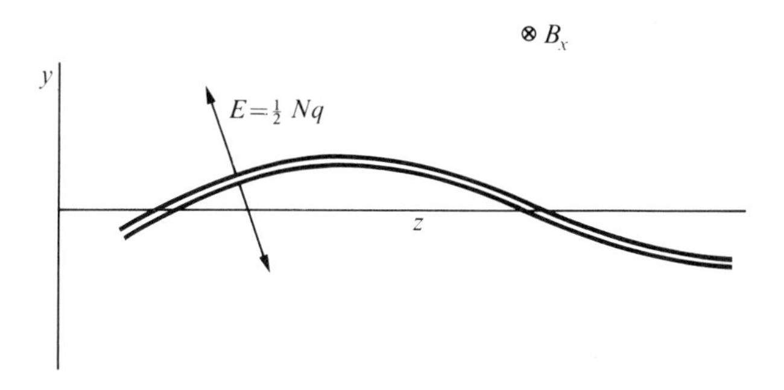

Fig.6.25. Rippled-sheet beam; the magnetic field is perpendicular to the paper and the electric field near the beam is perpendicular to it.

order to keep the analysis simple we assume that the frequency is small, so that only the mode corresponding to the zero-frequency mode in the zero-density beam is studied.

We look first at the steady state of the beam. Coordinates are defined such that the magnetic field is in the x-direction and the xz plane is a plane of symmetry. For a strip beam with charge density Nq per unit area there is a uniform z-velocity gradient in the y-direction with maximum velocity $E_y/B_x = \pm Nq/2\varepsilon_0$ at the edges. In our calculation we consider the perturbation of a charge originally at $y = 0$. Charges at the edge of the beam see a slightly different force from those at the centre and also have a small drift velocity. The electric field experienced by a charge will be taken as the average of the values at the top and bottom of the beam, $E = (E_+ + E_-)$. This is a reasonable approximation when the perturbation, though small, is large compared with the thickness of the beam.

The equations of motion of a charge originally at $y = 0$ may be written

$$\ddot{z}+\omega_c\dot{y} = \frac{qE_z}{m_0}$$
$$\ddot{y}-\omega_c\dot{z} = \frac{qE_y}{m_0} \,. \tag{6.217}$$

In the unperturbed state all these quantities are zero; introducing a harmonic perturbation

$$-\omega^2\tilde{z}+j\omega\omega_c y = \frac{q\tilde{E}_z}{m_0}$$
$$-\omega^2\tilde{y}-j\omega\omega_c z = \frac{q\tilde{E}_z}{m_0} \,. \tag{6.218}$$

Assuming $\omega \ll \omega_c$, for reasons explained earlier, and writing $E = \frac{1}{2}(E_+ + E_-)$, equation (6.218) may be written

$$\tilde{E}_{z+}+\tilde{E}_{z-} = 2j\omega B_x\tilde{y}$$
$$\tilde{E}_{y+}+\tilde{E}_{y-} = -2j\omega B_x\tilde{z} \,. \tag{6.219}$$

To proceed we need the equation of continuity and expressions for the fields in terms of the perturbation. First, the equation of continuity, by analogy with equation (6.4), is

$$jNk\tilde{z} = \tilde{N} \,. \tag{6.220}$$

In this expression, and also when deriving of the fields, we assume that $ky \ll 1$. From the geometry of Fig.6.25, and Poisson's equation the values of E may be seen to be

$$\tilde{E}_{z+}-\tilde{E}_{z-} = \frac{jkNq\tilde{y}}{\varepsilon_0}$$
$$\tilde{E}_{y+}-\tilde{E}_{y-} = \frac{\tilde{N}q}{\varepsilon_0} = \frac{jNqk\tilde{z}}{\varepsilon_0} \tag{6.221}$$

where the extreme right-hand side of the second equation makes use of equation (6.220). Eliminating z and y from equations (6.219) and (6.221)

$$\frac{\tilde{E}_{y+} + \tilde{E}_{y-}}{\tilde{E}_{y+} - \tilde{E}_{y-}} = \frac{2\omega B_x \varepsilon_0}{kNq}$$

$$\frac{\tilde{E}_{z+} + \tilde{E}_{z-}}{\tilde{E}_{z+} - \tilde{E}_{z-}} = \frac{\omega B_x \varepsilon_0}{kNq} \; . \tag{6.222}$$

These fields must now be matched to the boundaries. Initially we consider a system in free space so that the only alternating fields are those of the beam itself. The main contribution to the field at distances greater than y comes from the bunching rather than the rippling; for a sinusoidal perturbation with wavelength $(2\pi k)^{-1} \gg c/\omega$ the field decays away exponentially in the + and - y-directions, and $k_y^2 \approx -k_z^2$. Furthermore,

$$\tilde{E}_{z\pm} = \pm j\tilde{E}_{y\pm} \; . \tag{6.223}$$

Then $E_{z\pm}$ and $E_{y\pm}$ can be eliminated from the four equations (6.221) and (6.222) to yield

$$\frac{\omega}{k} = \pm j\frac{Nq}{2B_x\varepsilon_0} \tag{6.224}$$

The situation is thus inherently unstable, and the amplitude grows at a rate proportional to k. If the beam width is a, the factor in the r.h.s. may be written as $2E_{ya}/B_x$, where E_{ya} is the field at the edge of the unperturbed beam. This is just the relative shear velocity between charges on both sides of the layer. A diagrammatic explanation of how the instability builds up is given in Fig.6.26.

This effect in hollow beams has been described in detail by Kyhl and Webster (1956) and is illustrated by some beautiful photographs. They also point out that a similar effect occurs in hollow electrostatically focused beams of the type discussed in section 3.4.5. Similar behaviour has been reported in intense relativistic beams by Kapetanakos *et al.* (1973).

The method of analysis used here follows that of Buneman (1957). Gould (1957) analysed essentially the same problem for a beam with finite velocity in the z-direction; this is

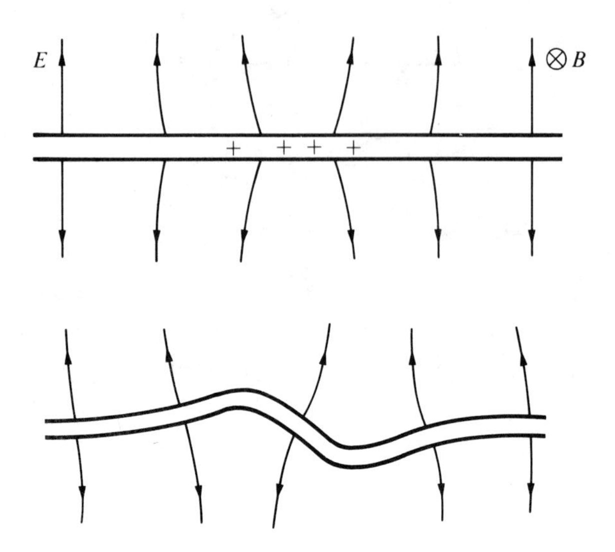

Fig.6.26. Diagram illustrating the growth of a perturbation on a sheet charge. A local excess charge density causes motion in such a direction as to cause buckling. Once this has occurred, the configuration is unstable and the perturbation grows.

appropriate if one is interested in the problem of matching the wave to a structure, as is done in transverse travelling-wave tubes. The method, which is in principle the same as that described in section 6.3.4 for longitudinal waves, will not be considered here.

When the beam has an overall drift velocity large compared with the shear velocity, the dispersion relation can be written

$$\omega = k\left(\beta c \pm \frac{\mathrm{j}Nq}{2B_x\varepsilon_0}\right) \tag{6.225}$$

where βc is the ratio of the external electric field in the y-direction to B_x. This external field, E_{y0}, is large compared with the self-field E_{ya} defined above. In terms of these fields equation (6.225) may be written

$$\omega = \frac{k(E_{y0} \pm \mathrm{j}E_{ya})}{B_x} \tag{6.226}$$

Equation (6.225) is the same form as for the negative-mass instability in an accelerator beam (section 6.3.4). Indeed, there are close relations between these two types of instability, explored by Neil and Heckrotte (1965). As indicated earlier much of the motivation for the study of this instability was directed towards understanding and developing the magnetron and other cross-field devices. A great deal of this work is described in the two-volume compilation edited by Okress (1961). Motivation for some of the later work was the 'Hipac' device mentioned in section 4.5.6, and some experimental results are described by Daugherty, Eninger, and Janes (1969a).

The effect has been analysed in cylindrical geometry in connection with experiments on non-neutral plasmas. Analysis and references are given in the book by Davidson (1974). Instabilities arising from longitudinal shear in cylindrical beams have been analysed by Rome and Briggs (1972). Analogies between crossed-field flow and inviscid shear flow in classical hydrodynamics have been developed in detail by Briggs, Daugherty, and Levy (1970).

6.R.4.7-13. Notes and references Much of the material described in these sections is scattered widely in the literature. Work until 1963 is covered in the text by Briggs (1964a); plasma waveguides are discussed at some length in a short book by Trivelpiece (1967), though this does not contain references later than 1957. More recent studies of waves on un-neutralized beams in a magnetic field are described in a paper by Theiss, Mahaffey, and Trivelpiece (1976), which contains an extensive list of references. The basic features of beam-plasma interactions are covered in some texts on plasma physics, but the very large amount of detailed work is not collected into reviews later than those already quoted in the appropriate sections.

6.5. Dynamic phenomena in bunched beams

6.5.1. Introduction The beams and plasmas considered in this chapter have all been uniform in the direction of propagation, with waves and instabilities representing a small linear perturbation. Many practical devices, on the other hand, operate with beams which are strongly bunched.

Travelling-wave tubes and backward-wave oscillators have been analysed in the linear 'small-signal' regime (section 6.3.9), whereas for high efficiency it is necessary to operate under non-linear 'large-signal' conditions. Although TWT's and BWO's are traditionally analysed in terms of waves, klystrons are normally tackled from the opposite extreme. Space charge is neglected to first approximation, and a purely ballistic theory is constructed. The electron energy on passing through the gap is modulated harmonically in time; this results in a beam structure downstream from the gap which is far from harmonic in time and does not vary periodically along the beam. Indeed, at some places the majority of the current flows in a small fraction of the time cycle. A prominent feature of this type of behaviour is the overtaking of particles slowed down in the gap by later particles which are accelerated there. This phenomenon, made use of in klystrons and bunchers designed to increase the acceptance efficiency of linear accelerators, is very straightforward to analyse when space-charge forces are neglected. When they are important, however, it becomes very complex, even in the approximation in which transverse variation is neglected. Many particular studies have been made using numerical methods, though a great deal of knowledge has come from systematic experiment. Numerous problems of this type are discussed in the book by Rowe (1965) in which both analytical and numerical approaches are used.

Most accelerators operate with bunched beams. In bunches of low density, where space charge and image effects can be neglected, the behaviour is determined by the properties of the betatron and synchrotron oscillations. The azimuthal extent of the bunch cannot exceed that determined by the range of phase stability calculated in section 2.8.2 and illustrated in Fig.2.20. If it is very much less than this, then the oscillations are linear and particles describe ellipses in phase space with the synchrotron oscillation frequency Ω_s (equation 2.171). As calculated in section 2.8.2 they also make radial excursions, so that a simple idealized bunch with no betatron oscillations would consist of an elliptical sheet of particles in the orbit plane. In addition, betatron oscillations are possible in both planes, and it is not difficult to visualize a 'reasonable' distribution of roughly ellipsoidal shape. If space charge is included, it is necessary to ensure that the distribution is self-consistent, and it is interesting to note in this connection that a distribution for an ellipsoidal beam analogous to that of Kapchinskij and Vladimirskij (section 4.3.3) does not exist. Nevertheless, various approximate models have been constructed; a discussion of this problem, and references to earlier work relevant to linear accelerators, is given by Gluckstern in the book edited by Lapostolle and Septier (1970). In such a bunch the presence of space-charge forces weakens the potential well for the particles and slows down the oscillation frequencies, except in the negative-mass regime in cyclic machines, where longitudinal space-charge forces increase the oscillation frequency. The negative-mass instability does not, however, occur; as the bunch becomes denser the velocities associated with the phase oscillations increase, and this decreases the tendency to instability. Image fields in smooth lossless walls shift the oscillation frequencies as in uniform beams; walls with loss or discontinuities produce more complicated effects, some of which are described later.

An analysis of the effect of space charge and smooth perfectly conducting walls on phase motion in a synchrotron has been given by Sørenssen (1975) in a comprehensive review which refers to earlier work on this problem. This paper contains also a detailed analysis of the passage through transition energy, where $1/m^*$ goes through zero from a positive to a negative value, resulting in a 'jump' in the stable phase position.

The discussion so far refers to a single bunch; when many bunches are present, significant interaction is sometimes possible through the effect of fields arising from wall currents when the vacuum chamber is not smooth and perfectly conducting. Indeed, such currents can cause interaction between parts of the same bunch as we see later. Not only can betatron and synchrotron oscillations cause change of the relative positions of particles within the bunch, but the whole bunch can oscillate in position or shape. This arises if the oscillations of individual particles are 'coherent' and the distribution of oscillation phase for a given amplitude among particles in the bunch is not uniform. It is evident that there are many degrees of freedom for coherent oscillations, both longitudinal and transverse; if coupling between bunches is taken into account, the general situation becomes very complicated indeed. A general analysis is hardly possible, but a number of idealized situations have been studied which give a good indication of the kinds of behaviour which are observed. This is a rather specialized problem which we do not treat in depth. Some of the effects will, however, be discussed in physical terms and some simple special situations analysed.

6.5.2. Wake fields In the discussion which follows we introduce a rather different physical viewpoint from that used before. In some ways it is easier to understand in a qualitative way the behaviour of bunched beams, but more difficult to make reliable calculations.

We now discuss the concept of a 'wake', already briefly introduced in section 6.4.3. If a bunch travels along a perfectly conducting tube currents are induced in the walls; for a bunch of length l in a tube of radius b such that $l \ll b/\gamma$ these occupy a length of wall of extent about b/γ in the neighbourhood of the particle; if on the other hand $l \gg b/\gamma$, the corresponding length is about l. In the limit of $\gamma \to \infty$ a bunch which represents a uniform current I_0 flowing for a time Δt induces a current $-I_0$ in the walls over a length $l = c\Delta t$; elsewhere the current is zero.

If now the wall has a finite resistivity, the wall current does not drop sharply to zero after the current has passed. This may be seen physically by regarding the finite current pulse as the sum of two component Heaviside step functions $I_0\mathrm{U}(ct-z)$ and $-I_0\mathrm{U}(ct-z-l)$, U being zero for negative argument and unity elsewhere. We examine first what happens if the wall is perfectly conducting. At $z = 0$ the first current component 'switches on' a wall current at time $t = 0$, and the second switches it off at $t = l/c$. When the wall has finite resistivity, however, the current set up by the first step does not remain constant after the step has passed; the field diffuses into the wall and its amplitude decays in time. The second step does not then cause cancellation, and a slowly decaying current in the wall, with which is associated a finite field within the tube, remains after the pulse has passed. This field has radial and longitudinal components which can act on subsequent pulses, or on the *same* pulse on subsequent revolutions in a cyclic machine. The precise form of this wake in realistic conditions, when for example γ is large but not infinite, is not easy to calculate precisely even in simple cylindrical geometry. The problem has been analysed by Morton, Neil, and Sessler (1966). Various formulae, with their range of validity explained, are derived. Here we merely quote the asymptotic forms, valid a long time after the bunch has passed. For a tube of inner radius b, with conductivity $\sigma\ \Omega^{-1}\mathrm{m}^{-1}$, and an infinitely thick wall, the longitudinal field E_z and the transverse magnetic field B_y arising from a displacement x_0 of the bunch N_0 particles from the axis may be written

$$
\begin{aligned}
E_z &= \frac{N_0 q \mu_0^{1/2}}{4\pi^{3/2}\sigma^{1/2} b t^{3/2}} \\
B_y &= \frac{-N_0 q_0 \mu_0^{1/2} c \pi^{3/2} y}{b^3 \sigma^{1/2} t^{1/2}}
\end{aligned}
\qquad (6.227)
$$

The electric field is of such a sign as to accelerate a following bunch, and the magnetic field to deflect it in the same direction as the original bunch was displaced.

In a circular machine with one bunch the effect of the wake field depends on the Q value. The wake fields add up to a forcing term in the equation of motion which causes a frequency shift together with growth or damping. With a finite number of bunches there are a number of collective modes, some of which are found to be stable and some unstable. This problem will be examined in more detail in the next section.

So far the wake associated with a smooth resistive wall

has been discussed. Another situation commonly encountered is that of a bunch passing a cavity. This might be an accelerating cavity in the machine or an 'accidental' cavity associated with changes in tube cross-section or a side-arm for example. Such a cavity in general has a spectrum of resonant frequencies, some of which are excited by the passing bunch. This excitation can be considered as a localized damped oscillatory wake, with amplitude, field configuration, and decay time dependent on the geometry. For bunches not long compared with the tube diameter, radiation can be produced which is not localized but which travels down the tube behind or ahead of the particle. If the radiation from an array of such obstacles travels with the same phase velocity as the bunch, a continuous energy transfer from the bunch to the field occurs and the effect may be termed Cherenkov radiation. Radiation into cavities or from obstacles produces a reaction force which varies along the length of the bunch, causing change of bunch shape if the bunch is not rigid. Many factors have to be taken into account in trying to estimate what happens in a particular situation.

These problems are very relevant to high-energy electron storage rings, where the charge is concentrated into short intense bunches. The analysis, however, is difficult even in relatively simple geometries, and experimental work is not simple. The behaviour of a point charge passing an array of cavities has been frequently studied theoretically; a numerical treatment which refers to earlier work (mainly analytical) is given by Keil (1972a).

In the next section we comment briefly on the earlier 'rigid-bunch' model in which chromaticity is neglected.

6.5.3. Rigid-bunch approximation As explained in the previous section the fields associated with wall interactions in an accelerator can be considered as the sum of the wake fields laid down by particles which have previously passed by. It is convenient to distinguish between 'slow wakes', by means of which one bunch influences subsequent bunches, and 'fast wakes', where the wake laid down by the leading particles influences those at the rear end. The problem of interaction between a number of hypothetical 'rigid' bunches interacting through the slow wake was analysed by Courant and Sessler (1966). For a single bunch interaction is through its own wake, and the convergent sum of decaying wakes laid down on previous transits is calculated. The motion is stable or not depending on the change of phase of betatron oscillation between transits; if I is the nearest integer to Q it is not difficult to derive the condition

$$\begin{aligned} &I + \tfrac{1}{2} > Q > 1 && \text{stable} \\ &I > Q > I - \tfrac{1}{2} && \text{unstable} \end{aligned} \qquad (6.228)$$

nor to calculate the growth rate making use of equation (6.227). For a train of equal bunches half the possible normal modes are stable and half unstable. For unequal bunch sizes or spacings more complicated rules may be found.

This rigid-bunch picture is, however, of limited interest, since the chromaticity, not considered in the single-bunch picture and effectively assumed to be zero, plays an essential part.

6.5.4. General treatment of bunched-beam instabilities In the previous section the behaviour of idealized 'rigid' bunches was discussed. When finite chromaticity is allowed for, internal motions are found which account for an important class of instability which is troublesome in storage rings. A precise theory of the behaviour of bunches, taking into account their internal structure and mutual interaction through wake fields, involves the solution of self-consistency problems of considerable complexity. Nevertheless, the introduction of well-chosen simplifying assumptions enables a useful theory to be constructed.

We now discuss some of the main features of the problem in a qualitative way†. We note first that, because of the finite chromaticity ξ, the betatron oscillation frequency of a particle depends on its energy and therefore varies as the

†The remainder of this section is closely based on a manuscript kindly supplied by F. Sacherer.

particle undergoes a synchrotron oscillation. As a result the betatron oscillation phase of the particle advances or lags behind that of the synchronous particle which does not undergo energy oscillations. If this relative phase is zero when the particle is at the head of the bunch, the maximum value χ is obtained at the tail and this then returns to zero as the particle returns to the head.

We first consider many particles with the same synchrotron oscillation amplitude arranged such that their betatron oscillation phases at time $t = 0$ vary linearly from zero at the head of the bunch to χ at the tail. This phase variation along the bunch will then be maintained with time. For example, as a particle leaves the tail, another always arrives with the correct accumulated phase shift χ relative to the head of the bunch. Thus the head and tail oscillate with the phase difference χ which is given by $\xi Q\omega_0\tau_b/\eta$, where ω_0 is the orbital revolution frequency, η is defined in equation (2.145), and τ_b is the bunch length in seconds. For electron machines χ is typically less than a radian, while for proton machines it can be much larger, and approaches infinity at transition energy ($\eta = 0$).

The fact that different parts of a bunch oscillate with different phases provides a new instability mechanism called the head-tail effect. For a rigid bunch, for which $\xi = 0$, the wake field laid down by the head of the bunch cannot drive the tail on the same revolution, since both wake field and bunch oscillate with the same phase and instability requires an out-of-phase component. Thus if the wake decays rapidly, before one revolution is completed, no instability results. However, for finite chromaticity the required phase shift is present. Instability occurs when the tail lags behind the head (χ negative). This occurs above transition energy for the normal machine chromaticity (negative) or below transition for positive chromaticity. (This contrasts with the unbunched beam behaviour where chromaticity helps prevent instability by increasing the Landau damping.)

This type of instability driven by short-range wakes was first discovered and analysed by Pellegrini (1969) and

Sands (1969). It is characterized by a dependence on chromaticity and lack of dependence on Q-value. Of course this non-rigid motion can also be driven by long-range wakes, so both ξ-dependence and Q-dependence can be present at the same time. A more recent and complete analysis is given by Sacherer (1972, 1974).

Other types of head-tail motion are also possible. The one considered so far, which reduces to rigid motion in the limit $\xi = 0$, is called 'mode zero'. For 'mode one' and $\xi = 0$, the head and tail of the bunch move in opposite directions and the centre is stationary; in other words, it represents a standing-wave pattern with one mode. Similarly, the higher modes are standing-wave patterns, approximately sine waves, with the number of nodes equal to the mode number. For these standing-wave patterns any two parts of the bunch move either in phase or 180° out of phase, and therefore no head-tail instability results. For finite chromaticity an additional phase shift occurs between any two parts of the bunch, but the amplitude of oscillation along the bunch is still given by the standing-wave pattern so the number of nodes remains the same. Experimental measurements of these patterns, taken from the signals observed on pick-up electrodes, are shown in Fig.6.27 (frontispiece).

Given the mode patterns and the coupling impedance, in this case the transverse impedance $Z_{\perp}(\omega)$ defined in equation (6.173), of the surrounding structures (vacuum chamber, cavities, bellows, etc.), it is straightforward to compute the growth or damping rate for the various modes (Sacherer 1972, 1974). Considerable effort is devoted to preventing these instabilities. For example, in electron storage rings it is essential to keep the chromaticity zero or slightly positive to prevent the head-tail effect. Instabilities due to long-range wakes are controlled by proper choice of Q-value, octupoles to increase Landau damping, radiofrequency quadrupoles to decouple the bunches, and active feedback systems.

Longitudinal instabilities can be treated in a similar way. There can be rigid motion of the bunches, 'breathing' motion in which the bunches expand and contract, and higher modes. In a normalized phase plane, where the particle trajectories are circular, a distribution with no dependence on phase is stationary. If a stationary distribution is displaced from the origin, dipole or rigid-bunch motion results, while, if instead the distribution is elongated or stretched, quadrupole or breathing motion occurs. In general, mode m has an m-fold symmetry of rotation in the phase plane.

If we consider the line density for the different modes, a familiar pattern emerges. Rigid motion is the sum of the stationary distribution $N_0(s)$ plus an oscillating perturbed density $N_1(s)$ with an excess charge at one end and a lack of charge at the opposite end, in other words, a sine-wave type pattern with one node at the bunch centre. Similarly, the perturbed line density $N_2(s)$ for the quadrupole mode has an excess charge density at both ends and a deficiency in the centre, or *vice versa*, with two nodes. The higher modes have more nodes and higher frequencies in their spectra. A resonant element in the vacuum chamber will excite primarily that mode whose frequency spectrum overlaps the resonator frequency. The growth rate can be found by summing the product of the mode spectrum with the appropriate values of the longitudinal coupling impedance $Z_{\|}(\omega)$. Some examples are given by Sacherer (1973), together with references to earlier work. Higher-order effects of the same class, dependent on 'longitudinal chromaticity', and asymmetries in the vacuum chamber are also possible, but have not yet been identified as being of practical importance.

At the time of writing there is still uncertainty about anomalous bunch-lengthening effects in electron storage rings, though it is probable that the physical nature of the instabilities of practical importance is now understood.

6.R.5. Notes and references The instabilities of bunched beams, analysed in this section, are of interest in high-current cyclic accelerators and in electron storage rings. References to the most important papers have already been given in the text.

It is convenient at this point to review the subject

of references to work on all types of instability in accelerators. A number of references have been given at appropriate points in the text, but review papers have not been discussed. Unfortunately many important papers appear as internal reports of the larger accelerator laboratories and are therefore not readily accessible.

Historically, work in this field began in the mid 1950's; a notable early analysis of the stability of beams is contained in the papers of Budker (1956). Work along similar lines was carried out both in the U.S.S.R. and U.S.A, and references to much of this early work in both countries may be found in reviews by Sessler (1966) and Kolomenskij and Lebedev (1970). In the later stages important contributions were made at CERN, Geneva, and in other European laboratories. The subject was again reviewed in 1972 by Sessler (1972), and many papers on instabilities, both theoretical and experimental, appear in accelerator conferences after this date.

6.6. Concluding remarks

In this chapter the properties of waves on beams have been related to the broader and more general field of waves in plasma. Although many of the problems relate to rather specific applications, some perhaps only of historic interest, they illustrate the main classes of linear behaviour of uniform and bunched beams in rectilinear and orbital configurations. The aim, as elsewhere in the book, has been to present the material in such a way that the essential physical behaviour is illustrated, in the full knowledge that this is not the most useful form for assisting in the solution of particular problems. Attention has centred on the explicit descriptions of various types of wave, both in terms of the form of perturbations imposed on the steady-state beam, and a discussion of the dispersion relations.

In practical applications the question of temporal or spatial growth is all important. In an isolated beam this can occur as the result of the negative-mass instability. It also can occur when a beam passes through or near a resistive environment, whatever the sign of the effective mass.

Further, for positive and negative mass, instability is induced respectively by an inductive or capacitive environment.

In the presence of two beams, a beam and a plasma, or a beam and a nearby propagating structure, coupling may occur, which again can give rise to growth. Simplifications which arise when this coupling is weak have been described. Such growth is required in microwave tubes, but must be discouraged in accelerators. This may be done either by feedback, a topic not discussed here, or by taking advantage of the Landau damping which occurs when the particles in the beam exhibit a spread in energy. These phenomena are well understood, though many of the practical situations in which they occur are complicated in detail, especially if the coupling is between more than two waves and is not weak.

Beyond the range of linear behaviour essentially new features appear, such as parametric coupling, limits to the growth of unstable configurations, and relaxation-type oscillations. The phenomena to be encompassed are too varied, too special, and often too scantily understood, however, for a compact all-embracing treatment to be given.

APPENDIX 1

Fields seen by a particle in a rotating frame in a uniform magnetic field

In a uniform magnetic field B_1 the forces on a moving particle may be written

$$F_r = m_0(\ddot{r}-r\dot{\theta}_1^2) = qB_1r\dot{\theta}_1 \tag{A1.1}$$

$$F_\theta = m_0(r\ddot{\theta}_1+2\dot{r}\dot{\theta}_1) = -qB_1\dot{r} \tag{A1.2}$$

where r and θ refer to cylindrical co-ordinates orthogonal to the magnetic field, and non-relativistic velocities in the r and θ directions are assumed. (If the particle has a relativistic velocity in the z-direction, as in a paraxial beam, m_0 can be replaced by γm_0.) The subscript 1 denotes that the angles are measured in the stationary frame of reference. In a frame rotating with angular velocity ω_f, for which the subscript 2 is used,

$$\theta_1 = \theta_2+\omega_f t. \tag{A1.3}$$

Substituting in equations (A1.1) and (A1.2) and writing Ω_L for $-qB_1/2m_0$ yields for the forces in the new frame of reference

$$F_{r2} = m_0(\ddot{r}-2r\dot{\theta}_2^2) = -m_0r\omega_f(2\Omega_L-\omega_f)-2m_0r\dot{\theta}_2(\Omega_L-\omega_f)\ . \tag{A1.4}$$

$$F_{\theta 2} = m_0(r\ddot{\theta}_2+2\dot{r}\dot{\theta}_2) = 2m_0\dot{r}(\Omega_L-\omega_f)\ . \tag{A1.5}$$

The force F_{r2} contains one component which is independent of the particle velocity in the rotating frame and one component perpendicular to the velocity. These may be identified with a radial electric field E_2 and axial magnetic field B_2 to give

$$B_2 = -\frac{2m_0}{q}(\Omega_L-\omega_f) = \left(1-\frac{\omega_f}{\Omega_L}\right)B_1 \tag{A1.6}$$

$$E_2 = \frac{m_0}{q}r\omega_f(-2\Omega_L+\omega_f) = -r\omega_f\left(1-\frac{\omega_f}{2\Omega_L}\right)B_1\ . \tag{A1.7}$$

The variation of B with ω_f is linear and that of E is parabolic. Particular cases of interest are

$$\omega_f = \Omega_L, \quad B_2 = 0, \quad E_2 = -\tfrac{1}{2}r\Omega_L B_1 \tag{A1.8}$$

and

$$\omega_f = \omega_c = 2\Omega_L, \quad B_2 = -B_1, \quad E_2 = 0. \tag{A1.9}$$

APPENDIX 2

Derivation of the paraxial ray equation from the principle of least action

The principle of least action states that for a system in which the Hamiltonian H is conserved

$$\delta \int_{x_1}^{x_2} \underline{\mathrm{P}}.\mathrm{d}\underline{\mathrm{l}} = 0 \qquad \text{(A2.1)}$$

where x_1 and x_2 are the co-ordinates of the end points of the path, $\underline{\mathrm{P}}$ is the canonical momentum, and δ represents a variation subject to the following constraints. The system is conservative, and all paths are physically possible, each with a different value of H; the end points x_1 and x_2 are fixed, but the *time* taken to traverse the path varies.

For a cylindrically symmetrical system, with electric and magnetic fields, there is only one component A_θ of the vector potential. The appropriate variational equation, with z chosen as the independent variable may be written

$$\delta \int_{x_1}^{x_2} (\underline{\beta}\gamma m_0 c + q\underline{\mathrm{A}}).\,\mathrm{d}\underline{\mathrm{l}} =$$

$$m_0 c\delta \int_{z_1}^{z_2} \{\beta\gamma(r'^2+r^2\theta'^2+1)^{1/2} + \frac{qA_\theta}{m_0 c} r\theta'\}\mathrm{d}z = 0 \qquad \text{(A2.2)}$$

where primes denote $\mathrm{d}/\mathrm{d}z$.

To proceed, we form the Euler-Lagrange equations corresponding to equation (A2.2)

$$\frac{\mathrm{d}}{\mathrm{d}z}\frac{\partial}{\partial r'}(\underline{\beta}\gamma m_0 c + q\underline{\mathrm{A}}) - \frac{\partial}{\partial r}(\underline{\beta}\gamma m_0 c + q\underline{\mathrm{A}}) = 0 \qquad \text{(A2.3)}$$

together with a similar equation with θ in place of r. The θ equation can be integrated immediately to give

$$\frac{\beta\gamma r^2\theta' m_0 c}{(r'^2+r\theta'^2+1)^{1/2}} + qA_\theta r = \text{const.} \qquad \text{(A2.4)}$$

Writing the equation in terms of $\dot{\theta}$ it is readily verified that this is Busch's theorem, equation (2.5) or (2.6), and that the constant is P_θ.

The first equation is rather more complicated:

$$\frac{\partial}{\partial z}\frac{r'\beta\gamma}{(r'^2+r^2\theta'^2+1)^{1/2}} - \frac{(r^2+r^2\theta'^2+1)^{1/2}}{2\beta\gamma}\frac{\partial}{\partial r}(\beta\gamma)$$

$$- \frac{\beta\gamma r\theta'^2}{(r'^2+r^2\theta'^2+1)^{1/2}} + \frac{q}{m_0c}\theta'\frac{\partial}{\partial r}(rA_\theta) = 0 \quad \text{(A2.5)}$$

Then θ' can be eliminated from equation (A2.4), and the resulting rather complicated expression may be simplified by writing

$$\eta = \frac{1}{\beta\gamma m_0c}\left(\frac{P_\theta}{r} - qA_\theta\right) = \frac{r\dot{\theta}}{\beta c} \quad \text{(A2.6)}$$

$$\lambda = \beta^2\gamma^2(1-\eta^2) \quad \text{(A2.7)}$$

to

$$r'' + \frac{1+r'^2}{\lambda^2}\left\{(1+r'^2)\frac{\partial\lambda}{\partial r} - r'\frac{\partial\lambda}{\partial z}\right\} = 0. \quad \text{(A2.8)}$$

This is an exact equation for the radial position of a particle, given A_θ and ϕ as a function of r and z; ϕ determines β and γ, which together with A_θ and the initial condition P_θ determine η and hence λ. The angular position θ is determined from equation (A2.4).

The paraxial equation (2.14) may readily be obtained from equation (A2.8) by neglecting r', taking $A_\theta \propto r$ (so that B and hence Ω_L is independent of r), and expressing λ in terms of β, γ, and Ω_L. The derivation here follows that given by Panofsky and Phillips (1962).

APPENDIX 3

Paraxial equation for a strip beam with curvilinear axes

We derive the paraxial equation for a strip curvilinear beam uniform in the y-direction, in the presence of fields B_y, E_x, and E_s. The method follows that of Kirstein *et al.* (1967), except that the treatment is relativistic and the co-ordinate notation is changed. This is shown in Fig.2.16; O is a point on a ray on the axis surface for which y is constant, and O' is distant x from O. Point P is separated by ds' and dx respectively in the s and x directions from O'. The centre of curvature of the ray axis is at C, so that CO = R.

The co-ordinate system of Kirstein *et al.* may be converted to the system used here by the transformations $r \rightarrow -x$, $z \rightarrow -y$, $s \rightarrow s$, $R \rightarrow R$. The metric for this set of co-ordinates is

$$dl^2 \quad = \quad dx^2 + dy^2 + (1+\tfrac{x}{R})^2 ds^2 \qquad \text{(A3.1)}$$

where $R(s)$ is the radius of curvature of the ray axis, taken as positive when the x-co-ordinate of the centre of curvature is negative.

The electric field is expressed in terms of its integral, the potential ϕ, and the following expansion about the ray axis is needed:

$$\phi \quad = \quad \phi^{(0)}(s) + x\phi^{(1)}(s) + \tfrac{1}{2}x^2\phi^{(2)}(s). \qquad \text{(A3.2)}$$

The equation of motion of a particle may be found from the Lagrangian. In order to construct this the product $\underline{\beta}.\underline{A}$ is required; it is not difficult to show that

$$\underline{\dot{r}}.\underline{A} \quad = \quad \tfrac{1}{2}B_y\{\dot{x}s - x\dot{s}(1+x/R)\} \qquad \text{(A3.3)}$$

so that the Lagrangian (equation 4.1) may be written

$$L = -m_0 c\{c^2 - \dot{s}^2(1+x/R)^2 - \dot{x}^2\}^{1/2} - q\phi$$
$$-\tfrac{1}{2}qB_y\{\dot{x}s - x\dot{s}(1+x/R)\}\ . \tag{A3.4}$$

The equations of motion follow directly by substitution into equation (4.2).

In the x-direction, with the usual paraxial assumptions, the equation of motion is

$$\gamma\ddot{x} + \dot{\gamma}\dot{x} - \frac{\gamma\dot{s}^2(1+x/R)}{R} + \frac{qB_y}{2m_0}\dot{s}(1+\frac{x}{R}) + \frac{q}{m_0}\frac{\partial\phi}{\partial x} = 0. \tag{A3.5}$$

Setting the sum of the terms of zero order in x to zero gives the condition that the axis surface contains a possible trajectory. Making use of equation (A3.2) to expand ϕ gives

$$-\frac{\gamma m_0\beta^2c^2}{R} - qB_y\beta c + q\phi^{(1)} = 0. \tag{A3.6}$$

In order to find the paraxial equation it is necessary to know the energy of a particle off the ray axis. The energy equation is

$$1 - \frac{q\phi}{m_0c^2} = \gamma = \frac{c}{\{c^2-(1+x/R)^2\dot{s}^2-\dot{x}^2\}^{1/2}}\ . \tag{A3.7}$$

Expanding about a trajectory on the axis surface, again using equation (A3.2), and also the relation

$$\dot{s} = \beta c + \dot{s}^{(1)}x \tag{A3.8}$$

yields

$$\dot{s}^{(1)}\beta c - \frac{\beta^2c^2}{R} - \frac{q\phi^{(1)}}{\gamma^3m_0} = 0. \tag{A3.9}$$

The paraxial equation, with time as independent variable, can be obtained by removing the zero-order terms from the equation of motion (A3.5) using equation (A3.9) to eliminate $s^{(1)}$.

It is convenient also to express the potentials in terms of derivatives along the ray axis. From the relation $\nabla^2\phi = 0$ in curvilinear co-ordinates

$$\phi^{(2)} = -\phi^{(1)}/R - \phi^{(0)\prime\prime}. \tag{A3.10}$$

Finally, from equation (A3.9) using equation (A3.2) to expand the potential, equation (A3.10) to eliminate $\phi^{(2)}$, equation (A3.9) to eliminate s, and the relation $\gamma = \gamma^{(0)} + q\phi^{(1)}x/m_0c^2$ to expand γ, we obtain

$$\ddot{x} + \frac{\dot{\gamma}}{\gamma}\dot{x} + \left[\frac{2\beta^2c^2}{R^2} - \frac{2q\beta cB_y}{\gamma m_0 R} + \frac{q^2}{\gamma^2 m_0^2}(B_y^2 + \phi^{(1)2}) - \frac{q}{\gamma m_0}\left\{\phi'' + \frac{\beta^2\phi^{(1)}}{R}\right\} - \frac{2q^2B_y\phi^{(1)}\beta}{\gamma^2 m_0^2 c}\right]x = 0 \tag{A3.11}$$

where the superscript (0) has been dropped.

This equation may be expressed in terms of s as the independent variable by using equation (2.24). In order to bring it into the same style as, for example, equations (2.13) and (2.94), the potentials are expressed in terms of the fields (equation 2.15) and ϕ is related to γ. In this form the equation is

$$\beta^2x'' + \frac{\gamma'}{\gamma}x' + \left\{\frac{2\beta^2}{R^2} - \frac{2q}{\gamma m_0c^2R}(\beta cB_y - \beta^2E_x) + \frac{\gamma''}{\gamma} + \frac{q^2}{\gamma^2m_0^2c^4}(c^2B_y^2 + \beta^2E_x^2 - 2\beta cE_xB_y)\right\}x = 0. \tag{A3.12}$$

This may be re-arranged to give equation (2.129).

APPENDIX 4

The effect of non-linearities on the period of a harmonic oscillator

If the restoring force for betatron oscillations is not proportional to displacement, then the displacement is not sinusoidal and the oscillation period is a function of amplitude. We consider here the effect on the amplitude and period of two simple departures from linearity. The perturbation is assumed to be of such form that the equation of motion is

$$\ddot{x} + Q^2\omega_0^2\left\{1 + \left(\frac{x}{x_0}\right)^n\right\}x = 0. \tag{A4.1}$$

The displacement x will be considered small compared with x_0 so that both amplitude and frequency are only slightly perturbed. For $n = 1$ the perturbation acts differently on the positive and negative cycles of oscillation, whereas for $n = 2$ the disturbance is symmetrical. To calculate the effect we solve equation (A4.1). Let the amplitude of oscillation, when $1/x_0 = 0$, be a; then the first integral of equation (A4.1) is readily found to be

$$\dot{x}^2 = Q^2\omega_0^2\{a^2 - x^2 + \frac{2}{(n+2)x_0^n}(a^{n+2} - x^{n+2})\}. \tag{A4.2}$$

When $x = 0$,

$$\dot{x} = Q\omega_0\left\{a^2 + \frac{2a^{n+2}}{(n+2)x_0^n}\right\}^{1/2}$$

$$\approx Q\omega_0 a\left(1 + \frac{a^n}{(n+2)x_0^n}\right). \tag{A4.3}$$

Compared with simple harmonic motion, the fractional increase of velocity $\Delta\dot{x}/\dot{x}$ at $x = 0$ to produce the *same* amplitude is $a^n/(n+2)x_0^n$. For the same value of $\dot{x}$ at $x = 0$, therefore,

$$\frac{\Delta a}{a} = -\frac{1}{n+2}\left(\frac{a}{x_0}\right). \tag{A4.4}$$

For $n = 1$, $\Delta a/a$ has the opposite sign to x_0 for positive half-cycles and the same sign for negative half-cycles. For $n = 2$, however, the amplitude of positive and negative half-cycles is symmetrically reduced as compared with simple harmonic motion. To calculate the change in period, we expand equation (A4.2)

$$\dot{x} \approx Q\omega_0 (a^2-x^2)^{1/2} \left(1 - \frac{a^{n+2} - x^{n+2}}{(n+2)(a^2-x^2)x_0^n}\right)^{-1} \qquad \text{(A4.5)}$$

from which, by integration,

$$Q\omega_0 t = \int \frac{1}{(a^2-x^2)^{1/2}} \left(1 - \frac{a^{n+2} - x^{n+2}}{(n+2)(a^2-x^2)x_0^n}\right) \mathrm{d}x \ . \qquad \text{(A4.6)}$$

Compared with a simple harmonic oscillation, the change in period is

$$Q\omega_0 \Delta t = -\int \frac{a^{n+2} - x^{n+2}}{(n+2)(a^2-x^2)^{3/2}\, x_0^n} \mathrm{d}t \ . \qquad \text{(A4.7)}$$

When n is odd, this alternates in sign between positive and negative half-cycles; for $n = 1$ the total Δt over a complete cycle is, to first order, zero. When $n = 2$, on the other hand, Δt is negative for both half-cycles. Evauation of the integral is straightforward but slightly tedious, yielding the final result

$$\frac{\Delta t}{t} = -\frac{\Delta Q}{Q} = -\frac{3}{8}\frac{a^2}{x_0^2} \ . \qquad \text{(A4.8)}$$

APPENDIX 5

Coupled-mode theory

Coupled-mode theory, which is essentially of the same form as perturbation theory used in quantum mechanics, was introduced in the study of the coupling of beams and structures by Pierce (1954b). It is set out in detail in the book by Louisell (1960), and more briefly by Johnson (1965).

If a number of linear systems are coupled together, then it is always possible to find a set of normal modes to describe the overall system which have co-ordinates consisting of linear combinations of the co-ordinates describing the component systems. If the coupling is weak, however, it is often more convenient to describe the behaviour in terms of continuous slow interchange of energy between the various component systems. The term 'slow' in this context implies that the amount of energy exchanged during one oscillation period is small compared with the total energy in the system. Although the theory may readily be formulated for an arbitrary number of coupled systems, we restrict consideration to the situation of present interest, namely two systems.

We are interested in the coupling between waves, which involves the variable k as well as ω. Let k_1 and k_2 be the propagation constants of the two systems in the absence of coupling. In the presence of coupling, let c_{12} and c_{21} be the coupling constants from mode 1 to mode 2 and *vice versa*. The coupled equations are then

$$\begin{aligned}\frac{\partial a_1}{\partial z} &= jk_1a_1+c_{12}a_2\\ \frac{\partial a_2}{\partial z} &= jk_2a_2+c_{21}a_1\end{aligned} \tag{A5.1}$$

together with similar relations for complex conjugates. We now look for a relation between c_{12} and c_{21}. This may be found by considering the steady-state power flow in the system, which in the absence of dissipation must be independent of z and t. The amplitudes are normalized in such a way that the power is given by

$$W = a_1 a_1^* \pm a_2 a_2^* \tag{A5.2}$$

where the lower sign implies that the second system carries negative energy. This can arise from a positive-energy wave with negative group velocity or *vice versa*. If the modes are coupled, the power flow in either of them changes with z,

$$\frac{\partial}{dz}(aa^*) = a_1 \frac{\partial a_1^*}{\partial z} + a_1^* \frac{\partial a_1}{\partial z} \,. \tag{A5.3}$$

Making use of equation (A5.1) and its complex conjugate gives for both modes

$$\frac{\partial W}{dz} = 0 = \{(c_{12} \pm c_{21}^*) a_1^* a_2 + (c_{12}^* \pm c_{21}) a_1 a_2^*\} . \tag{A5.4}$$

The phases of a_1 and a_2 are independent, so that $a_1^* a_2 \neq a_1 a_2^*$. Equation (A5.4) can therefore only be true if

$$c_{12} = \pm c_{21}^* \,. \tag{A5.5}$$

This is used in equation (6.117), where K is written for c_{12}.

APPENDIX 6

The energy associated with transverse waves on a filamentary beam

By considering a specific mechanism for the excitation of transverse waves on a filamentary beam, we are able to calculate the energy and show that its sign depends on whether the wave is fast or slow. For simplicity we assume a non-relativistic beam of N particles per unit length moving with velocity v along the axis of a focusing system. Coupling between particles is neglected, and the transverse oscillation frequency is ω.

Imagine now a short parallel-plate condenser moving along the beam with velocity V, as shown in Fig.A6.1. A wave

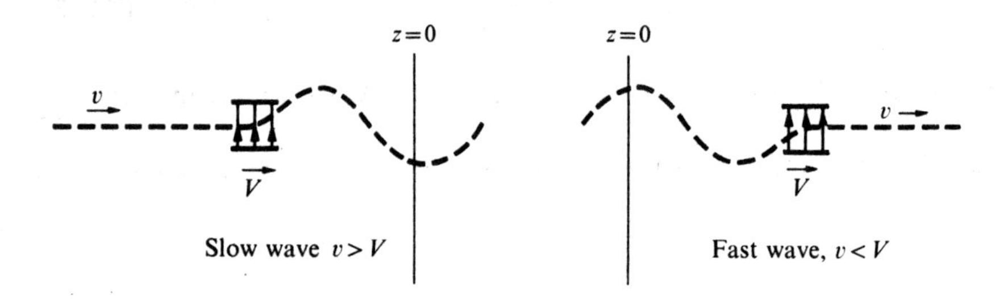

Fig.A6.1. The setting up of slow and fast transverse waves on a beam by means of a moving condenser.

is generated in front of or behind the condenser, according to whether v is greater or less than V. The phase velocity of the wave is equal to V, so that the two situations correspond to fast or slow waves. In order to calculate the wave energy we examine the situation in a frame in which the condenser is at rest. In this frame the energy of the beam particles is conserved, and the transverse oscillation energy and first-order velocity associated with it come from a second-

$$\dot{y}_2 = (v-V)\sin\theta_2$$
$$\dot{z}_2 = V+(v-V)\cos\theta_2 . \qquad \text{(A6.1)}$$

The change of energy per particle in the stationary frame is

$$\Delta U = \tfrac{1}{2}m(v^2-\dot{z}_2^2-\dot{y}_2^2)$$
$$= 2V(1-\cos\theta_2)(V-v) . \qquad \text{(A6.2)}$$

If now we assume that θ_c is small, then

$$\Delta U = \tfrac{1}{2}mV^2\theta_c^2\left(1-\frac{v}{V}\right) . \qquad \text{(A6.3)}$$

This is positive or negative according to whether V is greater or less than v.

For a condenser of given geometry, with fixed charge, θ_c is inversely proportional to $|V-v|$. Setting $\theta_c|V-v| = k$, equation (A6.3) becomes

$$\Delta U = \frac{mVk^2}{2(V-v)} . \qquad \text{(A6.4)}$$

If now the beam velocity is considered fixed and the condenser velocity is varied, $\Delta U \to \infty$ as $v \to V$. In this limit, the linearization breaks down since, for fixed k, θ increases to such an extent that it is no longer approximately equal to $\sin\theta$. When $v = V$ resonant interaction, of the type encountered in Landau damping, occurs.

It is interesting to note that for fast waves a stationary observer counts the crests and troughs in the order in which they were emitted, but for a slow wave they are seen in the reverse order. Thus, in a sense, slow waves have a negative frequency. The relevance of this observation in quantum systems, where $E = \hbar\omega$, is discussed by Musha (1964).

REFERENCES

Alfvén, H. (1939). On the motion of cosmic rays in interstellar space. *Phys. Rev.* 55, 425.

Allis, W.P., Buchsbaum, S.J., and Bers, A. (1963). *Waves in anisotropic plasmas.* MIT Press, Cambridge, Mass.

Amboss, K. (1969). The analysis of dense electron beams. *Adv. Electronics electron. Phys.* 26, 1.

Andrews, M.L., Davitian, H., Fleischmann, H.H., Kusse, B., Kribel, K.E., and Nation, J.A. (1971). Generation of Astron-type *E* layers using very high current electron beams. *Phys.Rev.Lett.* 27, 1428.

Ash, E.A. (1964). Limiting-current densities in the presence of a magnetic field. *J. appl. Phys.* 35, 298.

——— (1968). Electron Optics. In *Handbook of vacuum physics* (ed. A.H. Beck), vol.2., p.399. Pergamon Press, Oxford.

Ashkin, A. (1958). Dynamics of electron beams from magnetically shielded guns. *J. appl. Phys.* 29, 1594.

Atkinson, H.H. (1963). A new type of high vacuum gauge. *Tenth Nat. Vacuum Symp.*, p.213. American Vacuum Society, Boston, Mass.

Baker, F.A., and Hasted, J.B. (1966). Electron collision studies with trapped positive ions. *Phil. Trans. R. Soc.* 261, 33.

Bakish, R. (1962). *Introduction to electron beam technology.* John Wiley, New York.

Banford, A.P. (1966). *The transport of charged particle beams.* Spon, London.

Barnes, C.W. (1964). Conservative coupling between modes of propagation - a tabular summary. *Proc. inst. elect. electron. Engrs* 52, 64, and 52, 295. (correction).

Beck, A.H.W. (1958). *Space-charge waves and slow electromagnetic waves.* Pergamon Press, Oxford.

Beck, P.A. Belbeoch, R., Gendreau, G., and Leleux, G. (1967). Shifts in betatron frequencies due to energy spread, betatron amplitudes and closed orbit excursions. *Proc. 6th Int. Conf. on High Energy Accelerators, Cambridge Electron Accelerator Lab. Rep. No. CEAL 2000*, p.A63.

Bekefi, G. (1966). *Radiation processes in plasmas*. John Wiley, New York.

Benford, G., and Book, D.L. (1971). Relativistic beam equilibria. *Adv. plasma Phys.* 4, 125.

——— , ——— and Sudan, R.N. (1970). Relativistic beam equilibria and back currents. *Phys. Fluids* 13, 2621.

Bennett, W.H. (1934). Magnetically self-focusing streams. *Phys. Rev.* 45, 89.

Bermel, W. (1970). Elektronenbeschleunigung in einem Plasmabetatron mit und ohne hohem Leitungsstrom. *Z. Phys.* 237, 345.

Bick, J.H. (1965). The first-order theory of quadrupole lenses including the effects of space charge. *IEEE Trans. electron. Devices* 12, 408.

Birdsall, C.K., and Bridges, W.B. (1966). *Electron dynamics of diode regions*. Academic Press, New York.

——— and Whinnery, J.R. (1953). Waves in an electron stream with general admittance walls. *J. appl. phys.* 24, 314.

Blauth, E.W. (1966). *Dynamic mass spectrometers*. Elsevier, Amsterdam.

Blewett, J.P. (1946). Radiation losses in the induction accelerator. *Phys. Rev.* 69, 87.

Bobroff, D.L., Haus, H.A. and Kluver, J.A. (1962). On the small signal power theorem of electron beams. *J. appl. Phys.* 33, 2932.

Boersch, H. (1954). Experimentelle Bestimmung der Energieverteilung in thermisch ausgelosten Elektronenstrahlen. *Z. Phys.* 139, 115.

Bogdankevich, L.S., and Rukhadze, A.A. (1971). Stability of relativistic electron beams in a plasma and the problem of critical currents. *Sov. Phys.-Uspekhi* 14, 163.

Bohm, D., and Foldy, L. (1946). Theory of the synchrotron. *Phys. Rev.* 70, 249.

Born, M., and Wolf, E. (1965). *Principles of optics* (3rd edn). Pergamon Press, Oxford.

Boussard, D. (1970). Radiofrequency focusing in heavy ion linear accelerators. In *Linear accelerators* (eds. P.M. Lapostolle and A. Septier), p.1073. North-Holland, Amsterdam.

Braginskij, S.I. (1958). The behaviour of a completely ionized plasma in a strong magnetic field. *Sov. Phys. - JETP* 6, 494.

Branch, G.M., and Mihran, T.G. (1955). Plasma reduction factors in electron beams. *IRE Trans. electron. Devices* ED-2, 3.

Brewer, G.R. (1959). Some characteristics of a magnetically focused electron beams. *J. appl. Phys.* 30, 1022.

—— (1967a). High-intensity electron guns. In *Focusing of charged particles* (ed. A. Septier), vol. 2, p.23. Academic Press, New York.

—— (1967b). Focusing of high density electron beams. In *Focusing of charged particles* (ed. A. Septier), vol. 2, p.73, Academic Press, New York.

—— (1970). *Ion propulsion. Technology and applications.* Gordon & Breach, New York.

Briggs, R.J. (1964a). *Electron stream interaction with plasmas.* MIT Press, Cambridge, Mass.

—— (1964b). Transformation of small-signal energy and momentum of waves. *J. appl. Phys.* 35, 3268.

—— (1971). Two-stream instabilities. *Adv. Plasma Phys.* 4, 43.

—— Daugherty, J.D., and Levy, R.H. (1970). Role of Landau damping in crossed-field electron beams and inviscid shear flow. *Phys. Fluids.* 13, 421.

—— Hester, R.E., Lauer, E.J., Lee, E.P. and Spoerlein, R.L. (1976). Radial self-focusing of relativistic electron beams. *Phys. Fluids* 19, 1007.

—— and Neil, V.K. (1967). Negative-mass instability in a cylindrical layer of relativistic electrons. *J. nucl. Energy* 9, 209.

Brillouin, L. (1945). A theorem of Larmor and its importance for electrons in magnetic fields. *Phys. Rev.* 67, 260.

Bruck, H. (1966). *Accélérateurs circulaires de particules.* Presses Universitaires de France, Paris.

Budker, G.I. (1956). *CERN Symp. on High Energy Accelerators*, vol.1. p.68, CERN, Geneva.

—— (1967). An effective method of damping particle oscillations in proton and antiproton storage rings. *Sov. Atom. Energy* 22, 438.

Budker, G.I., Derbenev, Ya. S., Dikanksky, N.S., Kudelainen, V.I., Meshkov, I.N., Parkhomchuk, V.V., Pestrikov, D.V., Skrinsky, A.N., and Sukhina, B.N. (1975). Experiments on electron cooling. *IEEE Trans. nucl. Sci.* NS22, 1093.

Busch, H. (1926). Berechnung der Bahn von Kathodenstrahlen in axial symmetrischen electromagnetischen Felde. *Z. Phys.* 81 (5), 974.

Buneman, O. (1957). Ribbon beams. *J. electron. Control* 3, 507.

—— (1961). The RF theory of crossed-field devices. In *Crossed-field microwave devices* (ed. E. Okress), vol.1, p.367. Academic Press, New York.

——, Levy, R.H., and Linson, L.M. (1966). Stability of crossed-field electron beams. *J. appl. Phys.* 37, 3203.

Chen, F.F. (1974). *Introduction to plasma physics.* Plenum Press, New York.

Chodorow, M., and Susskind, C. (1964). *Fundamentals of microwave electronics.* McGraw-Hill, New York.

Christofilos, N.C. (1950). Focussing system for ions and electrons. U.S. Patent No. 2 736 799. Reprinted in *The development of high-energy accelerators* (ed. M.S. Livingston). Dover Publications, New York (1966).

—— (1958). Astron thermonuclear reactor. *Proc. 2nd U.N. Int. Conf. on the Peaceful Uses of Atomic Energy.* 32, 279. Geneva.

—— Condit, W.C., Fessenden, T.J., Hester, R.E., Humphries, S., Porter, G.D., Stallard, B.W., and Weiss, P.B. (1971). Trapping experiments in the Astron. *Proc. 4th Int. Conf. on Plasma Physics and Controlled Fusion Research*, p.119, IAEA, Vienna.

Chu, L.J. (1951). A kinetic power theorem. *Annual IRE Conf. on Electron Tube Research.* Durham, N.H.

—— and Jackson, J.D. (1948). Field theory of travelling-wave tubes. *Proc. Inst. radio Engrs* 36, 853.

Clemmow, P.C., and Dougherty, J.P. (1969). *Electrodynamics of particles and plasmas.* Addison-Wesley, Reading, Mass.

Collin, R.E. (1966). *Foundations for microwave engineering.* McGraw-Hill, New York.

Colonias, J.S. (1974). *Particle accelerator design: computer programs.* Academic Press, New York.

Connor, J.W., and Hastie, R.J. (1975). Relativistic limitation on runaway electrons. *Nucl. Fusion* 15, 415.

Corben, H.C. and Stehle, P. (1960). *Classical mechanics* (2nd edn). John Wiley, New York.

Courant, E.D., Livingston, M.S., and Snyder, H.S. (1952). The strong-focusing synchrotron - a new high energy accelerator. *Phys. Rev.* 88, 1190.

——— and Sessler, A.M. (1966). Transverse coherent resistive instabilities of azimuthally bunched beams in particle accelerators. *Rev. sci. Instrum.* 37, 1579.

——— and Snyder, H.S. (1958). Theory of the alternating gradient synchrotron. *Ann. Phys.* 3, 1.

Cox, J.L., and Bennett, W.H. (1970). Reverse current induced by injection of a relativistic electron beam into a pinched plasma. *Phys. Fluids* 13, 182.

Crawford, F.W. (1971). Microwave plasma devices - promise and progress. *Proc. Inst. elect. electron. Engrs* 59, 4.

Creedon, J.M. (1975). Relativistic flow in the high ν/γ diode *J. appl. Phys.* 46, 2946.

Cutler, C.C. (1956). Spurious modulation of electron beams. *Proc. Inst. radio Engrs* 44, 61.

——— and Hines, M.E. (1955). Thermal velocity effects in electron guns. *Proc. Inst. radio Engrs* 43, 307.

Daugherty, J.D., and Levy, R.H. (1967). Equilibrium of electron clouds in toroidal magnetic fields. *Phys. Fluids* 10, 155.

———, Eninger, J.E., and Janes, G.S. (1969a). Experiments on the injection and containment of electron clouds in a toroidal apparatus. *Phys. Fluids* 12, 2677.

——— ——— ———, and Levy, R.H. (1969b). The HIPAC as a source of highly stripped heavy ions. *IEE Trans. nucl. Sci.* NS-16, 51.

Davidson, R.C. (1974). *Theory of non-neutral plasmas.* Benjamin, Reading, Mass.

Davidson, R.C. and Lawson, J.D. (1972). Self-consistent Vlasov description of relativistic electron ring equilibria. *Particle Accel.* 4, 1.

Dawson, J.M. (1960). Plasma oscillations of a large number of electron beams. *Phys. Rev.* 118, 381.

———, and Whetten, N.R. (1969). Mass spectroscopy using RF quadrupole fields. *Adv. Electronics electron. Phys.* 27, 59.

Dow, W.G. (1958). Non-uniform d-c electron flow in magnetically focused cylindrical beams. *Adv. electronics electron. Phys.* 10, 1.

Dunn, D.A., and Self, S.A. (1964). Static theory of density and potential distribution in a beam-generated plasma. *J. appl. Phys.* 35, 113.

Enge, H.A. (1967). Deflecting magnets. In *Focusing of charged particles* (ed. A. Septier), vol.2, p.203. Academic Press, New York.

Faulkener, J.E., and Ware, A.A. (1969). The effect of finite ion mass on the stability of a space-charge-neutralized electron beam. *J. appl. Phys.* 40, 366.

Ferrari, L.A., and Rogers, K.C. (1967). Experimental study of runaway electrons in a plasma betatron. *Phys. Fluids* 10, 1319.

——— ———, and Landau, R.W. (1968). Observation of the negative-mass instability in a plasma betatron. *Phys. Fluids* 11, 691.

——— and Zucker, M.S. (1968). Stability of orbits in combined betatron and azimuthal magnetic fields. *Phys. Fluids* 12, 1312.

Field, L.M., Spangenberg, K., and Helm, R. (1947). Control of electron beam dispersion at high vacuum by ions. *Electl. Commun.* 24, 108.

Fink, J.H., and Schumacher, B.W. (1975). Characterization of charged particle beam sources. *Nucl. Instrum. Meth.* 130, 353.

Fleischmann, H.H. (1975). High-current electron beams. *Phys. Today* 28, 35.

——— and Kammash, T. (1976). Systems analysis of the ion-ring-compression approach to fusion. *Nucl. Fusion* 15, 1143.

Forsyth, E.G., Lederman, L.M., and Sunderland, J. (1965). The Brookhaven-Columbia plasma lens. *IEEE Trans. nucl. Sci.* NS-12, 872.

Frey, J., and Birdsall, C.K. (1966). Instabilitics in a neutralized electron stream in a finite-length drift tube. *J. appl. Phys.* 37, 2051.

Furth, H.P. (1965). Unstable precession under the influence of drag forces. *Phys. Fluids* 8, 2020.

Gabor, D. (1945). Dynamics of electron beams. Applications of Hamiltonian dynamics to electronic problems. *Proc. Inst. radio Engrs* 33, 792.

Garren, A.A. (1969). Thin lens optics with space charge. *Lawrence Berkeley Lab. Rep. No. UCRL-19313.*

——— Judd, D.L., Smith, L., and Willax, H.A. (1962). Electrostatic deflector calculations for the Berkeley 88-inch cyclotron. *Nucl. Instrum. Meth*. 18, 525.

Gewartowski, J.W., and Watson, H.A. (1965). *Principles of electron tubes.* Van Nostrand, Princeton, N.J.

Ginzton, E.L., and Wadia, B.H. (1954). Positive-ion trapping in electron beams. *Proc. Inst. radio Engrs* 42, 1548.

Glaser, W. (1952). *Grundlagen der Elektronenoptik.* Springer Verlag, Vienna.

Gluckstern, R.L. (1967). Space charge effects. In *Linear accelerators* (eds. P.M.Lapostolle and A. Septier) p. 827. North-Holland, Amsterdam.

——— (1970). Oscillation modes in two dimensional beams. *Proc. 1970 Proton Linear Accelerator Conf.* National Accelerator Laboratory, vol.2, p.133, Batavia, Ill.

Goldstein, H. (1950). *Classical mechanics.* Addison-Wesley, Reading, Mass.

Gould, R.W. (1955). A coupled mode description of the backward wave oscillator, and the Kompfner dip condition. *IRE Trans. electron. Devices*, ED-2, 37.

——— (1957). Space charge effects in beam-type magnetrons. *J. appl. Phys.* 28, 599.

Graybill, S.E., and Nablo, S.V. (1966). Observations of magnetically self-focusing electron streams. *J. appl. Phys. Lett.* 8, 18.

Green, T.S. (1974). Intense ion beams. *Rep. Prog. Phys.* 37, 1257.

——— (1976). Beam formation and space charge neutralization. *IEEE Trans. nucl. Sci.* NS-23, 918.

Greenway, T.J.L., and Hyder, H.R.M. (1971). The influence of extraction geometry on a negative helium ion source with an extended plasma. *Proc. Symp. Ion Sources and Formation of Ion Beams*, p. 277 Brookhaven National Laboratory, New York.

Grivet, P., and Septier, A. (1972). *Electron optics* (2nd ed). Pergamon Press, Oxford.

Groendijk, H., Vlaardingerbroek, M.T., and Weimer, K.R.U.(1965). Waves in cylindrical beam-plasma systems immersed in a longitudinal magnetic field. *Philips Res. Rep.* 20, 485.

Grunder, H., and Lambertson, G. (1971). Transverse beam instabilities at the bevatron. *Proc. 8th Conf. on High Energy Accelerators*, p. 308. CERN, Geneva.

Haeff, A.V. (1939). Space-charge effects in electron beams. *Proc. Inst. radio Engrs* 27, 586.

Hagedorn, R. (1957). Stability and amplitude ranges of two dimensional non-linear oscillations with periodical Hamiltonian applied to betatron oscillations in circular particle accelerators. *Rep. No. CERN 57-1*. CERN, Geneva.

Hahn, W.C. (1939). Small signal theory of velocity-modulated electron beams. *Gen. Elec. Rev.* 42, 258.

Halsted, A.S., and Dunn, D.A. (1966). Electrostatic and magnetic pinch effects in beam-generated plasmas. *J. appl. Phys.* 37, 1810.

Hammer, D.A., and Rostoker, N. (1970). Propagation of high current relativistic electron beams. *Phys. Fluids* 13, 1831.

Hardt, W. (1968). A few simple expressions for checking vacuum requirements in proton synchrotrons. *Rep. No. CERN ISR-300/GS/68-11*. CERN, Geneva.

Harker, K.J. (1957). Non laminar flow in cylindrical electron beams. *J. appl. Phys.* 28, 645.

Harris, L.A. (1952). Axially symmetric electron beam and magnetic field systems. *Proc. Inst. radio Engrs* 40, 700.

Harrison, E.R. (1958). On the space-charge divergence of an axially symmetric beam. *J. electron. Control* 4, 193.

Hawkes, P.W. (1966). Quadrupole optics. *Springer tracts in modern physics* (ed. G. Hohler), vol.42. Springer-Verlag, Berlin.

——— (1970). Quadrupoles in electron lens design. *Adv. electron. and electron. Phys.* Suppl. 7.

——— (1972). *Electron optics and electron microscopy.* Taylor & Francis, London.

Heitler, W. (1954). *The quantum theory of radiation* (3rd ed), Chap.5. Clarendon Press, Oxford.

Helm, R.H., and Loew, G.A. (1970). Beam breakup. In *Linear accelerators* (eds. P.M. Lapostolle and A. Septier), p.173. North-Holland, Amsterdam.

——— ——— and Panofsky, W.K.H. (1968). Beam dynamics. In *The Stanford two-mile accelerator* (ed. R.B. Neal), p.163. Benjamin, Reading, Mass.

Hermann, G. (1958). Optical theory of thermal velocity effects in cylindrical electron beams. *J. appl. Phys.* 29, 127.

Hines, M.E., Hoffman, G.W., and Saloom, J.A. (1955). Positive-ion drainage in magnetically focused electron beams. *J. appl. Phys.* 26, 1157.

Hübner, K., Ruggiero, A.G., and Vaccaro, V.G. (1969). Stability of the coherent transverse motion of a coasting beam for realistic distribution functions and any given coupling with its environment. *Proc. 7th Int. Conf. on High Energy Accelerators*, vol.2, p.343. Yerevan, USSR Academy of Sciences.

Hutter, R.G.E., and Harrison, S.W. (1960). *Beam and wave electronics in microwave tubes*. Van Nostrand, Princeton, N.J.

Jackson, J.D. (1960). Longitudinal plasma oscillations. *J. nucl. Energy* 1, 171.

—— (1963). *Classical electrodynamics*. John Wiley, New York.

Jepson, R.L. (1961). Magnetically confined cold-cathode gas discharges at low pressures. *J. appl. Phys.* 32, 2619.

Johnson, C.C. (1965). *Field and wave electrodynamics*. McGraw-Hill, New York.

Johnston, T.W. (1959). Nonlaminar flow in magnetically focused electron beams from magnetically shielded guns. *J. appl. Phys.* 30, 1456.

Jory, H.R., and Trivelpiece, A.W. (1969). Exact relativistic solution for the one-dimensional diode. *J. appl. Phys.* 40, 3924.

Kampen, N.G. van (1955). On the theory of stationary waves in a plasma. *Physica* 21, 949.

Kapchinskij, I.M. (1966). *Particle dynamics in linear resonance accelerators* (in Russian). Atomizdat, Moscow.

—— and Vladimirskij, V.V. (1959). Limitations of proton beam current in a strong focusing linear accelerator associated with the beam space charge. *Proc. Int. Conf. on High Energy Accelerators*, p.274, CERN, Geneva.

Kapetanakos, C.A., Hammer, D.A., Striffler, C.D., and Davidson, R.C. (1973). Destructive instabilities in hollow intense relativistic beams. *Phys. Rev. Lett.* 30, 1303.

Keil, E. (1972a). Diffraction radiation of charged rings moving in a corrugated cylindrical pipe. *Nucl. Instrum. Meth.* 100, 419.

Keil, E. (1972b). Intersecting storage rings. *Rep.No. CERN 72-14*. CERN, Geneva.

Keil, E. and Schnell, W. (1969). Concerning longitudinal stability in the ISR. *CERN Rep. No. CERN-ISR-TH-RF/69 48*. CERN, Geneva.

Kellogg, O.D. (1930). *Foundations of potential theory*. Frederick Ungar, New York. (Also Dover Publications, New York, 1953.)

Kerst, D.W., and Serber, R. (1941). Electronic orbits in the induction accelerator. *Phys. Rev.* 60, 53.

King, N.M. (1964). Theoretical techniques of high energy beam design. *Prog. nucl. Phys.* 9, 71.

Kino, G.S., and Gerchberg, R. (1963). Transverse field interactions of a beam and plasma. *Phys. Rev. Lett.* 11, 185.

Kirstein, P.T., Kino, G.S. and Waters, W.E. (1967). *Space charge flow*. McGraw-Hill, New York.

Klemperer, O., and Barnett, M.E. (1971). *Electron optics* (3rd edn.) Cambridge University Press, London.

Knauer, W. (1966). Diocotron instability in plasmas and gas discharges. *J. appl. Phys.* 37, 602.

Kollath, R. (ed.) (1967). *Particle accelerators*. (Transl. by W. Summer.) Pitman, London.

Kolomenskij, A.A., and Lebedev, A.N. (1966). *Theory of cyclic accelerators*. (Transl. by M. Barbier.) North-Holland, Amsterdam.

——— and ——— (1970). Collective effects in accelerators. *All-Union Conf. on Particle Accelerators*. (In Russian.) (Available in French as CERN transl. 70-7, CERN, Geneva, 1970).

Koshkarev, D.G. and Zenkevich, P.R. (1972). Resonance of coupled transverse oscillations in two circular beams. *Particle Accel.* 3, 1.

Kyhl, R.L., and Webster, H.R. (1956). Breakup of hollow cylindrical electron beams. *IRE Trans. electron. Devices* ED-3, 172.

Landau, L.D. (1946). On the vibrations of the electronic plasma. *J. Phys. USSR* 10, 25.

——— and Lifshitz, E.M. (1960). *Electrodynamics of continuous media*. Pergamon Press, Oxford. (Transl. from Russian.)

Langmuir, D.B. (1937). Theoretical limitations of cathode ray tubes. *Proc. Inst. radio Engrs* 25, 977.

Lapostolle, P. (1969). Les sources d'ions dans les grands accelerateurs. *Première Conf. Int. sur les sources d'ions*, p.165. INSTN, Saclay, France.

—— (1971). Possible emittance increase through filamentation due to space charge in continuous beams. *IEEE Trans. nucl. Sci.* NS-18, 1101.

—— (1972). Possible ion beam quality deterioration due to non linear space charge effects. *Proc. 2nd Int. Conf. on Ion Sources*, p.133. Osterreischische Studentgesellschaft für Atomenergie, Vienna.

—— and Septier, A.L. (eds.) (1970). *Linear accelerators*. North-Holland, Amsterdam.

Larsen, R.R., Altenmueller, O.A., and Loew, G.A. (1963). Investigation of travelling-wave separators for the Stanford two-mile linear accelerator. *Proc. Int. Conf. on High Energy Accelerators, Dubna*. Transl. ed. USAEC Division of Technical Information. Conf. No. 114, p.1044.

Laslett, L.J. (1963). *Proc. 1963 Summer Study on Storage Rings, Rep.No. BNL 7534*, p. 324. Brookhaven National Laboratory, New York.

—— (1967). Strong focusing in circular particle accelerators. In *Focusing of charged particles* (ed. A. Septier), vol.2, p.355. Academic Press, New York.

—— (1969). On the focusing effects arising from the self fields of a toroidal beam. *Lawrence Berkeley Lab. Rep. No. ERAN-30.*

—— (1974). "Stochasticity". *Proc. 9th Int. Conf. on High Energy Accelerators*, p.394. Stanford Linear Accelerator Center, Stanford, Calif.

—— Möhl, D., and Sessler, A.M. (1972). Transverse two-stream instability in the presence of strong species-species and image forces. *Lawrence Berkeley Lab. Rep.No. LBL 1072.*

—— Neil, V.K., and Sessler, A.M. (1965). Transverse resistive instabilities of intense coasting beams in particle accelerators. *Rev. sci. Instrum.* 36, 436.

—— and Resegotti, L. (1967). The space-charge intensity limit imposed by coherent oscillation of a bunched synchrotron beam. *Proc. 6th Int. Conf. on High Energy Accelerators, Cambridge Electron Accelerator Lab. Rep.No. CEAL 2000*, p.150.

Lau, Y.Y., and Briggs, R.J. (1971). Effects of cold plasma on the negative mass instability of a relativistic electron layer. *Phys. Fluids* 14, 967.

Lawson, J.D. (1954). Electron trajectories in strip beams constrained by a magnetic field. *Proc. Inst. elect. radio Engrs* 42, 1147.

——— (1959). On the classification of electron streams. *J. nucl. Energy C* 1, 31.

——— (1973). Simple models of solid and hollow relativistic electron beams with arbitrarily high current. *Phys. Fluids* 16, 1298. Comments by Hon-Ming Lai and correction. *Phys. Fluids* 19 1068.

——— (1975). Optical and hydrodynamical approaches to charged particle beams. *Plasma Phys.* 17, 567.

——— Lapostolle, P.M., and Gluckstern, R.L. (1973). Emittance, entropy and information. *Particle Accel.* 5, 61.

Lee, E.P. (1976). Kinetic theory of a relativistic beam. *Phys. Fluids* 19, 60.

——— and Cooper, R.K. (1976). General envelope equation for cylindrically symmetric charged particle beams. *Particle Accel.* 7, 83.

Lee, M.J. , Morton, P.L., and Mills, F.E., (1967). Throbbing beam instabilities in particle accelerators and storage rings. *IEEE Trans. nucl. Sci.* NS-14, 602.

Lehnert, B. (1964). *Dynamics of charged particles*. North-Holland, Amsterdam.

Levine, L.S., Vitkovitsky, I.M., Hammer, D.A., and Andrews, M.L. (1971). Propagation of an intense relativistic electron beam through a plasma background. *J. appl. Phys.* 42, 1863.

Levy, R.H. (1965). Diocotron instability in a cylindrical geometry. *Phys. Fluids* 8, 1288.

Lichtenberg, A.J. (1969). *Phase-space dynamics of particles*. John Wiley, New York.

Lindsay, P.A. (1960). Velocity distribution in electron streams. *Adv. Electronics electron. Phys.* 13, 182.

Livingood, J.J. (1961). *Principles of cyclic particle accelerators*. Van Nostrand, Princeton, N.J.

——— (1969). *The optics of dipole magnets*. Academic Press, New York.

Livingston, M.S. (ed.) (1966). *The development of high-energy accelerators*. Dover Publications, New York.

——— , and Blewett, J.P. (1962). *Particle accelerators*. McGraw-Hill, New York.

Loeffler, K.H. (1969). Energy-spread generation in electron-optical instruments. *Z. angew. Phys.* 27, 145.

Loew, G.A. (1967). Electron linac instabilities. *IEE Trans. nucl. Sci.* NS-14, 529.

Longmire, C.L. (1963). *Elementary plasma physics.* Interscience, New York.

Louisell, W.H. (1960). *Coupled mode and parametric electronics.* John Wiley, New York.

Lynden-Bell, D. (1967). Statistical mechanics of violent relaxation in stellar systems. *Mon. Not. R. astr. Soc.* 136, 101.

Manickam, J., Carr, W., Rosen, B., and Seidl, J. (1975). Convective and absolute instabilities in beam plasma systems. *Phys. Fluids* 18, 369.

Marcuse, D. (1972). *Light transmission optics.* Van Nostrand, New York.

McCorkle, R.A., and Bennett, W.H. (1971). Thin electrostatically self-focusing streams. *Plasma Phys.* 13, 1153.

McLachlan, N.W. (1947). *Theory and application of Mathieu functions.* Oxford University Press, Oxford.

McMillan, E.M. (1950). The relation between phase stability and first order focusing in linear accelerators. *Phys. Rev.* 80, 493.

Meer, S. van der (1961). A directive device for charged particles and its use in an enhanced neutrino beam. CERN 61-7, Geneva.

Meleka, A.H. (1971). *Electron-beam welding. Principles and practice.* McGraw-Hill, London.

Meltzer, B. (1956). Single component space-charge flow. *J. electron. Control* 2, 118.

Mihran, T.G., and Andal, K.B. (1965). The growth of peak velocity, noise and signals in O-type electron beams. *IEE Trans. electron. Devices* ED-12, 208.

Möhl, D., Laslett, L.J., and Sessler, A.M. (1973). On the performance characteristics of electron ring accelerators. *Particle Accel.* 4, 159.

Montgomery, D.C., and Tidman, D.A. (1964). *Plasma kinetic theory.* McGraw-Hill, New York.

Morton, P.L., Neil, V.K., and Sessler, A.M. (1966). Wake fields of a pulse of charge moving in a highly conducting pipe of circular cross section. *J. appl. Phys.* 37, 1375.

Moss, H. (1968). Narrow angle electron guns and cathode ray tubes. *Adv. Electronics electron. Phys.* Suppl. 3.

Mukhovatov, V.S. and Shafranov, V.D. (1971). Plasma equilibrium in a Tokamak. *Nucl. Fusion* 11, 605.

Mulvey, T. (1967). Electron microprobes. In *Focusing of charged particles* (ed. A. Septier), vol.1, p.469. Academic Press, New York.

———, and Wallington, M.J. (1973). Electron lenses. *Rep. Prog. Phys.* 36, 347.

Musha, T. (1964). Amplification of waves due to quanta with negative energy. *J. appl. Phys.* 35, 137.

Nagy, G.A., and Szilágyi, M. (1974). *Introduction to the theory of space charge optics.* Macmillan, London.

Neal, R.B. (ed.) (1968). *The Stanford two-mile accelerator,* p.1169. Benjamin, New York.

Neil, V.K., and Heckrotte, W. (1965). Relation between diochotron and negative mass instabilities. *J. appl. Phys.* 36, 2761.

———, and Sessler, A.M. (1965). Longitudinal resistive instabilities of intense coasting beams in particle accelerators. *Rev. sci. Instrum.* 36, 429.

Neugebauer, W. (1967). Equilibrium solutions for partially immersed relativistic electron beams. *IEEE Trans. electron. Devices* 14, 686.

Nezlin, M.V. (1968). Plasma instabilities and the compensation of space charge in an ion beam. *Plasma Phys.* 10, 337.

Northrup, T.G. (1963). *The adiabatic motion of charged particles.* Interscience, New York.

Okress, E. (ed.) (1961). *Crossed-field microwave devices.* Academic Press, New York.

Packh, D.C. de, and Ulrich, P.B. (1961). Brillouin flow in relativistic beams. *J. electron. Control.* 10, 139.

Panofsky, W.K.H., and Phillips, M. (1962). *Classical electricity and magnetism* (2nd. edn). Addison Wesley, Reading, Mass.

Parzen, G. (1961). Accelerators with a general magnetic field. *Ann. Phys.* 15, 22.

Pease, R.S. (1957). Equilibrium characteristics of a pinched gas discharge cooled by bremsstrahlung radiation. *Proc. phys. Soc.* B-70, 11.

Pellegrini, C. (1969). On a new instability in electron positron storage rings.(The head-tail effect). *Nuovo Cim.* 64A, 447.

Penrose, O. (1960). Electrostatic instabilities of a uniform non-Maxwellian plasma. *Phys. Fluids* 3, 258.

Pfeiffer, H.C. (1970). Experimental investigation of energy broadening in electron optical instruments. *Rec. 11th Symp. on Electron, Ion, and Laser Beam Technology*. San Francisco Press, San Francisco.

——— (1972). Basic limitation of probe forming systems due to electron-electron interaction. *Proc. 5th Annual Scanning Electron Microscopy Symp*. IIT Research Institute, Chicago.

Pierce, J.R. (1950). *Traveling-waves tubes*. Van Nostrand, New York.

——— (1954a). *Theory and design of electron beams* (2nd edn). Van Nostrand, Princeton, N.J.

——— (1954b). Coupling of modes of propagation. *J. appl. Phys.* 25, 179.

——— (1954c). The wave picture of microwave tubes. *Bell Syst. Tech. J.* 33, 343.

——— (1961). Momentum and energy of waves. *J. appl. Phys*, 32, 2580.

———, and Walker L.R. (1953). 'Brillouin flow' with thermal velocities. *J. appl. Phys.* 24, 1328.

Piwinski, A. (1974). Intra-beam scattering. *Proc. 9th Int. Conf. on High Energy Accelerators*, p.347. Stanford Linear Accelerator Center, Stanford, Calif.

Pöschl, K. and Veith, W. (1962). Generalized Brillouin flows. *J. appl. Phys.* 33, 1013.

Pötzl, H. (1965). Wellentypen in magnetisch fokussierten Elektronenstrahlen. *Ark. elekt. Ubertr.* 19, 367. English transl. Types of waves in magnetically focused electron beams. Available from AERE, Harwell, Transl. No. LB/G/3055.

Poukey, J.W., and Toepfer, A.J. (1974). Theory of super-pinched relativistic electron beams. *Phys. Fluids* 17, 1582.

Ramo, S. (1939). Space charge and field waves in an electron beam. *Phys. Rev.* 56, 276.

Redhead, P.A. (1967). Multiple ionization of the rare gases by successive electron impacts (0-250 eV). *Canad. J. Phys.* 45, 1791.

Redhead, P.A. (1974). Multiple ionization by sequential electron impact. *Jap. J. Appl. Phys.* 13, Suppl.2, 403.

Regenstreif, E. (1967). Focusing with quadrupoles, doublets, and triplets. In *Focusing of charged particles* (ed. A. Septier), vol.1. p.353. Academic Press, New York.

Reiser, M. (1971). First-order theory of electrical focusing in cyclotron-type two-dimensional lenses with static and time-varying potentials. *J. appl. Phys.* 42, 4128.

—— (1973). On the equilibrium orbit and linear oscillations of charged particles in axisymmetric $\underline{E} \times \underline{B}$ fields and application to the electron ring accelerator. *Particle Accel.* 4, 239.

—— (1977). Laminar-flow equilibria and limiting currents in magnetically focused relativistic beams. *Phys. Fluids*, 20, 477.

Robinson, K.W. (1958). Radiation effects in circular electron accelerators. *Phys. Rev.* 111, 373.

Rognlien, T.D., and Self, S.A. (1972). Interpretation of dispersion relations for bounded systems. *J. plasma Phys.* 7, 13.

Rome, J.A., and Briggs, R.J. (1972). Stability of sheared electron flow. *Phys. Fluids* 15, 796.

Roos, B.W. (1969). *Analytic functions and distributions in physics and engineering*. John Wiley, New York.

Rose, M.E. (1938). Focusing and maximum energy of ions in the cyclotron. *Phys. Rev.* 53, 392.

Rossi, B. (1952). *High-energy particles*, Chapter 2. Prentice-Hall, Englewood Cliffs, N.J.

—— , and Greisen, K. (1941). Cosmic-ray theory. *Rev. mod. Phys.* 13, 240.

Rowe, J.E. (1965). *Nonlinear electron-wave interaction phenomena*. Academic Press, New York.

Ruggiero, A.G., and Vaccaro, V.G. (1968). Solution of the dispersion relation for longitudinal stability of an intense coasting beam in a circular accelerator (application to the ISR). *Rep.No. ISR-TH/68-33*. CERN, Geneva.

Sacherer, F.J. (1968). Transverse space-charge effects in circular accelerators. *Rep.No. UCRL-18454*, Lawrence Radiation Laboratory, Berkeley.

—— (1971). RMS envelope equations with space charge. *IEEE Trans. nucl. Sci.* NS-18, 1105.

——— (1972). Methods for computing bunched-beam instabilities. *Rep.No. CERN/SI-BR/72-5.* CERN, Geneva.

Sacherer, F.J. (1973). A longitudinal stability criterion for bunched beams. *IEEE Trans. nucl. Sci.* NS-20, 825.

——— (1974). Transverse bunched beam instabilities - theory. *Proc. 9th Int. Conf. on High Energy Accelerators.* p.347. Stanford Linear Accelerator Center, Stanford, California.

Samuel, A.L. (1949). On the theory of axially symmetric electron beams in an axial magnetic field. *Proc. Inst. radio Engrs.* 37, 1252.

Sands, M. (1969). The head-tail effect : an instability mechanism in storage rings. *Rep.No. SLAC TN-69-8.* Stanford Linear Accelerator Center.

——— (1971). The physics of electron storage rings: an introduction. *Proc. Int. School 'Enrico Fermi', Course XLVI.* Academic Press, New York.

Scherzer, O. (1936). Über einige Fehler von Elektronenlinsen. *Z. Phys.* 101, 593.

——— (1936). Die schwache elektrische Einzellinse geringster sphärischer Aberration. *Z. Phys.* 101, 23.

Schoch, A. (1958). Theory of linear and non-linear perturbations of betatron oscillations in alternating gradient synchrotrons. *Rep.No. CERN 57-21.* CERN, Geneva.

Self, S.A. (1969). Interaction of a cylindrical beam with a plasma. I. Theory, II. Experiment and comparison with theory. *J. appl. Phys.* 40, 5217.

Septier, A. (1966). The struggle to overcome spherical aberration in electron optics. *Adv. opt. electron. Microsc.* 1, 204.

——— (ed.) (1967). *Focusing of charged particles*, vols. 1 and 2. Academic Press, New York.

Sessler, A.M. (1966). Instabilities of relativistic particle beams. *5th Int. Conf. on High Energy Accelerators.* Comitato Nazionale per l'Energia Nucleare, Rome.

——— (1972). Collective phenomena in accelerators. *Proc. 1972 Proton Linear Accelerator Conf.* p.291. *Rep.No. LA5115.* Los Alamos Laboratory. N.M.

——— and Vaccaro, V.G. (1967). Longitudinal instabilities of azimuthally uniform beams in circular vacuum chambers with walls of arbitrary electrical properties. *Rep.No. ISR 67-2.* CERN, Geneva.

Shevchik, V.N., Shvedov, G.N., and Soboleva, A.V. (1966). *Wave and oscillatory phenomena in electron beams at microwave frequencies* (Transl. from Russian, ed. B. Meltzer). Pergamon Press, Oxford.

Shohet, J.L. (1971). *The plasma state.* Academic Press, New York.

Sims, G.D., and Stephenson, I.M. (1963). *Microwave tubes and semiconductor devices*, p. 271. Blackie, London.

Siegman, A.E. (1960). Waves on a filamentary electron beam in a transverse-field slow-wave circuit. *J. appl. Phys.* 31, 17.

Sivukhin, D.V. (1966). Coulomb collisions in a fully ionized plasma. *Rev. plasma Phys.* 4, 93.

Smith, L. (1963). Effect of gradient errors in the presence of space charge forces. *Proc. Int. Conf. on High Energy Accelerators, Dubna.* Transl. ed. USAEC Division of Technical Information, Conf. No. 114, p.1232.

Sørenssen, A. (1975). Crossing the phase transition in strong-focusing proton synchrotrons. *Particle Accel.* 6, 141.

Spitzer, L. (1962). *Physics of fully ionized gases*, (2nd edn.) Chap.5. Interscience, New York.

Spong, D.A., Clarke, J.F., Rome, J.A., and Kammash, T. (1974). Runaway electrons in the Ormak device. *Nucl. Fusion.* 14, 397.

Steele, M.C., and Vural, B. (1969). *Wave interactions in solid state plasmas.* McGraw-Hill, New York.

Steenbergen, A. van (1965). Recent developments in high intensity ion beam production and preacceleration. *IEEE Trans. nucl. Sci.* NS-12, 746.

——— (1967). Evaluation of particle beam phase space measurement techniques. *Nucl. Instrum. Meth.* 51, 245.

Steffen, K.G. (1965). *High energy beam optics.* John Wiley, New York.

Stix, T.H. (1962). *The theory of plasma waves.* McGraw-Hill, New York.

Sturrock, P. (1951). The aberrations of magnetic electron lenses due to asymmetries. *Phil. Trans. R. Soc. A.* 243, 387.

——— (1952). The imaging properties of electron beams in arbitrary static electromagnetic fields. *Phil. Trans. R. Soc. A.* 245, 387.

——— (1955). *Static and dynamic electron optics.* Cambridge University Press, London.

——— (1958). Kinematics of growing waves. *Phys. Rev.* 112, 1488.

——— (1959). Magnetic deflection focusing. *J. Electron. Control* 7, 162.

——— (1960). In what sense do slow waves carry negative energy? *J. appl. Phys.* 31, 2052.

Sturrock, P.A. (1962). Energy and momentum in the theory of waves in plasmas. In *Plasma hydromagnetics. 6th Lockheed Symp. on Magnetohydrodynamics*, (ed. D. Bershader). Stanford University Press, Calif.

Sudan, R.N. (1975). Accelerators for the fusion program. *IEEE Trans. nucl. Sci.* NS-22, 1736.

Sutherland, A.D. (1960). Relaxation instabilities in high-perveance electron beams. *IRE Trans. electron. Devices* ED-7, 268.

Symon, K.R., Kerst, D.W., Jones, L.W., Laslett, L.J., and Terwilliger, K.M. (1956). Fixed-field alternating gradient particle accelerators. *Phys.Rev.* 103, 1837.

Theiss, A.J., Mahaffey, R.A., and Trivelpiece, A.W. (1977). Rigid-rotor equilibria and wave propagation characteristics of magnetically-confined plasmas. To be published.

Thomas, L.H. (1938). The paths of ions in the cyclotron. I. Orbits in the magnetic field. *Phys. Rev.* 54, 580.

Thonemann, P.C., and Cowhig, W.T. (1951). The role of the self-magnetic field in high current gas discharges. *Proc. phys. Soc.* 64, 345.

Trivelpiece, A.W. (1967). *Slow wave propagation in plasma waveguides*. San Francisco Press, San Francisco.

Trivelpiece, A.W., and Gould, R.W. (1959). Space charge waves in cylindrical plasma columns. *J. appl. Phys.* 30, 1784.

Veksler, V.I. (1967). Collective linear accelerator of ions. *Proc. 6th Int. Conf. on High Energy Accelerators*, p.289, CEAL, Cambridge, Mass.

Vlasov, A.A. (1945). On the kinetic theory of an assembly of particles with collective interaction. *J. Phys.USSR* 9, 25.

Walcher, W. (1958). Space charge neutralization in ion beams. In *Electromagnetic isotope separators and applications of electromagnetically enriched isotopes* (ed. J. Koch), p.275. North-Holland, Amsterdam.

——— (1972). Some remarks on emittance and brightness of ion sources and ion beams. *Proc. 2nd Int. Conf. on Ion Sources*, p.111. Osterreischische Studientgesellschaft für Atomenergie, Vienna.

Walker, L.R. (1955). Generalizations of Brillouin flow. *J. appl. Phys.* 26, 780.

Walsh, T.R. (1963). A normal beam with linear focusing and space-charge forces. *J. nucl. Energy* 5, 17.

Weber, C. (1964). The electron beam in a cathode-ray tube. *Proc. Inst. elect. electron. Engrs* 52, 996.

Webster, H.F. (1957). Structure in magnetically confined electron beams. *J. appl. Phys.* 28, 1388.

Wollnik, H. (1967). Electrostatic prisms. In *Focusing of charged particles* (ed. A. Septier), vol.2, p.163. Academic Press, New York.

Wu, T.Y. (1966). *Kinetic equations of gases and plasmas*. Addison-Wesley, Reading, Mass.

Yadavalli, S.V. (1958). Focusing in high voltage beam-type devices. *J. electron. Control* 5, 65.

Yonas, G., Poukey, J.W., Prestwich, K.R., Freeman, J.R., Toepfer, A.J., and Clauser, M.J. (1974). Electron beam focusing and application to pulsed fusion. *Nucl. Fusion* 14, 731.

Zimmermann, B. (1969). Energy broadening in accelerated electron beams by Coulomb interaction. *Rec. 10th Sym. on Electron, Ion, and Laser Beams*, p.297. San Francisco Press, San Francisco.

——— (1970). Broadened energy distributions in electron beams. *Adv. Electronics electron. Phys.* 29, 257.

Zworykin, V.K., Morton, G.A., Ramberg, E.G., Hillier, J., and Vance, A.W. (1945). *Electron optics and the electron microscope*. John Wiley, New York.

INDEX

f denotes that the topic is discussed in the pages following the one given

r denotes references to the topic given in the 'notes and references' sections.